A dictionary of
microbial taxonomy

A dictionary of microbial taxonomy

THE LATE S. T. COWAN, D.Sc.

Formerly Curator, National Collection of Type Cultures
Administrative Director, Central Public Health Laboratory, Colindale, and
Deputy Director, Public Health Laboratory Service England and Wales

EDITED BY L. R. HILL, D.Sc.

Deputy Curator, National Collection of Type Cultures

CAMBRIDGE UNIVERSITY PRESS

CAMBRIDGE

LONDON · NEW YORK · MELBOURNE

Published by the Syndics of the Cambridge University Press
The Pitt Building, Trumpington Street, Cambridge CB2 1RP
Bentley House, 200 Euston Road, London NW1 2DB
32 East 57th Street, New York, NY 10022, USA
296 Beaconsfield Parade, Middle Park, Melbourne 3206, Australia

First published 1978

Printed in Great Britain at the
University Press, Cambridge

Library of Congress cataloguing in publication data
Cowan, Samuel Tertius
Dictionary of Microbial Taxonomy
Includes bibliographic references
1. Microbiology – Classification – Dictionaries
I. Hill, Leslie Rowland, 1935– II. Title
QR9.C66 576'.01'2 77-85705
ISBN: 0 521 21890 X

Contents

The serious student of nomenclature, and the student who takes nomenclature seriously – these are two quite different people.

Preface

This dictionary started as a revision of *A Dictionary of Microbial Taxonomic Usage* (1968, Oliver & Boyd, Edinburgh), but it has been greatly enlarged and the scope widened. It is not intended only, or even mainly, for taxonomists (who should know all that is in it or else where to find the answers to their questions) but it is written for those whose work brings them to taxonomy, often reluctantly. Many are lost in a maze of rules, jargon and the legalistic pseudo-Latin of nomenclatural codes, while others look for meanings in scientific names. Some search for rules of classification, but of these there are none.

To make it generally useful, the dictionary includes words that may be met by general readers or by those seeking the usage of a word outside their own disciplines; thus, there are words or terms used in botany and zoology, and terms used in day-to-day microbiology; the first are intended for microbiologists, the second are for botanists and zoologists. There is much cross-referencing and duplication but these should make it easier for the user to find what he wants.

Laboratory workers often find taxonomic details, and writing about them, more difficult than the bench work itself. The dictionary should help taxonomists prepare reports and papers and, from lists of initials (under **abbreviations** and dispersed in the text), understand some of the acronyms and other initials used particularly by the international organizations. There are examples of Latin phrases (and their abbreviation) used in nomenclature, and an extended table of suffixes used to make substantival generic or specific names from the names of people.

It has been necessary to coin a term, Chester code, because the names found in the literature for that coding system (group number and numerical classification) are insufficiently distinctive, or are confusing in the light of more recent use of the term numerical classification. Terms used in numerical taxonomy have been dealt with by L. R. Hill; his contributions are distinguished by the appendage of his initials.

References are of three kinds: (1) *in the text*, usually to papers in journals, occasionally to books; (2) *at the end of an entry*, enclosed in square brackets; these are brief references to the relevant sections of the nomenclatural codes, and (3) *at the end of the dictionary*, to books of general taxonomic interest, or books and papers referred to under several catchwords or in the introductory chapters.

The writing of this book was made possible by help from many friends and former colleagues.

Queen Camel S. T. Cowan.

Editor's note

Late in 1976 Dr Cowan died. He had been working, in retirement, for some years on a book which, as he says in the Preface, arose from his earlier (1968) *A Dictionary of Microbial Taxonomic Usage*, but which as the work progressed developed into a new, much wider, book. When he died, he left a typescript that showed how far advanced he was towards completion. The introductory chapters were in an apparently first draft, but the bulk of this book, the actual entries, were already in their final form up to the letter N and the rest was already largely edited by him, awaiting retype.

I had become involved in 1975 to contribute some extra entries, specifically concerning numerical taxonomy, and some of these he had already included in the text. Originally, I had intended to contribute a greater number of entries than I finally have; I have deliberately reduced my contribution because they have to appear without his final approval.

It was natural, therefore, that I should be asked to edit and prepare for the press the incomplete manuscript. I have done this with as much care as possible and with respect for the sometimes very personal views Dr Cowan held. I apologize to readers, especially taxonomists of his generation who knew him well, for any defects my editing may have caused. I think, however, the true heretical spirit of Dr Cowan could never be dulled or blunted by anyone!

Kingsbury, London L. R. Hill

Abbreviations and conventions used in this dictionary

abbr.	abbreviation
adj.	adjective
BAC.	International Code of Nomenclature of Bacteria
bact.	bacteriology; in bacteriology; bacteriological usage
BOT.	International Code of Botanical Nomenclature
bot.	botany; in botany; botanical usage
ca.	*circa*
cf.	compare
Fowler	*A Dictionary of Modern English Usage*, by H. W. Fowler
n.	noun
pl.	plural
q.v.	which see
sp.	species (singular)
spp.	species (plural)
syn.	synonym
SOED	*Shorter Oxford English Dictionary*
US	United States of America; American usage. (See *American–English Usage*, by M. Nicholson)
v.	verb
VIR	proposed code for the nomenclature of viruses
vir.	virology; in virology; virological usage
VR	Rules of Nomenclature of Viruses, as amended 1975 (Fenner, 1976)
Webster	*Webster's Third New International Dictionary of the English Language Unabridged*
ZOO.	International Code of Zoological Nomenclature
zoo.	zoology; in zoology; zoological usage

Note: numbers in the text:

(1) superscript: number references which immediately follow the corresponding dictionary entry (references enclosed in square brackets),

(2) in cross-references: ...(**word**)... **1** refers to the first meaning given for that word, ...(**word**)... **2**, to the second meaning, etc.

Minirefs

BAC. Bacteriological Code, 1976 Revision
BOT. Botanical Code, 1972
VIR. proposed code for the nomenclature of viruses
VR. Rules of Nomenclature of Viruses
ZOO. Zoological Code, 1964

Dates in parenthesis indicate editions other than the above.
Numbers alone indicate Rules (in BAC., VR.) or Articles (in BOT., ZOO.).

Ad Advisory Notes (in BAC. only)
Ap Appendix
GC General Considerations (in BAC. only)
Gl Glossary (in ZOO. only)
N Note, appended to a Rule (or Article), or Recommendation
PR Principle
Pre Preamble (in BOT., and ZOO.)
Prov. Provisional Rule (in BAC. only)
R Recommendation
S Statute Article of ICSB (in BAC. only)
T Table

I

Codes of nomenclature for microbes

Microbiologists are drawn from several disciplines and are subject to rules and conventions outside their particular field. Each sub-discipline has its own jargon which can be learnt by its own practitioners, but other wide-ranging aspects such as abbreviations, terminology, and nomenclature may present unnecessary difficulty to those in other sub-disciplines. This dictionary deals particularly with the problems of names as these are applied to micro-organisms, but it does not consider the names themselves as these are dealt with in *Bergey's Manual of Determinative Bacteriology* and *Index Bergeyana* (for bacteria) and *Ainsworth & Bisby's Dictionary of the Fungi*. Common sense suggests that the same rules should apply to all kinds of organisms, large and small, and this was one of the ideals and hopes of R. E. Buchanan, who can be described as the father of bacteriological nomenclature. In practice microbiologists are subject to three (possibly four in the future) codes of nomenclature: the Bacteriological, the Botanical, the Zoological and (eventually) the Viral. These codes agree in principle but differ in detail; it is the detail that creates the difficulties and disputes, which in turn engender heat and discord among normally rational and objective observers of nature.

The early bacteriologists thought of bacteria as small animals and few attempts were made to apply the binomial nomenclature of Linnaeus to them. Some botanists, notably Ferdinand Cohn, possibly empire-building and anticipating Parkinson's First Law, claimed them as plants and, using the paraphernalia of nomenclature as it was developing in the early botanical codes, assumed jurisdiction over the names of bacteria. At that time, however, most of the discovered bacteria were the target of, and under the influence of medical and veterinary bacteriologists, by whom such things as nomenclatural codes remained unread. This led to a multiplicity of names, some vernacular and some in the doggerel-Latin of medicine. For the few who cared, numbers of synonymies grew greater and greater.

More interest was taken in these matters in America than in Europe, possibly because a smaller proportion of their bacteriologists evolved from pathology or engineering, and a larger proportion started life as botanists. The young SAB (Society of American Bacteriologists) set up a committee to advise on the classification and nomenclature of bacteria. Among its members was Buchanan who was at that time contributing a series of papers on bacterial classification and, significantly, starting to collect and record the names that had been used and applied to bacteria. Meanwhile, an American medical bacteriologist, D. H. Bergey, encouraged and supported by the SAB, began to compile his *Manual of Determinative Bacteriology*, and among his helpers was R. S. Breed, a thoroughly practical dairy bacteriologist.

Gradually botanical thought and influence filled the vacuum in bacterial systematics left by the indifference and lack of interest shown by medical bacteriologists. But it was not long before bacteriologists became restive under the limitations of the botanical code of nomenclature, and the main taxonomic achievement of the first congress of microbiology in 1930 was the establishment of a nomenclature committee, with Buchanan, Breed, and R. St John-Brooks taking a prominent part. From the microbiology congress in Paris they went to Cambridge, where botanists were holding their congress, and either by persuasion or demand Breed and Brooks became members of a botanical committee to deal with bacteriological nomenclature.

By 1936 thoughts had turned to the development of a separate bacteriological code, and in 1939 a drafting subcommittee was set up; in 1947 the proposed code was considered, revised, and approved by the microbiological congress, and the first bacteriological code was published in 1948. References to bacteria were not deleted from the botanical code until 1975.

The Bacteriological Code (BAC. in this dictionary) now deals with the names of bacteria and, if they are considered to be bacteria, with mycoplasmas. The title of the second (1958) version of the code indicated that the rules applied also to the names of viruses, but they failed to do this to the satisfaction of virologists. Attempts to formulate a virological code did not gain general acceptance, and it is not yet possible to indicate when the proposed virus code will be internationally approved.

The nomenclature of fungi and yeasts is subject to the Botanical Code (BOT.), but there is evidence that mycologists are becoming restless and are not content with the provision for exceptions, contained mainly in Art. 59, for the nomenclature of fungi. Neither do mycologists like the botanical insistence on preserved lifeless specimens as types of species, for with modern methods of preservation of cultures a type strain would often be more acceptable than the dried herbarium specimen demanded by BOT. Algae also are subject to BOT., and the insistence of a Latin diagnosis seems to indicate a tightening of botanical control.

The third code of nomenclature with which the microbiologist should be familiar is the zoological (ZOO.), and to it the names of protozoa should conform.

In this dictionary the three codes of nomenclature at present in use and the viral nomenclature rules are compared and brief references are made to the numbered rules (articles), recommendations, and notes so that the origins of statements can be traced. To find the relevant part in a code can be a time-consuming task for the codes are not well indexed, and it is hoped that these code references (*minirefs*) in the body of the dictionary will spare the reader fruitless hours of search.

2

Source material for taxonomy

The value of taxonomic research will depend on (1) the source material, (2) the thoroughness with which it is examined and documented, and (3) the analysis and comparison of the characteristics of the individual units. Of these, the first probably receives the least thought when the work is planned, but it is at least equally as important as the others; and it is certainly worthy of greater consideration than it receives because source material forms the base on which the subsequent characterization rests.

Source material may be of three kinds: (i) when the organism cannot be grown in artificial culture by techniques at present available it must be studied in its natural environment (as the leprosy bacillus in tissues) or in tissue cultures; (ii) specimens or samples from which micro-organisms will be isolated can be collected from individuals, objects, or localities, and (iii) cultures of micro-organisms from other laboratories and culture collections. The nature of the taxonomic task, and the type of organism involved, will often dictate which kind of source material is used; with (i) the characterization is obviously limited to morphology, staining reactions, and possible evidence of pathogenicity as shown by tissue reactions and the experimental inoculation of ground-up tissue into laboratory animals.

Sampling errors and unintentional duplication of culture must be looked for, recognized, and eliminated whenever possible. The same strain of a bacterium may be received from different laboratories or collections under different labels; few workers keep adequate records of the history of their strains, and collections, dependant as they are on the isolation and transfer histories supplied by donors of cultures, can seldom by any more helpful in this respect; at best, their evidence is mostly hearsay. Sampling errors arise when the source material consists of specimens; when these are from quite separate sources (different animal species, geographical areas, or environments such as water, food, or soil, excreta from different epidemics or epizootics) we can reasonably assume that they are independent and that their microbial floras will also be independent. However we should always enquire about possible associations, e.g. cattle whose faeces are being sampled may have grazed on fields from which soil samples are collected; when our inquiries are negative we are entitled to assume the independence of the samples. On the other hand when we collect soil to find out the number of species of streptomyces and their relative frequencies, we need to know whether the samples came from the same or different localities. Even when several soil samples are taken from one area it is unlikely that each sample will, on culture, yield the same microbial flora either qualitatively or quantitatively, and when more than one streptomycete strain is isolated from a sample the decision has to be made whether apparently similar organisms are to be treated as separate and individual strains or as one and the same strain. Where the specimens are from geographically separated localities, it is reasonable to assume that similar organisms are not the same strain.

This kind of problem also arises in using strains of apparently similar bacteria isolated from different individuals during the course of an epidemic or epizootic, and for taxonomic and epidemiological work a decision has to be made in the even more difficult case of strains isolated from the same individual on different occasions; are these to be accepted as different isolates of the same strain, or as different strains? The problem is not made any easier because the medical and veterinary bacteriologists use the word 'strain' for what other epidemiologists and mycologists would use 'isolate'. The answer to the problem cannot be determined in advance because the repeated examination of the excreta of one person may show the continued presence of apparently the same strain or it may show the presence of a strain similar to, but not identical with, a strain isolated from an earlier specimen. Certain bacteria are fairly easily transformed or transduced *in vitro*, and there is reason to believe that similar phenomena can occur *in vivo*. The practical solution to this problem seems to lie in assuming that the strains from the different samples are different, but only accepting for taxonomic work those strains which detailed tests show to have minor differences in characters, however minor these may seem. The taxonomist must beware of biasing the results of his analytical work by having too many specimens from the same kind of source, and he should diversify by collecting specimens from widely separated areas and from different kinds of situation, habitat, or host.

Cultures as source material for taxonomic work need just as careful thought and selection (generally the deliberate absence of selection, or randomization), and their collection can be considered under several headings.

(*a*) Own isolations from specimens received from others or from specimens collected by oneself. As many details as possible about the origin of the specimens should be obtained so that obviously similar (and possibly identical) specimens may be eliminated. In connexion with the isolation techniques to be used, it can never be assumed that a single culture medium will grow all the different kinds of bacteria in a specimen; many and varied media will be needed to isolate representatives of the diverse flora, and subcultures made from at least one of each different colony form present. To obtain a reasonably complete picture of a bacterial flora it will be necessary to incubate duplicate cultures at different temperatures.

(*b*) Cultures of strains isolated by other bacteriologists will often be used to enlarge the scope of the work, and whenever possible details of their isolation, origin (source material, time and place), and transfer from one worker to another should be obtained. It is surprising how few strains are used for a great deal of research carried out in laboratories throughout the world; the same strains are used for quite different kinds of research, so that when one starts to collect strains it is possible to receive several apparently different strains from different donors, only to find on close inquiry that they have all been distributed from one source, often a culture collection.*

(*c*) Departmental stock cultures and strains from culture collections are commonly (sometimes, too commonly) used by taxonomists who rely on their authenticity and place too much credence in the histories of the strains and in the identifications (determinations) made by others. It should not be necessary, but a caution is needed here; many of the cultures are old and have been submitted to repeated subculture; time may have wrought change, and these old cultures cannot be assumed to have the characters they had on isolation; many show degenerative changes, usually in the loss of one or more characters, but occasionally the change involves the acquisition of a character not present (or detected) on first isolation. Included in this group will be the strains of historic interest (such as Heatley's Oxford staphylococcus which has now almost lost the ability to produce coagulase), alleged descendants of classical strains, and many nomenclatural type strains.

Principle of pure cultures

It may seem trite to lay stress on the importance of pure cultures in taxonomic work, but because purity is too readily assumed, it is a principle that cannot be too often repeated or too strongly emphasized. In its simplest form an impure culture will show features of all the organisms present so that the characterization of a supposed strain or culture may well be so bizarre or exotic as to suggest a new species or even a new, higher taxon. The simultaneous growth of more than one organism in a medium will be encouraged when they grow synergically and discouraged when one

* In numerical taxonomy studies, it is recommended that some strains be deliberately duplicated by simply making two subcultures of each selected strain, and numbering them randomly. Duplicate strains can provide an internal check on the within-laboratory test reproducibility at the end of the study. LRH.

organism produces a substance antagonistic or inhibitory to the other; in neither event is the characterization of the 'organism' taxonomically useful because synergic growth may produce a summation of characters, while growth in the presence of an antagonist, even if it occurs, may be as spheroplasts or other aberrant forms with atypical characteristics.

Much time has been wasted by taxonomists making two unjustified assumptions; namely, that cultures received from other workers are pure, and that they have been correctly identified and named. However reputable the worker from when the cultures are received, all strains should be checked carefully for purity, and also for correctness of identification (determination). Cultures that we ourselves isolate are our own responsibility, and for the sake of our reputations need even more careful checking for purity, and repeated tests at intervals to ensure that purity has been maintained.

Some bacterial taxonomic work is theoretically unsatisfactory in that it is based on the characterization of cultures that cannot be proved to have arisen from a single bacterial cell; most workers rely on the assumption that a well-isolated colony on a thinly-spread plate developed from a single cell or, in the case of bacteria that do not separate readily, from a group or chain of cells. The assumption is itself based on the further assumption that the medium on which the colony develops is capable of initiating and maintaining growth of all the bacterial cells present in the inoculum. This last supposition clearly excludes growth (or cultures) derived from a colony on a selective or inhibitory medium unless it has been replated several times on non-inhibitory media, which means that it is non-selective and contains all the nutritional factors essential not only for the organism being studied but also for likely contaminating or synergic bacteria.

Some groups of bacteria are technically difficult to purify, so that their characterization is more than usually suspect. In this category are the strictly anaerobic bacteria as some of them will not tolerate much exposure to air, which makes single cell isolation, colony selection, and picking unusually difficult, and all these manipulations put the survival of the bacteria in jeopardy. Furthermore, most of the spore-forming anaerobes are actively motile and the growth swarms over the surface of the medium in plates incubated in the moist atmospheres of anaerobic jars. An effective method to prevent swarming is to increase the agar concentration to 6–8%.

Some aerobic bacteria such as *Proteus* species also swarm on the surface of nutrient agar but the phenomenon is more easily controlled because exposure to air during colony selection and picking is not lethal to the bacteria.

Detection of mixed cultures

Because of its importance in taxonomy, it will not be out of place to indicate here methods for checking and ensuring the purity of cultures, before leading on to the maintenance of purity by suitable methods for the preservation of both viability and characteristics of the bacteria. Experience gained with the old cultures of the National Collection of Type Cultures (NCTC), most of which had been kept as slope cultures at room temperature (12–15 °C in England) between subcultures, had shown that cryptic contamination with psychrophilic bacteria (and occasionally fungi) was not uncommon, and a vigorous policy of checking all strains was started in 1948 with the object of eliminating contaminating strains which, in some cases, had replaced the original strain.

The techniques involved are not difficult; in fact, quite elementary bacteriology is all that is needed to 'clean-up' the majority of contaminated cultures. A routine should be established and before any taxonomic work is undertaken, all strains, new and old, should be checked for purity by spreading each strain thinly over the surface of a good non-selective nutrient medium. It is inadvisable to discard these culture plates after inspection; if they are left at room temperature for a few days new colonies may appear which may be aerial contaminants picked up when the plates were inspected, or evidence of a cryptic contaminant. The presence of cryptic contamination by organisms that do not grow at the temperature at which the plates are normally incubated is seldom looked for, but is not uncommon. The first problem is to reveal the contaminant, the second to eliminate it. Incubation of duplicate cultures at different temperatures is the simplest way to detect contamination by thermophils, psychrophils, or psychrotrophs; at the higher and lower temperatures the wanted culture may not grow and the contaminant is then shown up in isolation.

Bacteria that have been isolated on inhibitory, differential, or selective media are likely to carry another kind of cryptic contaminant: one that was inhibited but not killed by the chemical put in the special isolation medium. A contaminant of this kind will probably grow on the nutrient (non-inhibitory) medium used to propagate the organism that is wanted. When it is known that a bacterium was isolated from faeces special care must be taken for the wanted and unwanted bacteria may be very similar in cultural characteristics; furthermore, multiple salmonella infections have been found in man, so that even a single colony may be a mixture of two organisms.

Maintenance of purity and preservation of cultures

Once a culture has been isolated from a single cell or is otherwise regarded as pure, every effort must be made to keep it in that state and to shield it from contamination. Because of continuous variation or mutation in bacterial cultures propagated and maintained by active subculture at regular intervals, a method that limits these changes should be adopted, and two are recommended: (1) drying by sublimation (freeze-drying, lyophilization in US) in which water vapour is removed from the frozen suspension without passing through the liquid phase; the dried powder is preferably sealed in a vacuum: (2) freezing a suspension in liquid nitrogen or in the vapour above liquid nitrogen. Sealed ampoules of the freeze-dried product can be stored for years in the dark at room temperature, but the deep-frozen suspension must be kept in the nitrogen vapour until it is wanted for transfer. Both methods involve a quick freezing of the bacterial suspension and in this part of the process some of the organisms die, but convincing evidence of selective killing has not been produced, and there is good reason to suppose that the survivors are representative of the original population.

Although apparatus for the large-scale preservation (or maintenance) techniques used by the culture collections is not likely to be available in all laboratories, quite simple equipment can be used for adequate drying of small numbers of ampoules of each bacterium; the methods described by Rhodes (1950) and by Annear (1956) need nothing more elaborate than an efficient evacuating pump, but the possession of a high-frequency spark tester (to check the vacuum) is an advantage.

The place of culture collections in taxonomy

A centre for bacterial taxonomy will inevitably develop a collection of cultures from the material used in its research programmes, and the taxonomists working there will, in their own interests, repeatedly check not only the purity of the cultures maintained but also the constancy of their characters. Although designed to control the typicality of the cultures, these checks will yield valuable evidence for or against the permanence or stability of characters. A culture collection in such a centre will be an asset to taxonomy for it can provide adequately characterized and checked cultures. Unfortunately it is not a *sine qua non* that the institute housing a culture collection should be a centre of taxonomic research, but the culture collection itself must be such a centre if it is to attract the right type of scientist to work in it.

The earliest culture collections were either private ventures, such as the Král Collection in Vienna and Prague, or were departmental collections of 'stock cultures' used mainly for teaching in university departments or research institutes (such as the Lister Institute) or museums (as the American Museum of Natural History). Little money was available for these collections and those responsible for their upkeep were seldom able to undertake any worthwhile research, all their energies being directed to the tiresome task of repeatedly subculturing the hundreds of strains in the collection, and, on request, making further subcultures to send to workers who had asked for them. The opportunities for errors of technique and the fallibility of human endeavour to create havoc in the collection were enormous: too few workers responsible for too many cultures; lack of facilities for checking the purity and characteristics of cultures, which usually meant that the identification (determination) of the donor had to be accepted (and many such identifications were subsequently found to be wrong); inadequate storage space at appropriate temperatures; human errors in labelling (or mislabelling) tubes and plates, including the transposition of figures in an accession number; technical errors in subculturing so that contamination occurred, or cultures were transferred to wrong tubes. This is an unwholesome list of opportunities for error that are sufficiently serious to invalidate any claims that may be made that strains from culture collections are reliable descendants of what was deposited by a donor many years before; every one of these errors has been committed more than once to my knowledge, and several of them by me or by staff who were trained to work in a collection. Only since the time when it became possible to avoid routine subculturing by drying the cultures (in theory about 1935 but in fact about 1950) has it been possible to rely on cultures from the larger collections, and then only from those collections which regularly undertake checks of viability and of cultural characters. Little or no reliance can be placed on the histories of cultures deposited before freeze-drying became the routine method of preserving newly-deposited cultures. Many strains deposited before that time had, on re-examination at a later date, so many characters that differed from those they were supposed to have that, in the better collections they were discarded, but in others were retained to cause further confusion among taxonomists, particularly typologists. Mention of the word typologist leads naturally to the consideration of the nature of strains maintained in collections. Some collections have the words 'type culture(s)' in their titles (as the National Collection of Type Cultures and the American Type Culture Collection), but few of the cultures kept are type strains in the sense used in the nomenclatural codes or by typologists. Most of the strains are carried under names which reflect the identifications given to them by the workers who deposited them. Only within the last thirty years, and only in some collections, have the identifications, and names, been checked and challenged, and a serious attempt made to rid bacteriology of its mislabelled cultures.

The aim of culture collections is to maintain, and make available for distribution to others, strains of bacteria that are representative (typical) of their taxa, or have features (characters, properties, products) of special interest. When the characters of a bacterium can be preserved by some process such as freeze-drying, the cultures serve as guides to the accuracy and completeness of descriptions made at the time of isolation. They can therefore be said to ensure 'the continuity between generations' of taxonomists (a phrase for which I am indebted to Dr Ruth E. Gordon). But in bacteria variation or mutation can take place so rapidly that cultures maintained even for a short time by subculture are no longer representative of the organism in its natural or wild state, and the specimens received after such maintenance are as likely to confuse as to enlighten the recipient.

Future taxonomists should not repeat the errors of the past, one of which was a blind acceptance of information provided by a collection catalogue or the older records of a collection. This warning against accepting cultures from collections at their face value is issued in all seriousness because I have met so many taxonomists who have based their arguments on characteristics of cultures obtained from so-called 'type culture' collections. It is a warning given on the basis of examining many hundreds of such cultures from my own and from other collections, and in the knowledge that taxonomic posterity would have been better served if all such type culture collections, founded before the days of satisfactory maintenance methods, had been destroyed by some cataclysm such as the Flood.

To rely on the authenticity of old cultures received from collections, and to endow them with special merit because they came from a 'type culture collection' is to place the whole of a piece of taxonomic work in jeopardy, and to expose it to violent and justifiable criticism on the grounds that the author(s) did not know what he was working with, and that he was basing his conclusions on false premises.

Summary in precepts

This chapter is intended to help taxonomists raise the standard of bacterial taxonomy and, with this in mind, it is proposed to enunciate a series of precepts, or practical rules of the art, the first of which will be:

Do not assume that cultures, alleged to be descendants of old or classical strains, have the characters of the original isolates.

Such old cultures will have been maintained for many years by repeated subculture, even if you receive them as freeze-dried ampoules; they may have historical interest, but their taxonomic worth is at best doubtful.

Taxonomic work should be based on cultures maintained from the outset by methods (such as freeze-drying or deep-freezing) which limit variation or mutation. The value of old cultures is limited to indications of character stability.

A second precept will state:

Do not rely too much on descriptions of organisms isolated before the era of pure cultures.

(about 1900 for most bacteria but some time after 1920 for anaerobes).

A third precept will be:

Do not rely on the purity of cultures received from others, or assume that the identification made by the donor (however good or senior he may be) is correct.

In other words, choose your reference strains carefully, rely on your own results in characterization and comparative studies, and, above all, use your common sense.

3
Philosophy of classification

The *Shorter Oxford English Dictionary* defines philosophy as 'the love, study or pursuit of wisdom or of knowledge of things and their causes'. More specifically, 'philosophy of' is defined as 'the study of the general principles of...(a) branch of knowledge, experience or activity', but the phrase is said to be 'less properly' used for 'the study of any subject or phenomenon'. From these definitions it is apparent that the heading of this chapter is either the improper use of 'philosophy of', or uses the word philosophy in the old and general sense of love, study, and pursuit of wisdom, or of knowledge of things and their causes. It is doubtful whether bacteriologists ever love, study or pursue wisdom for its own sake, but they show an avidity for knowledge of things and their causes, and it is with this sense of association of knowledge, causes, and effects, that this chapter is concerned. I shall not discuss metaphysical philosophy in which the essence of reality is questioned, but shall deal with philosophy as it concerns the scientist in general and the microbiologist in particular.

To the biologist, philosophy is a frame of mind that is attuned to the vagaries of living things and does not expect an experiment to be exactly repeatable, an attitude that is surprised by the expected, and astounded by the fulfilment of a forecast or prediction. Thus, if we say that a classification is logical we know that it is not a classification of biological units, whereas if we say it is expedient we will be ready to lay odds on its being the work of a biologist dealing with units of his own creation. Creation may not be the best word to describe what I mean, but since the units are mainly in the mind of the beholder, they are to some extent his creations. Because the philosophy of a taxonomist is a frame of mind (or an attitude determined by a frame of mind), it is an indication either of beliefs or perceptions, and may show elements of each; the first taxonomists believed that each kind of animal or plant was the result of a separate act of divine creation, but as time went on they perceived that living things did not 'breed true', and this observation cast doubts on the correctness of the belief. Unlike the librarian or the exact scientist, biologists were trying to classify infinitely variable units, and with each additional item of new knowledge a change might be forced upon even the best and most reliable classifications. The extent of the change could depend on the stubbornness of the taxonomist and on his willingness to recognize the fallibility of the best-laid schemes of mice and men.

If we accept the definitions of the *Oxford English Dictionary* (and no scientist would dare dispute its meanings of non-scientific words) we must recognize that there is not, and cannot be, a philosophy of classification, and that what we are about to discuss is the philosophy of the classifiers or taxonomists; there is not a single philosophy common to all taxonomists, who are individuals of independent mind, but there are many and varied philosophies. To some extent the philosophy will be dictated by (or under the influence of) the taxonomist's main occupation or, in other words, what he is paid to do, and from which he snatches time to indulge in taxonomy. When the occupation allows a wide discretion of interest, as enjoyed by those who are members of the staff of a university department, research institution, or government laboratory, the philosophy of the taxonomist will be related to his main interest, whether it be nomenclature, classification, or merely identification. I use the word 'merely' because some taxonomists regard identification as a lowly occupation and the least important task that a taxonomist has to do; this attitude overlooks the fact that without the call or need for identification (which is the practical day-to-day aspect of taxonomy) there would not be a demand for classification or nomenclature, and therefore no need to have taxonomists. However, it should not be thought that taxonomy is an example of Parkinson's First Law; the three parts of taxonomy as defined in this dictionary are consecutive and interdependent, and none of them is undertaken merely to fill the time available. A microbiologist's philosophy will depend on his approach to taxonomy itself, either as something dull, uninteresting, and time-consuming, or as a stimulating exercise which brings worthwhile rewards and is interesting enough to do for its own sake. The reluctant taxonomist is one whose interest lies outside the field but whose work has developed a taxonomic aspect, or has reached a stage at which progress is held up until a taxonomic problem has been resolved, and he will be less than enthusiastic for the subject. The professional taxonomist, on the other hand, must be enthusiastic, otherwise he would not be able to pursue successfully the repetitive benchwork and the laborious (and often tedious) literature searches.

Since few taxonomists philosophize for the sake of Philosophy (with a capital P), this chapter deals primarily with the attitude of taxonomists to their subject; few have had the urge or time to philosophize in print, so our discussion will reflect opinions expressed by the spoken word. Compared with botanists and zoologists, microbial taxonomists have been working at their subject for a relatively short time, and to the older practitioners must be regarded as mere amateurs at the game, but, we would hope, without the denigratory implication of 'enthusiastic amateurs'. This chapter, headed the Philosophy of classification, might have been better labelled the Philosophy of taxonomists, but, because my philosophy is based on an attitude that regards headings and names as meaningless but useful labels, I would not think of changing the title.

Philosophy of the taxonomist

The phrase 'philosophy of the taxonomist' summarizes in a few grand words, a man's views of mental approach to his work, which is taxonomy. Political and religious beliefs may play big parts in determining his attitude to life and to his fellow men, and decisive in shaping his general philosophy, but towards his work his attitude will be based on more tangible factors, which can be considered briefly under several different, but not exclusive, headings.

(1) *Worthiness of the subject*, which can be examined by asking a series of questions:

(*a*) is taxonomy worthwhile? And this must be considered at the same time as:
(*b*) is it important (or needed) in everyday microbiology; is it of value to the applied microbiologist or only academics who can afford to spend time on the theoretical aspects of the discipline?

It is inconceivable that an intelligent and trained person would spend his life working at something that he did not regard as worthwhile; every taxonomist's philosophy must start from the premise that taxonomy is a universally needed discipline, and this is borne out by experience because the practitioners of any applied branch, or research workers who use micro-organisms as tools, depend on the correctness of the taxonomist's work, whether it be the classification, the identification, or merely the labelling of the relevant micro-organisms. If he did not hold this view clearly before him, and also project it to others, the taxonomist would rarely complete any piece of taxonomic research for the very practical reason that it is unlikely that a non-taxonomist would have so exalted (or correct) a view of taxonomy; few would encourage him to continue, and fewer still would pay for taxonomic work. The end-product of taxonomy is almost always the result of a man's zeal, of time stolen from some applied microbiological problem that he is paid to investigate. Probably the most valuable and practical part played by taxonomists in the applied branches is to provide the detailed background and construct the various tables and keys that can be used for the identification of the micro-organisms. It goes without saying that the academically-minded worker is concerned with taxonomy, and that some knowledge of the subject is essential to the teacher. Which leads on to the next question.

(*c*) Does taxonomy, and its techniques, have a place in training microbiologists?

Taxonomy provides training and experience in working to a plan or system, and encourages the completion, in all its details, of work in which the problem may seem to have been solved. Because it discourages the use of short cuts, it minimizes the risks of overlooking some detail. In the characterization work so essential to the subject, it provides opportunities to compare different techniques; it nearly always indicates the importance of the standardization of techniques, with freedom to modify them so long as the standard methods are run in parallel. It teaches the beginner (and the experienced) the need to have adequate controls, and to evaluate his own results as critically as he would those of others.

(*d*) Does taxonomy help in communicating knowledge about microbes to others?

As part of a communications system, taxonomy applies not only labels but, by careful preparatory work, supplies a warranty that the labels are correct at that time. It is unfortunate that the phrase *at that time* is a necessary qualification, but, like price tags in times of inflation, microbial name changes are frequent with advances in taxonomic knowledge.

(2) *Usefulness of taxonomy* and the taxonomist must not be in doubt, and indeed cannot be doubted when seen in terms of the diagnostician (determinator) working in a clinical laboratory identifying the causal agents of infectious diseases, or as the guide of the epidemiologist, as the technical adviser to industrialists attempting to produce a microbial end- or by-product on a large scale, or isolating natural elements such as sulphur. Taxonomy is also of help to the agronomist and to the social scientist or economist in search of protein for the undernourished.

(3) *Reasons for being a taxonomist or for doing taxonomy* are provided by answers to questions such as:

(*a*) Why do we classify microbes?
(*b*) Why do we need to name them?
(*c*) Why is it important to distinguish between, or show the identical qualities of microbes?
(*d*) Why do we need to communicate and share our knowledge about microbes?

When a man first becomes a taxonomist he may treat these questions lightly but as he gets older (and more set in his ways) he regards them more seriously and his particular answers to the questions become major points in his philosophy. In old age, which most of us reach before we realize it, he becomes so deeply attached to the reasons that have satisfied him that he resents any attack made on them. The questions themselves read like the summary of an essay in which we are attempting to define taxonomy.

(4) *The relative values and importance of classification and nomenclature.* Differences here not only determine the philosophy but also the character of the taxonomist (using the words 'determine' and 'character' in their everyday, non-taxonomic, senses). Nomenclature is associated with, and largely dependent on the type method or type concept, in which a particular strain (or specimen) is chosen to be the type. Among zoologists, and perhaps among bacteriologists there is disagreement about what the 'type' is attached to; some think that the attachment of the strain (specimen) is to a particular name or epithet, others that the attachment is to a taxon (or taxonomic group) of a specified category or rank. It will be clear to the unprejudiced observer that this is a fundamental difference (discussed under **type**) which has a great influence on a taxonomist's philosophy. Nomenclature appeals to bibliophils, and to those with minds that, like a lawyer's, revel in the interpretation of rules or laws.

Classification, on the other hand, is much less concerned with the published literature, and as there are no laws of classification, a legally orientated mind is not necessary. The classifier is an acute observer of dynamic processes and he is the interpreter of his own observations. The nomenclator is a contemplator of literary problems and must be able to exercise judgement of the evidence before him.

(5) *Value (or weight) to be given to characters or organisms* is more than usually controversial and opinions (and consequently the holder's philosophy) may be extreme. At one end are views similar to those of Kauffmann who weight certain characters (or sets of characters) heavily, and take slight cognizance of others, thus biasing those that are weighted at the expense of those neglected or ignored. The other extreme, in which all characters are regarded as contributing equally to the whole, is attributed to Adanson but Mayr (1965) disputes this attribution and quotes passages from Adanson's writings which seem to show that he ignored trivial and inconsequential detail. The puzzle here is who (and how does he do it) determines which characters are trivial and inconsequential, or are the ones to be ignored those that are technically difficult or impossible to determine? The effects of Adansonian or Kauffmannian thought on a man's philosophy are

diametrically opposed, and there is little opportunity for compromise. Weighting, or not weighting, is intimately concerned in (6) below.

(6) *The assessment of relatedness* may be made by comparing the overall similarity of characters, or alternatively by the dependence on comparisons between a smaller number of selected characters which are regarded as being particularly important, and therefore considered better able to show relatedness. Once again there is a sharp division of opinion among taxonomists, not a little acrimony, and, for good measure, prejudice of a bitter kind not normally associated with scientists. Few men can be so objective that they are without prejudices, but different views on the assessment of relatedness lead to the taking of stances or positions on the subject and, unfortunately, may vitally affect judgement. As factors affecting the philosophy of a taxonomist, the questions of the evaluation of an organism's characters and the assessment of relatedness are of the utmost importance and, together with the points dealt with in the next two sections (evolution, and clumping versus splitting) determine those qualities of personal character that make taxonomists individualistic and aggressive protagonists of viewpoints; in short, they are the reasons behind the qualities that make a man what he is.

(7) *Place of evolution in taxonomy.* Although microbiologists do not deny the importance of evolution, they do question its place in the classification of microbes; almost nothing is known about the evolution of bacteria so that it is usually unprofitable to attempt a phylogenetic classification, but this lack of positive information does not exclude speculation. In practice, different approaches are seen in the views expressed by those who think that all classification schemes (even those of microbes) should be phylogenetic, and those who think only in terms of pragmatic systems; the first group is prepared to fill gaps in phylogenies with hypothetical taxa, but to the second there is an unacceptable difference between evidence that can be proved by experiment and speculation which is at present unproved and may, in fact, be unprovable.

(8) *Relative values of recognizing few or many different kinds of microbe.* (i.e. clumping versus splitting). On taxonomic problems bacteriologists can be broadly divided into two groups which are known as the lumpers (or clumpers) and the splitters, and these names (unlike those given to microorganisms) clearly describe the individuals, or rather, their mental approaches to a difficult problem in classification. Although there are probably more splitters on the western shores of the Atlantic than on the eastern, the correlation is not sufficiently good for the characteristic to be included in a list of differences between Americans and Europeans; the

difference is more likely to be the result of there being a greater number of microbiologists in the US and the earlier specialization of American workers. The tendency to become a splitter increases with specialization but the effect is not a simple one, and there are generally two opposing reactions. The massive accumulation of detailed information about a particular group of organisms, a narrowing of the field of work, and a limitation of reading to papers relating to the chosen group, puts the specialist in a unique position to see subtle differences between members of the group; at the same time he sees (or receives the impression) that other groups of organisms have far more similarities than differences; thus he inevitably becomes a splitter with regard to his own group, and a clumper of the groups that hold less interest for him.

Lumping and splitting are not confined to taxonomists or to those with a special interest in classification, the division is found among bacteriologists of all kinds because classification affects so many aspects of their work; it is also found among those who use bacteriology as a main source of information, such as clinicians who probably prefer lumping and epidemiologists who will benefit from splitting. This division of bacteriologists into lumpers and splitters, indicates the importance of taxonomy and amply justifies the time, often grudgingly given, spent in taxonomic pursuits.

(9) *Criticism and scepticism.* Taxonomy teaches its practitioners to approach each problem with an open mind but it does not encourage naivety; in fact the reverse is the case for all observations should be examined critically, and all theories and hypotheses need to be looked at with scepticism. A taxonomic hypothesis runs into deep water when it attempts to relate facts of microbial life as known today to evolution, and the 'evidence' for phylogenetic schemes of classification needs to be probed thoroughly and exhaustively.

A taxonomist's philosophy will seriously affect his faculty for criticism (especially of his own work) and scepticism, but the difference between individual taxonomists will be quantitative, for few will be found who either lack critical faculties or are so sceptical that they do not believe anything.

(10) *Moral issues* such as the control of life and living organisms, and the development of new and (?)advanced forms of life are not properly within the scope of a taxonomist's philosophy because taxonomy deals with organisms that exist or have existed. Papers have been written on hypothetical genera and species, and editors have accepted them for publication, but fortunately taxonomy does not concern itself with them, and any names proposed are without standing in nomenclature.

(11) *A means to spread knowledge* in a readily digestable form is essential to an effective communications system; in biology the taxa, or units being described, are most succinctly identified by a unique name or number, which although not itself descriptive, is fixed to a description, illustration, or specimen; this is nomenclature and to this subject different taxonomists apply quite different attitudes. Unlike classification, which is subjective but controlled by the application of common sense, nomenclature is supposedly objective and regulated by the rules and recommendations of the nomenclatural codes, rules which, though arbitrary, are intended to stabilize the names attached to bacteria, and to reduce the confusion that arises when names are changed. The rules stress priority of publication (except for viruses) and, at first sight, this should stabilize names but, in fact, it may require the abandonment of a universally used name if an older name (previously unused) is found during the course of a literature search. Some taxonomists delight in this form of logomachy, and in bringing these unknown names to the notice and annoyance of their colleagues. Actions of this kind have brought both taxonomists and nomenclature into disrepute, but should stop, for bacteria, after 1980 when new Approved Lists of Bacterial Names come into force.

In general, by their approach to nomenclature, taxonomists can be divided into three kinds: (i) those who acknowledge a Code and believe that its application should be rigid and essential in all cases; (ii) those who think that the application of the rules should be more flexible, applying them when there is a general consensus of opinion in favour or when it is convenient, but ignoring them when common usage demands or their strict application would be inconvenient; and (iii) those who think that names are unsatisfactory labels and that a code made up of numbers and/or letters can be both permanent and informative. This form of numerical coding of information about bacteria (which is quite old) is called numericlature and is different from the newer kinds of taxonomy in which comparisons of the characters of one bacterium are made with the characters of another and the similarity expressed as a number or percentage; this is numerical taxonomy and does not deal with the nomenclature of the taxa.

(12) *The need to systematize available knowledge* is obviously a requisite for the making of any classification of that knowledge, and this should be done as the information is collected by recording it in some way, and not leaving the details to the memory. Methods of recording are often highly characteristic of the worker, and may take the form of a laboratory shorthand which, although convenient, cannot usually be read by anyone else. For permanent record the information must be transferred to some more generally known medium, such as a paper in a scientific journal. The

idiosyncratic behaviour of a taxonomist in recording the results of his work may be the greatest weakness of his work, but there is a growing appreciation of the need to communicate and to co-operate with others.

The information recorded is useless unless it can be retrieved and understood; when it has been decoded it can be sorted and re-arranged in an orderly manner. This re-arrangement is another subjective process and different taxonomists may re-arrange the same material in several different ways, even when the aim of the re-arrangement is the same. To overcome these individualistic solutions and to remove the subjectiveness of the re-arrangement is one of the main arguments in favour of numerical taxonomy, in which the actual sorting and re-arrangement follows a pre-arranged programme. Numerical taxonomy is undertaken by those who do not weight particular kinds of character, and it is part of their philosophy that the taxonomist should detach himself from any privileged position he holds as a first-hand observer, and so diminish the subjectivity of biological systems of classification.

(13) *Co-operation among taxonomists.* There is an increasing tendency for taxonomists to co-operate in order to cover a wider field. It is often said that taxonomy cannot be undertaken by a committee and this is generally true, but co-operation in taxonomy is possible provided that it is joint action by workers in adjacent fields and there is not much overlapping of interests among members of the committee; in this way each member remains the expert and adviser on one section of the area covered by the committee. A good example of such co-operation is seen among those who work with members of the family Enterobacteriaceae, the individual genera of which are found as pollutants of water supplies, in soil, on the surface and in the tissues of diseased plants, in the intestinal tract and excreta of both healthy and diseased animals, warm- or cold-blooded; they are isolated by bacteriologists dealing with hygiene and public health, by clinical pathologists, veterinarians, and plant pathologists. In co-operative studies the individual workers are submerging, or at least modifying their individual views, and this should be to the ultimate advantage of all. But in studies of this kind there will be leaders and followers, and the philosophically rigid may get their way because they will not yield to the others. On the whole the system of taxonomic subcommittees set up by the International Committee for the Systematics of Bacteria has worked well and in most cases more has come out of these committees than was expected when they were set up.

4
Early history of bacterial classifications

⌊*Editor's note*: This chapter is only a fragment. Its original title was 'Development of bacterial classifications', but the apparently first-draft notes covered a subsection (*a*) called 'Historical Survey', which was complete up to criticisms of *Bergey's Manual*. The fragment has to end at that point for there were no indications of what other subsections were planned. Despite its incompleteness, and although much of the ground covered in this fragment can, so far as the facts are concerned, be found elsewhere, it was thought worthwhile to include it for the comments on general principles that Dr Cowan had made. LRH.]

Bacterial classification began, like its nomenclature, at least two centuries before we knew anything, other than that they existed, about bacteria; this quixotic situation arose because there was a gap of about 200 years between the first sighting of bacteria by van Leeuwenhoek and their isolation and study by Pasteur and his contemporaries. In the seventeenth century, van Leeuwenhoek drew and described the bacteria he found in association with other animalcules; Bulloch (1938) said of him 'Not only did he discover bacteria but he described and figured all the morphological types known to-day, viz. cocci, bacteria, and spiral forms...' and, in doing so, he undoubtedly (and unwittingly) made the first classification of what we now call bacteria. But, although he has been called the first bacteriologist and the creator of bacteriology, neither van Leeuwenhoek nor those who came immediately after him distinguished the bacteria from the other animalcules, and it was in an early classification of the infusoria that bacteria first appeared in a formal classification; even then they were not put into a separate group or class, but were included with genera of infusoria; indeed, some genera consisted of species that were probably bacteria, together with species that were protozoa.

According to Buchanan (1925), who devoted 94 pages of his *General Systematic Bacteriology* to the history of bacterial classifications up to 1923, the first organisms that we would now accept as bacteria to appear in a classification did so in 1773, in Mueller's classification of Vermes. Buchanan accepted *Vibrio* Mueller as a bacterium, and because *V. lineola* came first on Mueller's list, Buchanan designated it as type species. However, Hugh

(1964) believed that although some of Mueller's species of *Vibrio* were bacteria, others, including *V. lineola*, were infusoria and argued that *Vibrio* Mueller must be regarded as a genus of infusoria. This is a brief and relatively simple example of the confusion that has arisen from accepting a starting date for nomenclature (and automatically for classification) before the nature of bacteria were known, a confusion that continued until the International Committee on Systematic Bacteriology (ICSB), in its wisdom, changed the starting date and withdrew the names of unrecognizable bacteria, whether pictured or described.

It is not my intention to discuss the early classifications of bacteria except insofar as they involve general principles; readers who are interested in early classifications can find the outlines of the various schemes in Buchanan (1925), in shorter versions in the first three editions of *Bergey's Manual of Determinative Bacteriology* (1923, 1925, 1930), and, from the historical angle with many fascinating sketches of the individuals concerned, in Bulloch's *The History of Bacteriology* (1938).

The earliest classifications were based on morphological characters, and the bacteria were regarded as animals, even to the extent that they were supposed to have stomachs! However, in 1823 Bizio set a problem not yet solved to the satisfaction of all bacteriologists and mycologists. Bizio investigated the cause of the miracle of bleeding polenta, which occurred regularly in some churches when polenta (a doughy bread) was left exposed to air. He describes how 'the red color appears within twenty-four hours if the atmosphere is damp enough. Indeed, later than this time other types of molds appear which effectively impede the red coloring' (translation, Merlino, 1924). The organism described by Bizio, to which he gave the name *Serratia marcescens*, could have been a yeast or a bacterium, and the confusion caused by the nomenclatural acceptance of the name *Serratia marcescens* adds weight to the arguments of those who would deny recognition of taxa (and their names) first 'described' before, at the earliest, the Pasteurian era, or, better before the era of pure cultures or, best of all, the date in the future now established by ICSB (namely, 1980).

The changed appearance of the polenta, and the production of pigment by the organism studied by Bizio, introduced two new kinds of character to supplement morphology, and added an entirely new element (the effect produced by a micro-organism on the substrate) to the features that formed the bases of classification. The point of present interest is that while the first half of the nineteenth century saw the first stirrings of an awareness of microbial life, most of this was due to the painstaking work of the early microscopists, many of whom were more interested in the physics and mechanics of their instruments than in what they observed, and only a few were consciously exploring the mysteries of life among objects of small size.

Only the crudest culture techniques were available, so that a formal classification before the Pasteurian era could be based only on: (i) morphology; shape, size, movement of cells, (ii) pigment production, usually on (unsterilized) potato, and (iii) observable effect(s) on some substrate(s). Even pathogenicity was denied the earliest taxonomists, for at this time the microbial origin of infectious disease, or contagion as it was called, was only a hotly-disputed theoretical possibility.

It was Cohn, a botanist, who suggested that Ehrenberg's family of Vibrionia (the bacteria) should be regarded as members of the plant, rather than the animal, kingdom, and Näegeli who created the group Schizomycetes for the bacteria, in the plant kingdom. (The name Schizomycetes (fission fungi) has persisted up to the seventh edition of *Bergey's Manual*, 1957.) Since the nomenclature of plants and animals developed differences in detail, this transfer had some effect on the nomenclature of bacteria, which for a long time then followed the rules laid down by botanists. To this day the Bacteriological Code is closer to the Botanical than to the Zoological Code of Nomenclature.

Cohn was aware that the inoculum of a culture might be composed of more than one kind of bacterium, and he regretted his inability to isolate single bacterial cells, and to observe them over a period of time under different conditions. He thought that the only characteristics available to him for classification were the size and, within limits, the shape of the cells, and their combination or aggregation into groups. Chains posed a problem because the length was variable, and the individual units making up the chain were small and difficult to measure. Because they reproduced asexually, Cohn did not think that bacterial genera had the same significance as the genera of higher plants, and he adopted a principle used by mycologists, whereby each form that showed big differences from others was to be regarded as a genus. Since species were distinguished by even smaller differences, he was conscious that several of these 'species' might be variant forms or descendants of one parent cell but, in spite of his reservations, he thought that bacteria could be separated into good and distinct species. At the time he discussed the question of morphologically-identical but physiologically (or pathologically) different species, and while he was hesitant to treat them as 'good' species, he thought that they deserved separate recognition as physiological or pathological races.

Bulloch regards perusal of Cohn's *Untersuchungen über Bacterien* (1872) as like 'passing from ancient history to modern times' and since his classification of 1872 is outstanding for its time, I shall give some details of how species were separated, using *Micrococcus* as my example.

Tribus I, Sphaerobacteria, had one genus for which Cohn used Hallier's name *Micrococcus* but amended the description of characters to exclude various moulds. The description of the genus, good for its time, reads in free translation (by H. E. Ross):

Colourless or faintly coloured cells, very small, spherical or oval, which by transverse division form short rosary-like filaments with two or more components (Mycothrix, Torula form), or form many-celled families (colonies, balls, heaps), or unite in slime (Zoogloea, Mycoderma form), they are non-motile.

Because micrococci were difficult to classify by differences in shape and size of the cells, but were relatively easy to separate by their physiological activities, Cohn arranged them in three groups: (1) chromogenic, (2) zymogenic, and (3) pathogenic.

(1) *The chromogenic cocci* were divided into two subgroups:
(*a*) in which the pigment was insoluble; these included
1. *Micrococcus prodigiosus* (Note 1, below);
2. *M. luteus* which formed a thick yellow film on the surface of liquid medium and a yellow sediment; on potato, yellow colonies.

(*b*) pigment was soluble:
3. *Micrococcus aurantiacus*, yellow colonies on egg-white medium;
4. *M. chlorinus*, yellow or sap-green pigment on egg-white medium;
5. *M. cyaneus*, elongated cocci, said to be non-motile (Note 2, below);
6. *M. violaceus*, elliptical cells, non-motile, larger than *M. prodigiosus*; violet-blue colonies (Note 3, below).

(2) *Zymogenic cocci*:
7. *M. ureae*, breaks down urea;

(3) *Pathogenic cocci*
8. *M. vaccinae*, small cocci isolated from vaccine lymph;
9. *M. diphtericus* (*sic*) (Note 4, below);
10. *M. septicus*, isolated from various sources;
11. *M. bombycis*, chain-forming coccus from diseased silk-worms.

Note 1. Cohn had Ehrenberg's strain of *Monas prodigiosus*.
Note 2. *M. cyaneus* was probably *Pseudomonas aeruginosa*.
Note 3. *M. violaceus* was probably *Chromobacterium violaceum* or *C. lividum*, but possibly *Micrococcus violagabriellae* Castellani.
Note 4. *M. diphtericus* (*sic*) was probably the coccal form of *Corynebacterium diphtheriae* var. *intermedius*.

In the absence of good cultural methods it is not surprising that Cohn's genus *Micrococcus* includes more than cocci; those species that we now know to produce rods as the characteristic form also produce shorter rods (Cohn's elongated cocci and ellipsoidal cells) and even coccal forms. What is more surprising, indicative of an unexpected lack of critical faculty among taxonomists, is the continued apparent acceptance of Cohn as the father-figure of what we now recognize as *Micrococcus*; the attribution is a nomenclatural one and is limited to the name *Micrococcus*, but bacteriologists not familiar with the peculiarities of the rules of nomenclature and the vagaries of taxonomic practice, can be excused for thinking that the taxonomists continue to accept Cohn's (1872) *circumscription* of *Micrococcus*, which is a different matter.

Buchanan's (1925) review of the history of bacterial classification was written exclusively from the viewpoint of a nomenclaturist, and the inadequacy of a description accompanying the publication of a name, or the impracticability of a dichotomous key were not matters that unduly concerned Buchanan in this or any of his publications. This concentration on nomenclature and lack of criticism – except on points of nomenclature – are the only weaknesses of a remarkable compendium of information on names and schemes of classification.

From Zopf's 1885 classification onwards there was a tendency to put spore-forming rods into *Bacillus* or *Clostridium* and in his first classification (1890) Migula put all the non-sporing rods into *Bacterium*. Buchanan draws attention to Sternberg's inclusion of all rod-shaped forms (except those with fusiform spores, which went into *Clostridium*) in the genus *Bacillus*, and comments adversely on this practice by medical bacteriologists of including both sporing and non-sporing rods in the same genus. A later practice common to medical bacteriologists was to modify Migula's inclusion of all non-sporing straight rods in *Bacterium*, by restricting that genus to Gram-negative rods.

Thus far, the pure culture era has not been reached, but Buchanan's statement that 'During the year 1893 the only new generic name proposed was *Achromatium*...' seems to be unintentionally ironical, and preceded Perkin's masterpiece 'The discovery of the principles of pure-culture study resulted in such a sudden burst of investigation that it was a lost month in which a new organism was not described, catalogued, and laid away, very frequently in the wrong grave' (in Jordan & Falk, 1928).

In his 1894 classification Migula restricted *Bacterium* to non-motile straight rods, another amendment to the circumscription of this controversial genus, and *Bacillus* lost its spore requirement and gained peritrichous flagellation as an essential feature. In other words, for the straight rods, Migula changed the emphasis from spore formation to flagellation, and the site of flagella insertion became a matter of taxonomic importance: polarly flagellated rods were all placed in *Pseudomonas*.

In spite of technical difficulties and the misleading results obtained by light (and even sometimes electron) microscopy, flagellation had earned a surprisingly high regard from taxonomists. Both spore formation and flagella location appear in the 1895 classification of Fischer, but Buchanan remarks that 'No species were known for several of the genera...' and the classification seems to have been at least partly theoretical; this is contrary to perhaps the only principle of classification, namely, that which is classified (and named) exists and has been observed! In his 1897 classification Fischer puts non-motile spore-formers in *Bacillus* but also included in the same subfamily, Bacilleae, such organisms as *Bactrinium pyocyaneum* whose spore-forming potential was non-existent or unknown. In his third scheme of 1903, Fischer ignored sporulation, put non-motile rods in *Bacillus*, and motile ones in *Bactrinium*, *Bactridium*, or *Bactrillium*.

The first of Lehmann & Neumann's classifications appeared in 1896 and Buchanan (1925, p. 51) described it as 'one of the most satisfactory that has been proposed'. In it all non-sporing straight rods were put into *Bacterium* and all sporing rods into *Bacillus*; it was, in fact, a sound, lumper's classification, angled towards the medical bacteriologist. In the second edition, published in 1901, the Hyphomycetes were removed from an appendix to the main classification and were renamed Actinomycetes.

A Manual of Determinative Bacteriology by F. D. Chester appeared in 1901; in his Preface Chester describes the book as being composed of tables for identification, not for classification. His contribution to the *Bacterium-Bacillus* kaleidoscope was to ignore spore formation, put non-flagellated rods into *Bacterium*, peritrichously flagellated rods into *Bacillus*, and polarly flagellated rods into *Pseudomonas*. Chester's (1905) later work was directed towards simplifying classification by the use of coded characters (see **Chester code**).

Kendall (1903) was perhaps more conscientious than many of his contemporaries; after putting non-motile rods in *Bacterium* he created a new genus *Bacterius* for rods motile by flagella of uncertain or unknown insertion site. For all this honesty, Buchanan *et al.* (1966) ruled against the validity of the generic name on the ground that 'No species were named.' However, Kendall had anticipated such a step for he said 'The state of bacterial knowledge is such at the present time [1903] that any attempt to define a species is entirely out of the question' and, in a footnote to the name *Bacterius*, he added 'This group will disappear as the knowledge of the flagellation becomes more complete.'

Several bacteriologists who were dealing with small groups or were making specialized classifications began to introduce pathogenicity as a major criterion into their schemes; an example of a limited scheme is that made by Winslow & Rogers (1906) in their extension of their 1905 scheme for the Coccaceae; one of a more general scheme, in connexion with a textbook on hygiene, is that of Flügge. Buchanan complained about

Flügge's nomenclature, and well he might, for Flügge used trinomials and committed other nomenclatural atrocities that would never enter the mind of a botanist; but we should be tolerant when a bacteriologist (and especially a medical one) was ignorant of the Botanical Code of Nomenclature, and failed to anticipate the approval 40 years later of a bacteriological code. Flügge was a pioneer in that he brought the Gram reaction into taxonomy; Gram had described his staining method in 1884 but it did not appear to have influenced the thinking of the early bacterial taxonomists. But having introduced it, Flügge can be criticized for then putting both Gram-positive and Gram-negative organisms in *Micrococcus* (most species were Gram positive but he also included the gonococcus and meningococcus) and in *Bacterium* (again mostly Gram negative but some Gram-positive species such as the tubercle, diphtheria and Swine erysipelas organisms).

Orla-Jensen's classification had an important place in the history and development of bacterial classifications; he broke new ground by introducing bacterial physiology to a general scheme. Some earlier workers such as Winogradsky had used physiological characters for small groups of sulphur bacteria, but it was Orla-Jensen who had the imagination and skill to apply the techniques of the biochemist to the problems of general microbial taxonomist. Although he is best remembered for his approach to taxonomy by the physiological processes and metabolic products of bacterial growth, Orla-Jensen did not in fact ignore the older criteria of taxonomy; he used morphology extensively, and also tinctorial reactions and source. He had little respect for orthodox names or a legalistic approach to their usage and priority; he could be described as a nomenclatural anarchist, richly deserving Buchanan's rebuke for 'many times substituting new names for older and apparently perfectly valid terms'. Like other reformers, his intentions were excellent; he tried to make generic names informative by adding suffixes such as *-coccus* and *-monas* to indicate shape or flagellation (when known) of the organisms; the stem of the name also was informative in many cases, as in *Denitrimonas*, a genus of active denitrifiers, and *Propionibacterium*, a genus of bacteria that produce propionic acid.

A major step forward in the classification of bacteria was taken by Buchanan himself in a series of papers published between 1916 and 1918 in the first of which he made the suggestion that the Society of American Bacteriology (SAB) Committee that was sitting at that time, should not only consider the classification of bacteria but also make a classification of the chemical changes brought about by bacteria. In retrospect, this suggestion was an attempt to combine the best of orthodoxy with Orla-Jensen's approach to classification. In one of these papers, he proposed the generic name *Pfeifferella* for the taxon that included the glanders bacillus as the type; later he found that this allocation was an overlooked clerical error, and that the name was intended for the genus which included Pfeiffer's influenza bacillus! But the name had been validly published and appeared to be legitimate; when Becker (1951) showed that it was a later homonym of a genus of protozoa, Buchanan himself was most active in having *Pfeifferella* Buchanan 1918 declared a *nomen genericum rejiciendum* (Opinion 14). The preliminary report of the SAB Committee (Winslow *et al.* 1917) was published in the same journal and in the middle of the time span (1916–1918) covered by Buchanan's papers. The Committee, of which Buchanan was a member, was undoubtedly influenced by this relatively young but strong-willed taxonomist who was to become universally known and respected, and who dominated bacterial taxonomy, and particularly nomenclature, for the next fifty years.

Whether the classification and nomenclature included in the final report of the SAB Committee (Winslow *et al.* 1920) is to be regarded as the work of that Committee or of Buchanan, it, too was an important landmark in bacterial taxonomy, and would be a worthy contender to replace Linnaeus's *Species Plantarum* of 1753 as a more realistic starting date for bacterial nomenclature. Although there are brief definitions of the genera included in the classification there is, in the Introduction to the Report, a statement to the effect that members of the Committee were in accord with the Botanical Society of America which 'urged that the application of generic names should be determined by type species rather than by attempts at generic characterization'. Thus was allowed to slip a splendid opportunity to publish a modern classification of bacteria accompanied by a series of adequate descriptions of the genera and the designation of their type species; only the type designations were made in the Report of SAB Committee.

A final comment on the older classifications is appropriate; later taxonomists have tried to identify as bacteria some of the organisms described in classifications of infusoria; the generic names *Monas* and *Vibrio* have the longest lives but it is extremely doubtful whether all the organisms given these names were bacteria. In my view it is a waste of time to try to find useful bacteriological information from observations made before bacteria were clearly distinguished from algae, filamentous fungi, and yeasts, and I believe that we shall not lose anything by ignoring all work before the pioneer work of Pasteur. After Pasteur, and particularly after plating techniques were improved so that there was a reasonable chance that pure cultures were obtained, bacteriological records and descriptions began to have a meaning, but for some groups at least, the era of pure cultures did not start until the 1920s, which reinforces the significance of the time of publication of the SAB Committee's final Report (1920).

Bergey's Manual

The publication of *Bergey's Manual of Determinative Bacteriology* in 1923 was a landmark in bacterial taxonomy not because it brought anything radically new (it represented a touched-up version of the scheme put forward by Winslow *et al.* in their 1917 and 1920 reports), but because it presented in readily accessible form a kind of directory of bacteria for the use of non-biologically minded scientists who found in bacteria useful sources of enzymes. It also provided universities and schools with a textbook that treated the classification of bacteria more seriously and at considerably greater length than the introductory chapter of a book on botany, in which bacteria were probably the most troublesome and difficult subjects. But as a book for use in a general diagnostic laboratory (which was its avowed aim) *Bergey's Manual* found little acceptance, and it will be worthwhile to spend a little time on the criticism levelled at the early editions.

The first and most voluble criticism was that *Bergey's Manual* introduced a new nomenclature which was resented, disliked, and, outside the US, regarded as an American product. Often referred to as the American nomenclature, it was the target of ill-feeling by nationalistically-inclined Europeans, and the generator of anti-Americanism among otherwise pacific bacteriologists; in short, it engendered the heat and ill-temper that a new or controversial nomenclature always does in conservatively-minded Europeans. The criticism was, of course, absurd, unjust, and hasty, and should have rebounded on its makers; the nomenclature was mainly ferreted out of the old literature by Buchanan and, as bacteria were regarded as plants, applied according the the rules of the Botanical Code of Nomenclature. The critics, on the other hand, adopted what they regarded as a simpler nomenclature – and so it was, for all rod-shaped bacteria had the generic name *Bacillus* (though a few used *Bacterium* for some ill-defined groups) – but as a nomenclature it was neither descriptive, nor was it a system, and it was sterile. The nomenclature of the first *Bergey's Manual* followed Buchanan's (1916–1918) proposal generally but the intestinal bacteria of *Bacterium* were subdivided on some of the lines, but not all, suggested by the Europeans Castellani & Chalmers (1919) and introduced names which often aroused even more criticism than the 'American names' of *Bergey's Manual.*

The second criticism of *Bergey's Manual*, justified even up to the 7th (1957) edition, was that as a means for the identification of unknown bacteria it was impracticable, that the keys were badly constructed and introduced bizarre characters or qualities that were unlikely to be tested or even be testable; in short, that it was almost useless in a diagnostic laboratory. The keys were wide open to criticism on other counts; many of them were of the kind 'character *x* not known', or the character depended on the demonstration of a negative state (if that is possible, such as a lack of pathogenicity for a particular species of plant or animal); sometimes source entered into the key, at other times the key hedged as in 'acid from sucrose not reported'. In particular, the keys to the higher categories were so drawn up that they formed an almost impenetrable barrier to the recognition of a taxon in a category above genus. It was accepted by most users that it was essential to know enough bacteriology to place an unknown bacterium in its genus before consulting *Bergey's Manual*, and then (and not always then) the Manual might be of help in identifying a culture. [In fairness, and for those who do not know, the above criticism of the effectiveness of *Bergey's Manual* identification keys refers to those in the text of the Manual. Both the 7th (1957) and the current 8th (1974) editions (but not, strangely, the Shorter 8th edition, 1977) included as a supplement 'A key for the determination of the generic position of organisms listed in the Manual', by V. B. D. Skerman, which keys go a long way towards compensating for the inadequacies of keys in the manual itself. LRH.]

These are severe criticisms of a manual that has as its titular purpose the provision of aid in the identification of bacteria; it is criticism that has decreased with succeeding editions, and it is the hope of the editors and contributors to the 8th edition that it is no longer justified. The criticism, or some of it, can be answered in the statement that the early editions of the book were compilations from the literature; little of it was written from first-hand experience with the organisms; successive editions have seen the introduction of more and more contributors from specialized fields who are familiar with the species about which they have written. Moreover, the editors have become more selective and have deleted descriptions that are so vague that the organisms would not be recognized if they were isolated and studied by modern methods.

A

a-, an-. Without, lacking, unable to do something or produce a specified product. Prefix attached to the name of a character, as achromogenic, aflagellate, asporogenous, anaerogenic.

-a. 1. Suffix added to a personal name ending in *-e*, *-i*, *-o*, *-u*, *-y*, *-er* to form a new generic name (bact.,[1] bot.[2]).

2. A personal name ending in *-a* can be used to form a generic name by adding to it *-ea* (BAC.,[1] BOT.[2]) as *Silvaea* from Silva or *-ia* (ZOO.[3]) as *Silvaia*.

3. A personal name ending in *-a* can be used to form a specific epithet (name); the addition to the name differs with the discipline and also with the sex of the person whose name is to be used. BAC.:[1] add *-i* or *-e* to the name of a man (e.g. *shigae* from Shiga) or *-e* to the name of a woman (e.g. *almeidae* from Almeida). BOT.:[4] add *-e* to the name of a man or woman. ZOO.:[5] add *-i* to the name of a man (e.g. *brodai* from Broda), or *-ae* to the name of a woman, after elision of the final *-a* for euphony[6] (e.g. *dattae* from Datta).

[1, BAC.Ap9B. 2, BOT.R73B. 3, ZOO.ApD37. 4, BOT.R73C. 5, ZOO.R31A. 6, ZOO.ApD18.]

abbreviations. Writing about taxonomy involves much repetition of scientific and other names, and authors resort to abbreviation of words that are often repeated. Some abbreviations are easily recognized for what they are, and their meanings are clear, as DNA (or ADN in French) for deoxyribonucleic acid, but others are simply shortened forms of words, as BAC., intended for use only in one paper or book.

1. *Generic names.* In nomenclature there is often a desire to shorten a generic name and the abbreviation adopted may be dictated by **editorial whim** or **journal style**. Editors of bact. journals generally accept use of the initial letter of the generic name and prefer this to use of the first three or four letters of the name (*Sal.* or *Salm.* for *Salmonella*) (Cowan, 1954, *Science*, **120**, 1103). Authors should never use an abbreviation at the first mention of a generic name and at subsequent mention use one only when they are satisfied that its use will not cause ambiguity; in all cases of doubt spell out generic names. The *CBE Style Manual* recommends abbreviation of generic names unless omission of the full name could lead to misunderstanding. BAC.[1] gives official approval to the use of initial letter abbreviation of a generic name.

2. *Qualifying phrases, indications.* Abbreviations of Latin words and phrases may be appended to the names of taxa and are usually printed in roman. The following list includes those used most frequently:

ab.	*aberratio*	aberration
auct. non	***auctorum non***	not (of) this author
comb. nov.	***combinatio nova***	new combination
emend.	***emendavit***	emended
excl. gen.	*excluso genere*, ***exclusis generibus***	excluding genera
excl. sp.	*exclusa specie, exclusis speciebus*	excluding species
f. sp., f. spp.	***forma specialis, formae speciales***	special form(s)

gen. n.*	***genus novum***	new genus
mut.	*mutatio*	mutation
mut. char.	***mutatis characteribus***	changed characters
nom. ambig.	***nomen ambiguum***	ambiguous name
nom. confus.	***nomen confusum***	confusing name
nom. cons.	***nomen conservandum***	conserved name
nom. nov.	***nomen novum***	new name
nom. nud.	***nomen nudum***	bare name
nom. rejic.	***nomen rejiciendum***	rejected name
p. p.	***pro parte***	in part
pro syn.	***pro synonymo***	as a synonym
sp. ep. cons.	***epitheta specifica conservanda***	conserved specific epithets
sp. ep. rejic.	***epitheta specifica rejicienda***	rejected specific epithets
sp. incert. sed.	***species incertae sedis***	species of uncertain taxonomic position
sp. inq.	***species inquirenda***	unidentified species
sp. n.*	***species nova***	new species
var. n.*	*varietas nova*	new variety

3. *Name of author*. BOT.[2] advises abbreviation of the names of authors when they are appended to scientific names, i.e. in author citations. Prefixes and nobiliary particles are deleted unless they are an inseparable part of the name; thus J. B. P. A. Monet Chevalier de Lamarck becomes Lam.

A monosyllabic name may be cut down to the first consonants; E. M. Fries becomes Fr. In polysyllabic names the first syllable and the first letter of the second are used (the first two when both are consonants); thus Richard becomes Rich. To distinguish two botanists with the same surname the first (given) names or initials may be used; Robert Brown becomes R. Br. and Alexander Brown becomes A. Br. Names of certain well-known botanists are given even greater abbreviation; Linnaeus becomes L. and de Candolle becomes DC.

Author citation is used less frequently in bact. and zoo. but when used the names of the author(s) should not be abbreviated.

4. *Journal names*. Titles of scientific publications are sometimes given in abbreviated form. At one time there was great variation in abbreviation; this was part of **journal style**, but greater uniformity is now apparent and most English-language biological journals use one of two systems: (i) *World List of Scientific Periodicals* updated in *New Periodical Titles*, published by Butterworth, or (ii) an International List based on American Standards Association Z 39.5–1969 and British Standard 4148: 1970 (O'Connor & Woodford, 1975). The systems differ considerably in the manner by which the abbreviations are formed: *World List* uses initial capitals only for nouns, adjectives are shown with lower case letters; there is often greater shortening of words in *World List* than in the International List. It is difficult to summarize the principles of abbreviation

* The Latin form (gen. n., etc.) is preferred by editors of botanical journals; the English form (n. gen., etc.) by editors of other kinds of journal (*CBE Manual*, 49).

used, and the original lists should be consulted; it is inexcusable to guess at the abbreviation.

A stimulus to uniformity is the increasing use of computers for storing and retrieving scientific information; the International List has been modified and extended so that it is now bespoke for computers. The history of efforts to standardize abbreviations for journals is well described by Williams (1968). See also **name–date system**.

Dr A. F. B. Standfast, an editor of considerable experience, questioned the value of abbreviating journal titles. His view was that both time and space factors affected the value of abbreviation. The amount of space saved is trivial and if the reader has to puzzle over the meaning of an abbreviation the object is defeated; clarity and time-saving are on the side of full journal titles; the *Journal of General Microbiology* abolished abbreviation of journal titles in 1970.

5. *Names of organizations.* Abbreviation of the names of scientific societies, research councils, committees, international organizations, culture collections, and government departments often take the form of a string of initials, sometimes made up into pronounceable words, as Unesco (see **acronym**), but in the majority the abbreviation is used in written work and in speech as a sequence of letters such as NCTC, ATCC, or CMI. A selected list of such initials that concern microbiologists, or may be found in biological literature follows:

AC– Advisory Committee of...
ASM American Society for Microbiology
ATCC American Type Culture Collection
CBS Centraalbureau voor Schimmelcultures
CCM-O Commonwealth Collections of Micro-organisms
CDC Communicable Disease Center; now Center for Disease Control
CMI Commonwealth Mycological Institute
COMCOF Committees, Commissions and Federations (of IAMS, see below)
EB– Executive Board of...
EBIAMS Executive Board of the International Association of Microbiological Societies
EC– Executive Committee of...
ECBO European Cell Biology Organization
EMBO European Molecular Biology Organization
FAO Food and Agriculture Organization
GIAM Global Impacts of Appplied Microbiology (series of conferences)
IAMS International Association of Microbiological Societies
IAPT International Association for Plant Taxonomy
IBPTN International Bureau for Plant Taxonomy and Nomenclature
ICCC International Culture Collection Conferences (series of conferences)
ICNB International Committee on the Nomenclature of Bacteria (now ICSB, see below)
ICNV International Committee on the Nomenclature of Viruses (now ICTV, see below)
ICRO International Cultural Research Organization

ICSB	International Committee on Systematic Bacteriology
ICSU	International Council of Scientific Unions
ICTV	International Committee on the Taxonomy of Viruses
ICZN	International Commission on Zoological Nomenclature
IMI	Imperial Mycological Institute (now CMI)
ITZN	International Trust for Zoological Nomenclature
IUBS	International Union of Biological Sciences
MSG	Microbial Systematics Group (of SGM)
NCDO	National Collection of Diary Organisms
NCIB	National Collection of Industrial Bacteria
NCMB	National Collection of Marine Bacteria
NCPPB	National Collection of Plant Pathogenic Bacteria
NCTC	National Collection of Type Cultures
NCYC	National Collection of Yeast Cultures
NIH	National Institutes of Health
SAB	Society of American Bacteriologists (before 1964); Society of Applied Bacteriology
SGM	Society for General Microbiology
SSZ	Society for Systematic Zoology
UKFCC	United Kingdom Federation for Culture Collections
UKNC	United Kingdom National Committee (of the Commonwealth Collections of Microorganisms)
UNEP	United Nations Environment Program
WDC	World Data Center (of WFCC, see below)
WFCC	World Federation for Culture Collections
WHO	World Health Organization
WIPO	World Intellectual Property Organization (amongst other activities, concerned with patents involving micro-organisms)

Note: For a full list of culture collection abbreviations see: Pridham, T. G. (1974), *Micro-organism Culture Collections: Acronyms and Abbreviations*, Agricultural Research Service, US Department of Agriculture, 41 pp. ARC–NC–17, June 1974.

6. *Methods, tests.* Initials or other short forms are often used for the names of tests and techniques; a few examples are:

CAMP test	Acronym from Christie, Atkins, Munch-Petersen (see *Aust. J. exp. Biol. med. Sci.* **22**, 197)
CF	complement fixation; clumping factor activity of staphylococci
FA	fluorescent antibody
HPO	high pressure oxygen = hyperbaric oxygen
IMViC	Indole, Methyl red, Voges–Proskauer, Citrate utilization
ONPG	Test for β-galactosidase
VP (V–P)	Voges–Proskauer test for acetylmethylcarbinol

[1, BAC.Ad. 2, BOT.R46A.]

aberrant form. An organism that deviates in one or more ways from the norm. A bacterium of undoubted identity may show one or more unusual characteristics, e.g. a strain of *Salmonella typhi* that is oxidase positive but otherwise is typical would be an aberrant form. Often the aberrant state will persist and it must be assumed that the normally stable genotype has undergone a gene change to produce a phenotype which we regard as aberrant. The change may be substantial and involve more than one gene. Aberrant forms may be produced artificially, as in the development of L forms by growth in the presence of antibiotic.

absence of a rule. When a rule applicable to a particular problem or difficulty cannot be found, established custom, where this can be ascertained, should be followed.[1] The Commissions or General Committee appropriate to the discipline may be asked to interpret rules and presumably would consider the desirability of a new rule if one were needed.[2]

[1, BOT.Pre; ZOO.80. 2, BAC.S8c; ZOO.78.]

absorb, absorption. Terms used in antigenic analysis to indicate removal of specific antibody by union with homologous antigen. We say that antibody is adsorbed to the specific antigen; when the antigen is particulate (as in a bacterial suspension) the complex combines with similar antigen–antibody complexes and eventually flocculates (agglutinates). An antiserum can, in this way, be freed from a specific antibody by adding a sufficient amount of the corresponding (homologous) antigen to the antiserum; after removal of the flocculated particles the antiserum is described as absorbed.

American usage differs, see *CBE Style Manual*, 1972, p. 206.

abstract. Précis or summary of the contents of a paper. ZOO.[1] recommends that authors should not publish a name in an abstract issued before the paper in which the description appears, and the advice is equally applicable to other disciplines. In some journals an abstract follows immediately after the title and by-line; in that event the name of a new taxon or combination should be spelt out without abbreviation of the generic name, but phrases that qualify the author citation may be shortened. In bact. a new name appearing in the abstract of a paper presented at a meeting is not regarded as being validly published.[2]

[1, ZOO.ApE23. 2, BAC.25b(1).]

abstracting journals. Serials devoted to the publication of abstracts of papers, e.g. *Biological Abstracts*. Some abstracts are written by the author of the original paper; a new name for a taxon should not be published for the first time in such an abstract.[1]

[1, BOT.R29A.]

acapsulate. A morphological state in which a capsule in absent. In some bacterial species loss of capsule is associated with loss of virulence or pathogenicity, and can be shown to be due to loss of a surface antigen (e.g. *Streptococcus pneumoniae*); in others (*Bacillus anthracis*) virulence is not necessarily dependent on capsule formation. The acapsulate state may lead to difficulty in identification.

acceptance for publication. The date on which a paper is received at an editorial office or of its acceptance for publication is often printed at the beginning or end of the paper. This information does not have any significance in determining the priority of a name, which is dependent on the **date of publication** of the journal.[1]

[1, BAC.26b; BOT.45; ZOO.10; 11; 21.]

acceptance of a name. Agreement on a name; recognition of a name. A name accepted by an individual is the one that he thinks is, and treats as, the correct name of a taxon; certainly an author who proposes a name should accept it in that sense.[1] But a name accepted and used by the majority of workers in a discipline is not necessarily the **correct (valid)** name of a taxon. A name that has such *general acceptance* could be the subject of an appeal to a Commission or Committee for conservation[2] so that stability of nomenclature (of that taxon) will be assured. While the matter is being considered by the Commission **existing usage** should be followed.[3]

An **accepted name** in zoo. is the equivalent of a **conserved name** in bact. and bot.

[1, BAC.28b; BOT.34. 2, ZOO.4(a). 3, ZOO.80.]

acceptance. Statements of a biological nature that are helpful and supply a need but which cannot be shown to be true; they are believed, therefore accepted, but cannot be proved. Originated by Woodger, acceptances were elaborated by Blackwelder who thought that they had taken a commanding position in discussions of what are called principles.

The five acceptances are: (1) that older (classical) taxonomy was based on morphology and type specimens; (2) that so-called biological species are more meaningful than ordinary species based on comparative data; (3) that the species is different in nature from other taxa, and is more important; (4) that a classification must be based on phylogeny and taxonomists should study the origins of their taxa; (5) that a natural classification can only be a phylogenetic one.

accepted name. 1. To have any standing in nomenclature a name must be accepted by an author when he proposes (publishes) it; both BAC.[1] and BOT.[2] have a rule, similarly worded, to deny valid publication to names not accepted by their authors, and ZOO.[3] says that a name proposed conditionally is not available.

It takes a good imagination to conjure up the fantasy of a serious worker – and taxonomists are serious people – proposing a name that he does not accept. Perhaps it is the word proposed (or proposal) that causes more difficulty than the telepathy needed for determining acceptance by the author, for names that have been suggested have, by other authors been interpreted as proposals, which they obviously were not. Such interpretations reinforce the stress put on being definite given in **propose, -al**.

In contrast to BOT.,[2] BAC.[1] will not recognize as validly published (i.e. will not accept) a name published with a question mark even if the name is accepted by the author.

2. The Zoological Commission may be asked to consider the application of rules to a particular name and, in the interests of stability, may exempt that name from the rules; such a name is accepted, and corresponds to a conserved name in bact. and bot., and is published in an Opinion.[4] It will be entered in an Official List.

[1, BAC.28b. 2, BOT.34. 3, ZOO.15. 4, ZOO.78(b); (f).]

acceptor. Term applied to a strain that is able to act as a recipient and incorporate DNA from another strain in a transformation experiment.

accession number. Number given to a strain (culture) when it is accepted by a microbial **culture collection**, sometimes called the catalogue number.

The name attached to a culture may change after more detailed examination of its characters, or with changing views on classification, but the accession number is never altered. It is more stable than any name, even when the name is formed in accordance with all the rules of nomenclature applicable at that time.

-aceae. Suffix used in bact.[1] and bot.,[2] and attached to the stem of a legitimate name of an included genus to form a family name. Although the Latin suffix is a plural adjective qualifying Plantae (bot.) or Procaryotae (bact.), in nomenclature the family name formed by it is taken as a plural noun. The singular, which is seldom, if ever, used, ends in *-ea*.

BAC.[1] states that the stem should be that of the type genus but there are exceptions such as Enterobacteriaceae, a conserved family name whose type genus was designated as *Escherichia* (Opinion 15). The genus *Enterobacter* Hormaeche & Edwards is neither the type genus nor the name-bearing stem (**basionym**).

[1, BAC.9. 2, BOT.18.]

acetobacter. 1. In lower case roman, a common name for bacteria able to produce acetic acid.

2. In italic with initial capital, a generic name.

achromogenic. Failure or inability to produce a pigment.

acid–base composition of DNA. See **base composition**.

acid-fast. Ability to resist decolorization by mineral acids. Normally applied to bacteria that have been stained by Ziehl–Neelsen's method, in which steaming carbol fuchsin is applied to stain organisms that are not readily stained by simple dyes. The acid decolorizer may be mixed with alcohol; this makes decolorization more speedy and clears up the background, but it does not add to the taxonomic value of the staining method.

acidophil(e), -ic, *n.* and *adj.* Organisms that prefer acid conditions (pH in the range 4–6) for growth. Moulds are acidophilic and often contaminate acid solutions. Of the bacteria, lactobacilli which grow in the pH range 4–6, are unusual; most bacteria prefer slightly alkaline conditions for growth.

aciduric. Acid-tolerant; capable of resisting acid.

acronym. A word made up of the first letters of several words, such as Unesco from United Nations Educational, Scientific and Cultural Organization. Unlike the mnemonic, which is a memory aid, the acronym is a form of shorthand and is intended to be a time-saver; when the meaning of the letters has to be worked out by the reader it can be a time-waster, and if acronyms are to be used, abstruse ones should be avoided.

In bact. IMViC is formed from the names of (1) a product tested for (indole), (2) a pH indicator (methyl red), (3) the name of one of the originators of a test (Voges & Proskauer) for acetylmethylcarbinol, and (4) a chemical substance (citrate) utilized by some bacteria; the lower case *i* is added to make a pronounceable word. Muschin & Bassett added a fifth test and substituted capital E (from Eijkman, whose test in modified form, bear his name) for the lower case *i*, to make IMVEC. See also **sigla**.

actinomycete. A general term applied to rods that are Gram positive and show true branching. The plural adds the letter *s* and is printed in roman; *Actinomycetes* has been used as a generic name by Krasilnikov.

Adansonian, Adansonism. Terms applied to a method in which equal weight is given to every character of an organism used in the construction of a classification. In microbiology, perhaps more than in the macrobiologies, a serious attempt is made to use all the ascertainable characters in arriving at a classification, but we should always bear in mind that the characters we find and use are those most easily determined and are only a part of the whole.

As a principle of classification equal weighting is generally attributed to Adanson, a French naturalist who lived 1727–1806, but Mayr thinks that we malign him by doing so, and he quotes Adanson as saying that we should ignore minute and superflous characters as useless burdens on the memory. See **weighting of characters**.

adapt. To make a modification favouring survival under new (perhaps disadvantageous) circumstances.

adaptation. 1. Used to describe the state of an organism that is modified to take advantage of (and sometimes to survive in) a new environment or circumstance; adaptation may be shown by the appearance of an induced enzyme system (cf. **mutation**).

2. The process of becoming adapted.

adaptive, *adj.* To indicate modification brought about by reacting to an environment. Formerly used in relation to enzymes but now replaced by **inducible**.

adaptive radiation. Zoo. term for a multiplicity of forms each adapted for a different mode of life.

adequate. Sufficient, enough to satisfy. In taxonomy most often used in relation to numbers and to descriptions; the questions asked are: what are adequate numbers? And what is an adequate description? These are probably answered best by critics who comment on the inadequacy of numbers (of specimens or characters) and the inadequacy of descriptions.

What number of strains is sufficient on which to base a species or other taxon of low rank? Taxonomists have seldom applied the wise medical tenet which says that one should never base a new disease on a single case, and it is possible to found a species on one strain or specimen. This may be permissible but it is seldom wise; it explains why, in such circumstances, serious workers hesitate to create species, and prefer to describe a 'group' to which an identifying number or vernacular name is attached.

How many characters should contribute to the making of a species (or other taxon) is a question asked most often of numerical taxonomists, and the answer must surely be 'as many as possible' and certainly not less than 40 or 50, though Sokal & Sneath (1963) thought that there was little to be gained by having more than 100.

Descriptions in classical taxonomy often rely on fewer characters, generally chosen because they have been found to have particular value in distinguishing the taxon from others of similar (but not identical) kind. Although taxonomists tend to be long-winded in their writings, their prolixity seldom extends to their descriptions; neither is it analytical or self-critical enough to ensure adequacy of the description.

adjective. In the formation of generic names earlier editions of BAC.[1] and BOT.[2] advised against using adjectives as nouns, but many names of this kind have been validly published, some before the first edition of the appropriate code, e.g. *Alcaligenes* Castellani & Chalmers 1919.

Adjectives may be used as specific epithets, and they should agree grammatically with the generic name. Buchanan, in his annotations to the 1958 edition of BAC. (pp. 36–7), lists eight different kinds of adjective used as epithets, ranging from the simple Latin (*Staphylococcus aureus*) to a participial adjective formed from a past participle (*Clostridium malenominatum* – the badly-named clostridium); but the best known in bact. are those made from place names (*Brucella melitensis*) or from names of people (*Lactobacillus pasteurianus*). Buchanan was ambivalent about ordinal adjectives (rejected by BAC.[3] and BOT.[4]) and accepted *Bacillus tertius* Henry.

ZOO.[5] does not comment on the use of adjectival forms for generic names; for species-group names those of geographical origin are preferred in adjectival form, as *ohioensis* from Ohio.

[1, BAC.(1958)R5a(4). 2, BOT.R20A(f). 3, BAC.52(2). 4, BOT.23. 5, ZOO.ApD22(a).]

adopt. An author is said to adopt a name when he uses an older legitimate name (epithet) to replace a name (epithet) that has been rejected. Provided that there are no other difficulties in using the epithet, he may adopt one previously used in an illegitimate combination;[1] it is then treated as a new epithet.

[1, BOT.72, 72N.]

adsorb, adsorption. Specific adsorption occurs in antigenic analysis and other serological manipulations when antigen and antibody combine (see **absorb, absorption**).

Non-specific adsorption takes place on bacterial filters and depends on the electric charges of the filter and the material being filtered.

Advisory Notes. BAC.[1] provides a chapter of miscellaneous technical suggestions intended to facilitate the putting into practice the theory expounded in the Rules. LRH.

[1, BAC.Ch.4.]

-ae. Suffix used to form a specific epithet[1] or species-group name[2] from the surname of a woman, as *gordonae* from Gordon, except when the woman's name ends in *-a*, when *-e* is added, as *dattae* from Datta.

[1, BAC.Ap9B; BOT.R73C. 2, ZOO.R31A; ApD18.]

aerobe, -ic. An organism requiring oxygen; will grow in air on the surface of a solid medium (without the addition of reducing substances to lower the oxidation–reduction protential); in liquid media strict aerobes need molecular oxygen.

aerogenic, *adj.* Describes an organism that produces detectable gas during the breakdown of carbohydrate.

aerotolerant. 1. Term applied to an organism that normally grows under anaerobic conditions but is also able to grow to a small extent in the presence of air.

2. Growth in air at normal atmospheric pressure.

affinity. 1. Numerical taxonomists use the term for overall similarity between the characters of an unknown and those of known taxa. To give some quantitative expression the term is qualified by the words absolute and relative.

Absolute affinity indicates that the characters of the unknown agree at a specified degree (or percentage similarity) with those of only one taxon, and if the percentage is high enough it suggests identity (perhaps at the rank of species). *Relative affinity* indicates that the characters of the unknown do not reach the degree of similarity regarded as identity with any of the taxa used in the comparison, but the lesser degree reached may suggest relationships at another level which, in terms of a hierarchy, will be of higher rank.

Both Haas and Mayr strongly disapprove of the use of affinity for similarity.

2. In the macrobiologies affinity is widely used for relationship, but here Haas thinks that it should be limited to genetic relationships.

agamospecies. Species of animal or plant that do not reproduce sexually; populations of uniparental organisms.

agglomerative vs **divisive strategies.** In numerical taxonomy, hierarchical structure can be obtained *either* by successive linking together of units into higher ranking units until all are comprised in one unit (the agglomerative, or 'building from below' strategy), *or* by successive division of a unit into smaller and smaller units (the divisive strategy). Agglomerative strategy is more frequently used in bact. numerical taxonomy and incorporates more easily the concepts of **polythetic** taxa than divisive strategies which have been more used in ecological contexts, or for deriving **monothetic** taxa. LRH.

agglutination. The visible clumping of bacteria sensitized by specific antibody; particles combine with others and form large masses which then fall to the bottom of the tube and form a deposit.

Bacterial suspensions used in agglutination tests are sometimes incorrectly described as agglutinating suspensions; they should be described as agglutinable.

aggregated species. Term used by Colman for taxometrically produced groups (of streptococci) which could be further subdivided, generally on serological grounds, into subunits or subspecies. Among the indifferent or greening (viridans-like) streptococci one aggregate contained representatives of four different Lancefield serological groups (*A*, *C*, *F*, and *G*). Aggregated species were not genetically isolated for DNA from *S. milleri*, *S sanguis*, or *S. viridans* could all be accepted in transformation experiments by *S. milleri*.

The term aggregate species as used by Colman is not equivalent to the botanical **species aggregate** except that each is a grouping created for convenience.

aims. Codes of nomenclature aim at providing a stable method of naming taxonomic groups, of avoiding and rejecting names that may cause error, that are ambiguous, or may *throw science into confusion.*[1]

[1, BOT. Pre.]

alcohol-fast. Resistant to decolorization by alcohol. Formerly thought to have some value in distinguishing between species of mycobacteria, but it is now recognized that alcohol-fastness is not a reliable distinguishing character.

-ales. In bact.[1] added to the stem of the name of type genus to form the name of an order. The singular, *-alis*, is not used.

In bot.[2] *-ales* in added to the stem of the name of a family to form the name of an order, but it is not essential for the ordinal name to be formed from the name of a family as the principles of priority and typification are not mandatory for bot. taxa above family.[3]

ZOO. does not deal with names of orders.

[1, BAC.9T. 2, BOT.(1975)17. 3, BOT.(1975)16.]

aliquot. One of several equal parts of a whole. A word borrowed from chemistry and much abused by biologists. An aliquot of a solution or suspension is one part of a fraction that divides the whole volume into equal parts without leaving a residuum; thus 25 ml is an aliquot of a litre but 12 ml is not.

Generally better to use the word sample rather than aliquot for a representative part of the whole.

alkalophil(e), -ic. Preferring alkaline conditions for growth (pH values greater than 7). Bacteria tend to be slightly alkalophilic, and some species (e.g. *Vibrio cholerae*) are extremely tolerant (alkaloduric) of high pH values (8–9).

allele. The heritable nature of a character is determined by a pair of genes each of which is an allele; when the two alleles are alike the character is reproduced without alteration (homozygous reproduction) but when they differ one is said to be *dominant* and the other *recessive*, and the character is not always reproduced in the progeny (heterozygous reproduction).

allocate. In identification, the placing of the specimen being identified into a taxon. LRH.

allochronic. Occurring at different time periods, e.g. contemporary and fossil specimens.

allopatric. Forms or species of a genus that do not occur in the same area. There may or may not be differences between the species, and the differences may be quantitative rather than qualitative. They may, in fact, be conspecific. A term not used in microbiology.

allospecies. Species that make up a **superspecies**. The proposer of this term, Amadon, also proposed that the name of the superspecies should be enclosed in square brackets and inserted between the generic name and the allospecific epithet, as *Accipiter [gentilis] meyerianus*. Not used in microbiology.

alphabetization. The arrangement of words, names, and patronymics by the alphabetical sequence of letters in the word or name. This is usually a simple matter but difficulty may arise with prefixes and nobiliary particles indicative of canonization and sainthood. Some of the problems are discussed under **personal name 1** and **3**, but the alphabetical list of reference authors' names needs comment here.

The first point to be made is that a universally accepted method of preparing such a list does not exist. O'Connor & Woodford in their European ELSE–Ciba Guide for Authors say first find out the journal style and then follow it. Their main emphasis is on being consistent, and if a journal style is not apparent the options are: (1) if they are known, use the forms of the countries in which the different authors live, or (2) adopt a system that is consistent in itself, e.g. all names beginning with Van or van would be listed under V, regardless of whether a capital or lower case V is used by the author. The *CBE Style Manual* describes American practice in detail and is particularly useful because it contains many examples, especially of foreign names that have not been Americanized. But again lack of uniformity is apparent, and some statements of journal style have to be qualified by 'usually', 'a few journals' or 'most of them'.

Probably the commonest question in English-speaking countries is how to alphabetize Mac, Mc, and M'. M. D. Anderson in *Book Indexing* (Cambridge University Press, 1971) recommends treating them as if each was spelt Mac, and the order of the next letters of the name determines the sequence in the alphabetical list; this gives McAnally, M'Andrew, Mackenzie. The MacLeod, McLeod, M'Leod sequence must be determined by the initials of the individuals; this usage is well known and understood as it is used in British telephone directories.

St in a surname such as St John-Brooks is treated as if spelt Saint John-Brooks and is listed under S; when it is part of a forename, as in R. St John Brooks, it should be listed as Brooks, R. St. J.

Knights take their places in a normal alphabetical sequence, but a baron or life peer is listed by the first part of his name, e.g. Lord Stamp of Shortlands would be listed as Stamp, Lord.

alpha classification. The earliest stage in the development of a classification, usually based on morphology. Turrill, who used this term, thought that a classification passed through a series of stages the earliest of which he called alpha; as techniques and knowledge developed the classification would be improbed towards the ideal (or omega) classification. However, that ultimate stage can never be reached while evolution continues.

alphanumeric, *adj.* Applied to a code composed of letters, numbers, or both letters and numbers.

alteration of characters. Changes in the diagnostic characters or circumscription of a taxon that do not exclude the type (strain or specimen) of a name do not require any change of the name or of the author citation.

If the change is considerable, its nature may be indicated by a **qualifying phrase** appended to the **author citation** (which should be to the author who made the change[1]); many phrases are listed in BOT.[2] When the alterations are great enough to change the nature of the taxon and the type is no longer included, the name of the redefined taxon must be changed, and the name retained for the part that now includes the type.

[1, BAC.35. 2, BOT.47, R47A.]

alternative names. 1. Two (or more) names proposed by an author (or group of authors) for the same taxon. After 1 January 1953 such names proposed simultaneously are not considered to be validly published in bot.,[1] unless they are names for the perfect and imperfect states of the same fungus;[2] but see **simultaneous publication of names**.

2. By the rules a genus can have only one correct name, but another name based on a different classification of the taxon may be better known. When an author wants to show this, he sometimes puts the better-known name in brackets between what he believes to be the correct generic name and the specific epithet, as *Achromobacter* (*Acinetobacter*) *anitratus*. This is misleading because it is the method used to indicate the name of a subgenus, and suggests that *Acinetobacter* is a subgenus of *Achromobacter*, which is not what the author intends; he intends to show it as an alternative generic name. ZOO.[3] advises against this usage; BAC. does not mention the possibility of this combination in these circumstances, presumably because the Judicial Commission, when it revised the code, did not think that any author would want to show an alternative name; but the need arises at times and the difficulty occurs in some manuscripts.

The usual taxonomic practice is to list the **synonyms** of the generic name, but as the list may be long, an author may not want to do this. A practical method that indicates a single alternative generic name is to equate it with the correct generic name and enclose it in brackets, as in *Achromobacter* (= *Acinetobacter*) *anitratus*. But if this method (which is quite unofficial) is adopted proof-reading must be vigilant.

[1, BOT.34. 2, BOT.59. 3, ZOO.R44A.]

amend, amendment, *v.* and *n.* The act and result of making an alteration, not necessarily to correct a fault or error (cf. **emend**). In the circumscription of a taxon the list of characters may be amended, and if this is indicated after the name of the organism the Latin form, ***emendavit,*** or its abbreviation, emend., is used. In such a case a change of name of the taxon is not needed unless the amendment or the amended circumscription no longer fits the type of the taxon (which would then be said to be excluded, taking the name with it), or the amended taxon is transferred to or united with another taxon of the same rank, or by a change in rank.[1]

[1, BAC.37b; BOT.51.]

amendment of rules (articles) of the codes. 1. *Bact.* A proposal for amendment must be made by the Judicial Commission, but any bacteriologist may initiate a proposal through the Editorial Secretary of the Commission.[1] If approved by ten or more Commissioners the proposal is submitted to ICSB, and if accepted by 70% of members voting it is approved and then reported to the next plenary meeting of ICSB.[2]

2. *Bot.* Amendment of BOT. can only be made by members attending a plenary session of an International Botanical Congress voting on a resolution moved by the Nomenclature Section of the Congress. Proposals for amendment are prepared by the General Committee, and the decisions taken on them are prepared for publication by the Editorial Committee. Between Congresses the work of the Section of Nomenclature is carried out, proposals received and reviewed, by the various committees of IAPT.[3]

3. *Zoo.* Amendment to ZOO. can only be made by a Zoological Congress acting on a recommendation from the Commission. This recommendation must first be approved by a nomenclature section of the Congress, and must go forward with its blessing.[4]

[1, BAC.1b. 2, BAC.S8c. 3, BOT.Div.III. 4, ZOO.87.]

American Type Culture Collection. See **ATCC**.

amorphous. Without shape or organized structure.

ampersand (&). See **and**.

amphimixis. Reproduction by a sexual process.

amphitrichate, -ous. Terms used to describe a single polar flagellum at each end of a rod-shaped bacterium.

anaerobe, -ic. Terms used to describe an organism that has the capacity to dispense with oxygen, which it cannot use. Usually an anaerobe is incapable of growing on the surface of a solid medium exposed to air, and indeed may perish on even a brief exposure to atmospheric oxygen. But, as defined above, an anaerobe is an organism that does not use oxygen even if it can grow in air, and the term 'facultative aerobe' should not be used.

Many aerobes are facultative anaerobes, i.e. they can grow in the absence of atmospheric oxygen, though maybe not as vigorously as in air.

anaerogenic, *adj*. Describes an organism that does not produce visible gas when it breaks down carbohydrate. Should not be used in connnexion with organisms that do not attack carbohydrate, for which the term is asaccharolytic.

anagenesis. Term used by some zoologists for the transformation of species; see **speciation**.

anagram. The rearrangement of the letters of a word to form another word. Anagrams may be used in the formation of scientific names and ZOO.[1] gives as an example *Milax* from Limax. Such an arbitrary arrangement of letters should be treated as an indeclinable noun.[2]

An anagram of a personal name may form the basis of a generic name; it is to be regarded as different from the original name.[3]

[1, ZOO.ApD41. 2, ZOO.ApD26. 3, BOT.R73B.]

analogy, analogous. Features that cannot be traced back to the same feature in a common ancestor of two or more organisms. Analogy is the antonym of **homology**.

ancestor. One who had gone before; in taxonomy a taxon from which others are thought to have descended. *Not* a higher rank in a hierarchical system.

ancestor, ancestral tree. Terms taken from human genealogy for part or the whole of a dendrogram or graphic representation of an evolutionary classification. This use presents a false analogy.

ancestral character. A character that does not appear to have changed from that seen in the ancestral form of the organism. Preferred by Mayr (1969, p. 213) to the term **primitive** which implies simplicity, for what evidence there is suggests that simplification is the result (not the start) of evolutionary processes.

and, et, ampersand (&). In references use of the word *and* (*et*) or the ampersand depends on **journal style**, and before a paper is prepared the author should look at a recent issue of the journal of his choice, and advise his secretary of the form to be followed. This is another small reason why a paper should not be written before the journal has been chosen, and is all part of the general advice to an author to choose his journal and write expressly for it.

BOT.[1] allows use of the ampersand in author citations; examples in BAC.[2] show both the word *and* and the ampersand. ZOO.[3] uses *et* in the French text and the ampersand in the English.

[1, BOT.R46B. 2, BAC.34cEx; Prov.A1Ex. 3, ZOO.51(c).]

annotation. Comment or explanatory note. All the codes are annotated to some extent and give examples to help explain the workings of the rules and recommendations. The 1958 printing of BAC. was particularly well annotated by R. E. Buchanan, though the title page did not give him the credit he deserved. The annotations were unofficial and were not put formally before or approved by the Nomenclature Committee, but members of the Judicial Committee received copies and were able to comment on the original draft. These annotations were some of Buchanan's finest work and not only explained but stimulated interest in the dull and legalistic code.

anonymous publication. In zoo. a name published anonymously after 1950 is not **available.**[1] When a name was published anonymously but its author was known, his name, when cited, should be enclosed in square brackets to indicate the original anonymity.[2]

The other codes do not have rules on anonymous publication, but it would be in the spirit of the codes to reject the names.

[1, ZOO.14. 2, ZOO.R51A.]

anoxybiont, -ic. Bacteria not able to use atmospheric oxygen for growth.

antibiogram. Pattern of sensitivity to various antibiotics; sometimes included in the characterization of a micro-organism. Information on the sensitivity of an organism to antibiotics that can be used therapeutically may be useful to the clinician but has little value in taxonomy; organisms isolated from treated patients are the survivors of any antibiotic used in treatment and may have developed resistance.

On the other hand, information on sensitivity or resistance to non-therapeutic substances such as lysozyme or the pteridine derivative, 0/129, can be used in identifying a bacterium. See also **resistogram.**

antibody is produced by cells of the reticulo-endothelial system in response to an antigenic stimulus; it is a modified form of globulin and is detected by mixing the serum of the immunized animal with the antigen; if homologous antibody is present the type of reaction seen will depend on the physical state of the antigen. When the antigen is particulate antibody will agglutinate the particles (often the bacterial bodies); when the antigen is in solution (e.g. capsular polysaccharide) the antigen–antibody complex may be soluble in excess of one or other reagent, but in certain *optimal proportions* it is seen as a precipitate; the specific antigen–antibody complex may also be shown by a non-specific indicator such as the sensitized cells of a complement fixation system.

Writing about antibodies sometimes causes difficulty for authors, as in an example seen in a manuscript: 'removed anti-4, 12 antibody from an anti-O serum'. The anti-. . . anti- difficulty can be avoided by deleting the first anti- to read 'removed 4, 12-antibody. . .' and the 'anti-O serum' can, with advantage, be changed to 'O-antiserum'.

anticipate, anticipation. A name proposed on the assumption that the taxon to which it is applied will eventually be accepted; such a name is not regarded as validly published and has no status in nomenclature.[1]

[1, BAC.28b(2); BOT.34.]

antigen. A substance which, when introduced parenterally into the animal body, stimulates the reticulo-endothelial system of the animal to produce antibodies. The antigen may be particulate – as a whole bacterium – or in solution.

The taxonomic interest in antigens is that they may add highly specific elements to the characterization of a micro-organism. Each bacterium is made up of many antigens, some of which are more effective than others in stimulating the production of antibodies, so that the antibody response cannot be taken as a quantitative index of the antigens inoculated. Motile bacteria have flagella (H) in addition to somatic (O) antigens, and these are enumerated in the antigenic analysis of an organism. Some bacteria have capsular (K) antigens which may mask the O antigens.

antigenic analysis of a bacterium is an immunological method of characterizing the physico-chemical nature of the organism, particularly the surface. It is carried out *in vitro* by testing the organism against specific antibodies, and *in vivo* by inoculating the organism into an animal and later testing the serum of the animal for antibodies. The first of these methods is the kind of simple test carried out to identify an organism by its antigenic structure; the second method detects the antigenicity (i.e. the ability to stimulate antibody production) of the antigen. Antigenic analysis must take account of the antigens of the different anatomical parts of the organism (e.g. the flagella and the somatic surface) and the different kinds of **variation** that are associated with changes in these surfaces. S→R variation is a change in the somatic (O) antigens, and phasic variation affects the H antigens of the flagella. A motile bacterium with H and O antigens may produce non-motile variants, which have only O antigens.

antigenic formula. A method of expressing the antigenic structure, or the surface antigens, of a bacterium. Used most frequently in describing the relations of the enteric bacteria, the antigens of which are shared widely and in different combinations.

In the Salmonella group the antigens are expressed in the following sequence: O antigens (originally by roman numerals but since 1955 by arabic numerals), these are followed by a colon [:]; flagellar antigens of the first phase (originally called the specific phase), colon or double ended arrow [↔], flagella antigens of the second (originally group) phase. The flagellar antigens may be expressed as lower case letters, by the letter z with an inferior number, as z_{23}, or by arabic numerals. The specificity of the antigens is chemical; the different O antigens are different polysaccharides.

Examples: *Salmonella paratyphi-A* *1*, 2, 12: a: –
S. paratyphi-B *1*, 4, [5], 12: b: 1, 2
S. georgia 6, 7: b: e, n, z_{15}

In Salmonella groups A, B, and D the O antigen 1 (which is associated with **converting phage**) is indicated in the antigenic formulae by underlining the figure 1 in manuscripts, and by printing it in italic as shown in the formulae for *S. paratyphi-A* and *S. paratyphi-B*.

The presence or absence of antigen 1 in Salmonella groups A, B, and D does not affect the naming of the serotype, but the presence of converting phages in other groups may cause the serotype name to be changed. Thus phage ϵ_{15} will convert antigens 3, 10 to 3, *15*, so that *S. anatum*, 3, 10: e, h: 1, 6 becomes 3, *15*: e, h: 1, 6 and the name is changed to *S. newington*. Note that the antigen *15*, but not the antigen 10, is printed in italic.

Apart from antigens lysed by converting phases, other antigens may be variable in their presence in a particular serotype; antigens of this kind are shown within square brackets, as in *S. stanleyville*, which is represented by the formula:

1, 4, [5], 12, *27*: z_4, z_{23}: [1, 2].

Antigenic formulae and the names applied to the hundreds of salmonella serotypes are not governed by BAC.,[1] but the code[2] would like the word serotype to be replaced by serovar.

[1, BAC.5d. 2, BAC.Ap10BT.]

apomixis. Reproduction partly or wholly avoiding the normal sexual processes. Not used much in bacteriology because sexual reproduction of bacteria is the exception rather than the rule. In mycology it is used for the development of unfertilized sexual cells. Species definitions among apomictic populations are necessarily based upon phenetic similarities.

apomorph. A character that has changed during evolution from the form seen in the ancestor; Mayr prefers the term character-state; a derived character.

a posteriori. Reasoning from effect to cause, i.e. inductively.

application of names. The use of names in nomenclature. The proper application of a name is determined by the **nomenclatural type.**[1]
[1, BOT.PreII; 7.]

application of rules (articles). The rules of nomenclature apply in bact.[1] to names of taxa within the ranks of class to subspecies (variety); in bot.[2] from division to form (but not *forma speciales*); and in zoo.[3] from family-group (i.e. superfamily) to species-group (subspecies).
[1, BAC.5b; d. 2, BOT.4; 4N. 3, ZOO.1; 35(a); 45(a).]

apposition. The grammatical use of this word is the placing of one word beside or parallel with another. Used in this way the expression *a substantive in apposition*[1] in connexion with a specific epithet means that the epithet is a noun following another noun (the generic name). Buchanan (in *Bergey's Manual*, 1957, 7 edn, 20) said that a noun in apposition was an explanatory noun and gave as an example *Actinomyces scabies* (the scurf or scab actinomyces); he added the comment that this method of naming was uncommon in bact.
[1, BAC.12c(2).]

appropriate name. The name of a taxon does not have an enhanced standing in nomenclature because it is appropriate; the only significant test of a name's status is its legitimacy (bact., bot.) or validity (zoo.) A name may not be rejected or replaced because it is inappropriate.
[1, BAC.55; BOT.62; ZOO.18(a).]

Approved Lists of Bacterial Names. With the revision of BAC. published late in 1975 an entirely new way of fixing the priority of names was introduced to nomenclature; the new rules will not be fully implemented until 1 January 1980, and only afterwards will it be possible to judge the success of the method in stabilizing nomenclature. Note 3 to Rule 28a says that after 1 January 1980 search for validly published names and descriptions will not be needed, and that *Approved Lists of Bacterial Names will form the foundation of a new bacterial nomenclature and taxonomy*. A summary of the new sections, with comments in square brackets, follow but authors are advised to consult the original, particularly Rules 24–32 and Provisional Rules A1 and A2 on the publication of names, and Rules 33–36 and Provisional Rules B1–B4 on the author citation of names.

Priority of publication will start from 1 January 1980, by which date the Approved Lists of Bacterial Names should have been published in ***IJSB***. The names in these Lists will have been validly published before 31 December 1977, been *assessed by the Judicial Commission with the assistance of taxonomic experts*, approved by **ICSB**, and published in *IJSB*.[1] [It is not clear how the 'assessment' will be made, or what guidelines will be given to the taxonomic experts. Possibly, but this is not stated, current usage will be taken into account, as well as adherence to the rules of nomenclature. Since objective synonyms (different names based on the same type) will be allowed in the Lists, the freedom of taxonomists to classify as they wish will be assured.]

Names validly published under BAC. (as revised) between 1 January 1978 and 1 January 1980 will be added to the Approved Lists, but afterwards no further additions will be made. All names on Approved Lists will be assumed to have been validly published on 1 January 1980, which will become the starting date for the nomenclature of bacteria.

Of names published before 1 January 1980 only those in an Approved List will have any further standing in nomenclature; other names will not be made **rejected names** (*nomina rejicienda*) but will be available for **reuse** in the naming of new taxa.

Each name on an Approved List will be given the author citation, a reference to an effectively published description and, whenever possible, information on the type and, for species and subspecies, the location and designation of the type strain in a culture collection.

The Approved Lists may contain objective synonyms [but apparently not subjective synonyms, though this is not stated] but appearance in a List will not confer official approval.

When two names (objective synonyms) on the Lists are competitive, priority shall be determined by the date of the original valid publication.

The Judicial Commission may remove a name from an Approved List and place it on a list of rejected names.[2]

[1, BAC.24aN1. 2, BAC.24c.]

a priori. Reasoning from cause to effect, i.e. deductively.

apud, *abbr.* ***ap.*** In the work of. May be used in **author citations** when a name is published by one author in a book or other work of another author, though it is more usual to connect the two names by the word **in**.

arbitrary formation of names. Scientific names may be formed capriciously, without regard to derivation, but they should be treated as Latin words. All codes of nomenclature accept arbitrarily-formed names,[1] but there are restrictions such as those forbidding the formation of a species-group names from words related by a conjunction or the inclusion of a sign that cannot be spelt out in Latin;[2] examples are *rudis planusque* and *?-album*.

Letters not found in classical Latin may be used,[3] linked letters (**ligatures**) should be separated,[4] **diacritic signs** are suppressed,[5] and **diaeresis** signs are removed; though tolerated by BAC. and BOT. this archaic sign should be avoided in scientific names. An arbitrary combination of letters used as a species-group name should be treated as an indeclinable noun.[6]

[1, BAC.10a; 12c; BOT.20; 23; ZOO.11(b). 2, ZOO.11(g)(iii). 3, ZOO.11(b)(i). 4, ZOO.ApE3. 5, ZOO.27. 6, ZOO.ApD26.]

archetype, -(ic)al. A theoretical or assumed ideal type; the ideal type of the **typologist**. Its use in taxonomy arises from the Platonic philosophy in which archetypal meant the idea present in the divine mind before the creation of the world and life.

arrange, arrangement. 1. The act of placing or displaying objects in some order. In making a classification we try to do this in a logical way so that units that resemble each other are close together, and those that are unlike (or dissimilar) are separated.

2. The result of **1** is an arrangement or system (e.g. a system of classification).

arrangement of references. References in the body of a paper are collected and listed at the end; they are usually arranged in one of two systems: (1) alphabetically by the names of authors (see **name–date system**), or (2) by numbers (*a*) in sequence of references in the paper, (*b*) in sequence of an alphabetical list of authors (see **literature citation 2**).

When writing a paper the only safe method to avoid errors from last-minute changes is to use the name–date method; this is just as, or even more, important when the numbered sequence is to be used in the final manuscript.

arrangement of references to the codes. In this Dictionary references to the codes are made so frequently that they are dealt with at the end of each entry. The system used is simple and, I hope, self-explanatory. Those who have difficulty in deciphering the code should see **miniref** or the key on p. *xii*.

arthrospore. 1. A fungal spore produced by the fragmentation of a hypha into several separate 'spores'.

2. Used by some older bacteriologists for rods that did not produce endospores. This usage is now rare and its meaning obscure.

Article. BOT. and ZOO. label their rules of nomenclature Articles, and these must be followed; they correspond to Rules in bact. Recommendations and Examples attached to Articles are of lesser importance, and when they conflict with the Article, it is the Article that should be given the greater weight (*Reg. Veget.* **56**). However, the advice given in BOT. Recommendation 24B is so much more sensible than that of the Note in the fourth paragraph of Art. 24 that the Note should be ignored.

-arum. Suffix added in zoo.[1] to form a species-group name from the personal name (singular) of women (plural); presumably it applies only when all the women have the same name. BOT.[2] uses the suffix *-iarum* to form a specific epithet from the name of women, but BAC. does not make any recommendations for forming scientific names or epithets from the personal names of several authors or workers. [1, ZOO.R31A. 2, BOT.(1975)R73C.]

asaccharolytic. Term used to describe organisms that do not attack carbohydrates, alcohols, or polysaccharides (collectively labelled 'sugars'). Should not be used for bacteria such as *Brucella* spp. that fail to show acid production when grown in carbohydrate-containing media, but which nevertheless attack the sugar.

ascospore. Tough-skinned spore (of fungus) which contains a haploid nucleus.

asexual. Not differentiated into sexual forms.

asexual reproduction. Division of the microbial cell without preliminary combination of sexual forms or cells.

asporogenous. A variant of a spore-forming bacterium in which the spore does not develop; this failure may arise from intrinsic (i.e. genetic) or extrinsic (unsuitable culture medium, lack of essential growth constituents) factors.

ATCC. Initials of the American Type Culture Collection, 12301 Parklawn Drive, Rockville, Maryland 20852, USA. This is a collection of bacteria, viruses, fungi, and yeasts, and covers a wide range of interests.

The ATCC, started in 1925 under its present title, is a direct descendant of the Bacteriological Collection of the American Museum of Natural History, founded by C.-E. A. Winslow. The first curator of the ATCC was G. H. Weaver, and its first location was Chicago.

atrichous, atrichate, *syn.* aflagellate. Adjectives denoting the absence of flagella.

attribute. 1. *n.* Character state, character, quality, feature.

2. *v.* To refer to the author of a name, to the discoverer of a method.

atypical. Not typical or characteristic of what it is supposed to represent; an abnormal form. An adjective not much used in taxonomy because of the special meaning of the words *type* and *typical* as 'of the type'; in that sense atypical would mean 'not of the type', which is next door to nonsense. Assuming that all strains of a species are born equal, atypical would apply to all strains except the one more equal than others that is designated the type.

The term 'atypical mycobacterium', to which justifiable objection has been taken, is meaningless when it is applied to an organism that has all the characters typical of the genus, but presumably lacks the special characters that distinguish the type strains of the different species of the genus. It seems probable that a user of the term is not able to identify and name the organism beyond the level of genus.

auctorum non, *abbr.* **auct. non.** Not of the (this) author; used in the author citaton of a **misidentification**. The abbreviated form is placed between the misapplied specific epithet and the name of the original author, and this is followed by the bibliographical reference to the misidentification.[1]

A more straightforward way of handling this situation is to use ***sensu*** before the name of the author who made the misidentification, followed by non and the name of the original author of the name.

[1, BOT.R50D.]

authentic strain. Vernacular term for a strain of a species obtained from the author who described the species or from another worker who has made a special study of the species, genus, or group. From this strain may be produced a culture in freeze-dried form, a sample of which can be returned to the original author or supplier of the strain who, after checking purity and characters, can authenticate the dried product.

An authentic strain or culture is without standing in nomenclature;[1] nevertheless it can be much more useful to the taxonomist (and especially the future taxonomist) than the type strain of a species kept by repeated subculture.

[1, BAC.19.]

author. 1. A writer; in taxonomy, the writer of an account of work, experimental, bibliographical, or philosophical. Papers on taxonomy are often written by men with years of experience at the bench who have, by the nature of their work, become armchair scientists and critics; the critical faculty needs to be applied to their own work as rigorously as to the work described in the literature cited. Probably more than in any other branch of microbiology, taxonomy makes great demands on its writers, who must present their descriptive work clearly and concisely, and discuss the implications of this subjective science with unusual objectivity.

Young workers may get some help in the preparation and arrangement of their papers in the entries headed **writing for publication** and **writing to be read**,

and every author will benefit from reading O'Connor & Woodford's (1975) *Writing Scientific Papers in English*.
2. A person who gives a new name to a taxon is the author of that name. In the process known as **author citation** the name of the author of a scientific name is printed in contrasting type after the scientific name.
3. New names or new combinations may be proposed in papers written by several authors; when there are more than two authors the name of the first author may be spelt in full and the names of the others indicated by *et al.* See **author citation 3**.

author citation. 1. The practice of indicating the authority (authorship) of a scientific name by printing in a different type the name of the author who proposed the name. Full citation requires that the author's name should be followed by the year in which the name (or transfer) was proposed and the page number in the main text (not to the summary or abstract if this precedes the text) on which the name was proposed.[1] The page reference is helpful to the author checking a reference in a long paper or chapter in a book; without it the year indicates only the priority of the name. Incorrect author citation does not affect the validity of publication of a new combination.[2]

Provisional rules appear in BAC. for dealing with author citation of names in an **Approved List**, and also for the **reuse** of names not in an Approved List.[3] These provisional rules will not become effective until 1980, and they may be changed in the light of experience.

Author citation is a purely nomenclatural device and refers only to the author of a name; it does not necessarily indicate the person who discovered or described the organism (taxon) though it may do so. As long as the **type** is included an alteration of the definition of a taxon does not require the citation of the name of the author who revised the definition. However, if in the revision the type is excluded, a new name must be given to the redefined taxon and the name of the revising author should be cited, while the original name is retained for the residue that contains the type.

ZOO. is ambivalent on author citation. Under the heading of 'authorship' the name of the author is said not to *form part of the name of a taxon and its citation is optional*,[4] but when the name of a genus (or a taxon of lower rank) is cited, the name of the author and the date should be cited at least once in each publication.[5]

Author citation of names of ranks above genus need not be made in zoo.[6] but is usual in bact. and bot. Author citation of a binomial (binomen) should follow the scientific name without any intervening punctuation. See also **name change**.
2. BAC. and ZOO. do not suggest that an author's name should be abbreviated but BOT.[7] encourages this practice unless the names are very short, as Lam. for Lamarck and DC. for De Candolle. When there are two authors their names may be joined either by the ampersand or the word and (*et*).[8]
3. In bact. the suggestion was made that when there were more than two authors, one or at most two of them should be designated as the author of a new name proposed by them. This suggestion seems undesirable because authors, like cabinet ministers, usually accept collective responsibility for the work of their colleagues. BAC.[9] recommends that with three or more authors the citation should be made either by giving the names of all the authors, or by the name of the first-named author followed by *et al.*; BOT.[8] also recommends the use of *et al.*

[1, BAC.33bN2. 2, BOT.33N2. 3, BAC.Prov.A1–B4. 4, ZOO.51(a). 5, ZOO,ApE10. 6, ZOO.ApE9. 7, BOT.R46A. 8, BOT.R46B. 9, BAC.AdB(1).]

authority, authority citation. The authority for a name is the author whose name appears in the **author citation**.

autoecious. Term used to describe rusts that complete their life cycle on more than one host (cf. **heteroecious**).

autograph. A document in the handwriting of the author. An autographed description when reproduced by some mechanical or graphic process, e.g. lithography, and distributed (offered for sale, exchanged, or given) is effectively published in bot.[1] if before 1 January 1953, but not in bact. or zoo.
[1, BOT.29.]

automatically established name. See **autonym**.

automatically typified name. The name of a taxon above the rank of family is automatically typified when the name is ultimately based on the name of an included genus.[1] One taxon of each rank must include the type genus, and the names of the ranks are formed by the addition of the appropriate suffix to the stem of the name of the type genus.[2] In bot. the principle of priority need not be applied to names of taxa above family; in bact.[3] priority applies to names up to the rank of order.
[1, BOT.(1975)10; 16. 2, BAC.21a. 3, BAC.23a]

automatic tautonymy. The name of a subdivision of a taxon above family that contains the type (and thus has an automatically typified name) must itself bear a name based on the same generic name, with the ending appropriately modified.[1] Example: Pseudomonadaceae (based on *Pseudomonas*); pseudomonadineae; Pseudomonadales.[2]
[1, BOT.(1975)16. 2, BAC.9T.]

automation has already had considerable impact on taxonomy, especially microbial. On the one hand, machines are used more and more to collect data, and computers used to process data, either for classification and identification purposes. This has increased widely the taxonomic work that is now feasible; see **man versus machine controversy**. LRH.

autonym. An automatically established name arises when a taxon is divided and the part that includes the type of the higher (divided) taxon is not named by the author.[1,2] Thus if *Bacillus cereus* is divided into subspecies, one of them (that containing the type strain) will automatically be named *B. cereus* subsp. *cereus*, even if the author did not designate it.

The place of autonyms in establishing the priority of a name, and the definition of the word *autonym* are to be reconsidered by the Editorial Committee responsible for BOT. (*Taxon*, 1976, **25**, 170).
[1, BAC.39; 40. 2. BOT.19; 22; 26.]

autotroph, -ic. An organism that obtains its energy from inorganic sources or from light (cf. **heterotroph**).

auxanography. A technique described by Beijerinck in 1889 by which essential nutritional requirements were detected. A plate of medium deficient in some essential growth factor was heavily inoculated with the test organism; addition of the deficient factor to a small area allowed growth to occur in that area. Pontecorvo called this *positive auxanography*. The method described by Heatley for the assay of penicillin in which diffused penicillin inhibited growth of the test organism was the reverse procedure, *negative auxanography*.

auxiliary character. In identification, a character not used in a key, but included at the end of a pathway as supplementary information to confirm that the identification is correct. LRH.

auxotroph. Organisms that are variants deficient in synthesizing ability and need growth factors or amino acids not required by the **wild-type**, or **prototroph**.

available. A terms used in zoo. for a name that satisfies Art. 11 of ZOO.[1] which requires, among other things, that it shall be in Latin or in Latin form, that the author uses binomial nomenclature in the publication that contains the new name, and that he abides by the conventions used by zoologists in forming names for taxa of different rank. After 1930 a name must be accompanied by a description or a reference to a previously published description or diagnosis, and a genus-group name must be fixed by the designation of its type-species.[2]

As defined above the term available is the equivalent of **legitimate** in bact. and bot.

A zoo. species-group name remains available even when it is combined with an invalid or unavailable genus-group name.[3]

[1, ZOO.10. 2, ZOO.13. 3, ZOO.17(3).]

axenic, *adj.* Applied to a pure culture; an axenic culture is a single species, growing in a medium free from other living organisms.

B

Baarn collection. A world-famous collection of fungi; associated with the yeast collection at Delft. The address is Centraalbureau voor Schimmelcultures, Oosterstraat 1, Baarn, Netherlands.

BAC. Abbreviation used in this dictionary for **bacteriological code**, the full title of which has changed several times since the original code was approved in 1947. Unless qualified the references are to the 1976 revision, published in 1975.[1]

For the phrases 'in bacteriology' and 'bacteriological usage' the abbreviation used is bact. or Bact.

[1, *Note.* Some confusion may arise between the title page statement *Bacteriological Code 1976 Revision* of the edition published in 1975 and Rule 1a which refers to this as the 1975 Revision. Apparently the discrepancy is to meet requirements of US copyright law.]

bacillus. 1. In lower case roman type means a rod-shaped bacterium. *Pl. -i.*

2. In italic with initial capital, a generic name for a group of spore-forming aerobic bacteria of which the nomenclatural type is *Bacillus subtilis*, and the type strain NCTC 3610, ATCC 6051.

bact. In roman (not italic) type, with or without a capital B, the abbreviation used in this dictionary for bacteriology; in bacteriology; bacteriological usage.

-bacter. Suffix used to form many generic names for bacteria. By an Official Opinion (no. 3)[1] of the Judicial Commission it is to be treated as of masculine gender.

[1, BAC.Ap5.]

bacterial nomenclature is regulated by the **Bacteriological Code**, in this dictionary referred to as BAC., and by no other code. Before the original BAC. was approved in 1947 (it was published in 1948) bact. nomenclature had followed BOT.; recent revision of that code has deleted references to bacteria.[1]
[1, BOT.(1975) PrIFo.]

bactericide, -cidal. Substance or quality that kills bacteria; distinguished from **bacteriostat**, a substance that inhibits multiplication but does not kill the bacteria. Note the differences in the spelling of the roots, bacteri- and bacterio-.

bacteriocin. Inhibitory substance produced by bacteria; a strain able to produce one of these antagonistic substances is bacteriocinogenic. The antagonistic substance produced by one bacterium may be effective against only some strains of a species, against all strains of that species, or it may inhibit the growth of strains of several species; when the antagonistic action forms a constant pattern the phenomenon clearly has taxonomic interest. Bacteriocins produced by *Escherichia coli* and other coliforms are known as **colicins**.

Chemically, bacteriocins are a heterogenous group of macromolecular substances, ranging from simple proteins to particles of high molecular weight which resemble phages. Bacteriocins can be transmitted from one cell to another by conjugation or by transduction.

Bacteriological Code. Short title of the *International Code of Nomenclature of Bacteria* approved at the First International Congress of Bacteriology, Jerusalem, 1973. The short title had been in use for many years as the common (and unofficial) name of the Code approved by the Fourth Congress for Microbiology in 1947, and was made official in 1953 at the Fifth Congress.

The Code includes Principles, Rules, Recommendations, Appendixes, and the Statutes of the International Committee on Systematic Bacteriology and of the Bacteriology Section of the International Association of Microbiological Societies. The Statutes describe the means by which the rules can be altered and the procedures for the conservation and rejection of names. Although the rules cannot be enforced, the observance of both Rules and Recommendations is a self-discipline which, if followed by all bacteriologists, would reduce nomenclatural difficulties. Such observance might be made a principle of a Code of Ethics for taxonomists such as is found in the Zoological Code.

In this dictionary abbreviated to BAC.

bacteriophage, (phage). A bacterial virus; a virus parasitic on bacteria. The action of a bacteriophage may be highly specific (i.e. acting on only one kind of bacterium) or may be broad; the specific phages are used for phage typing. Broad-spectrum phages may be made more specific and even adapted to bacteria they did not originally lyse. Phage particles multiply in the host (susceptible) cell, and are released when the cell wall bursts, the lysis of the cell. Temperate phages have the ability to reproduce and be transmitted in the non-susceptible cell of a **lysogenic** strain by a phage precursor or prophage.

bacteriostat, -static. A substance or quality that inhibits the multiplication of bacteria, but does not kill them.

bactériotron. See **taxotron**.

bacterium, *pl.* **-ia. 1.** A micro-organism with a rigid cell wall and an elastic cell membrane; often unicellular but may be divided by one or more septa. Usually multiply by binary fission but under certain circumstances a form of conjugation, with fusion of nuclear material, occurs between pairs of the organisms. Not dependent on chlorophyll.

2. A generic name of a group of bacteria; at one time limited to Gram-negative, non-sporeforming rods (coliforms) but debased to the category of a dump heap to include all non-sporeforming rods. To avoid continuation of such indiscriminating use the Nomenclature Committee approved an Official Opinion (no. 4, revised)[1] which declared *Bacterium* to be a rejected generic name, and also rejected the family name, Bacteriaceae, derived from it.

[1, BAC.Ap5.]

bar diagram. Histogram prepared from % **S** values, which are plotted on the ordinate and the organisms on the abscissa. The % **S** values are represented as solid columns; the dips between columns indicate gaps between different taxonomic groups. The shape of the histogram may resemble a series of skyscrapers; hence the alternative name of skyline diagram.

base composition of DNA. The proportions of the four bases, guanine (G), cytosine (C), thymine (T) and adenine (A) that occur in DNA. The base composition can be expressed in several ways, but because of the equivalence of G to C, and A to T (see **base pair**), it is usually expressed as a percentage of G+C to total bases (% GC, moles % G+C); less frequently as A+T to total bases.

Correlation of DNA base-composition and taxonomic relations was first suggested in 1956: bacteria with similar ratios may have the same sequences (and so form in-vitro molecular hybrids) or the same ratio could arise by chance. Thus, while similar base compositions may or may not indicate relatedness, differences in base ratios are positive evidence of lack of relationship. Note the synonymy of base-composition and base ratios. LRH.

base pair. Bases which pair in double-stranded DNA or RNA. The pairs are guanine (G)–cytosine (C) and adenine (A)–thymine (T, replaced by uridine, U, in RNA). In each pair, one base is on one strand of the nucleic acid, the other base on the other strand. This is the Watson–Crick model of nucleic acids. A common misconception is that it is the hydrogen bonding between base pairs that holds the double helix structure together; the major forces contributing to the stability of double-stranded nucleic acids are vertical stacking forces, while the specificity of the base pairing ensures the two strands are in register. LRH.

basic, basal. Fundamental, foundation. BOT.[1] and the older editions of BAC. name the species as the basic category or rank to which every organism belongs. Some species may be split into subspecies (varieties) or other rank, but subspecific ranks are not recognized in all species.

[1, BOT.2.]

basionym,[1] **basonym.**[2] The name-bearing or epithet-bearing synonym which occurs in a new combination; e.g. *Bacillus coli* Migula is the basonym of *Escherichia coli* (Migula) Castellani & Chalmers.[3] A new combination is not validly published unless the basionym is referred to, giving the author, publication and date.[1] Reference to *Index Kewensis*, the *Index of Fungi*, or any work other than the original publication would not be considered adequate.[4]

[1, BOT.33. 2, BAC.34a. 3, BAC.Ap5, Op.15. 4, BOT.33N1.]

batch culture. Term used by biochemically-inclined microbiologists for a culture growing in a volume of unchanged medium. Contrasts with a **continuous culture** growing in a **chemostat** in which fresh medium is slowly but continuously added, and medium is withdrawn at the same rate to maintain a constant volume and, hopefully, a constant concentration of nutrients.

battery. A collective noun applied to a series of similar things, e.g. a battery of tests, battery of features. An alternative and preferred term is 'set'.

BCCM-O. Initials of the British Collections of Micro-Organisms, later changed to Commonwealth Collections of Micro-Organisms.

Bergey, David Hendricks, 1860–1937. Founder of *Bergey's Manual of Determinative Bacteriology*, the first edition of which was published in 1923. Bergey was Professor of Bacteriology at the Medical School of the University of Pennsylvania. In 1915 he was President of the Society of American Bacteriologists, and it was the Society that encouraged him and at first sponsored the Manual. In 1936 Bergey set up a Trust to ensure continuation of his manual by applying royalties from each edition to the preparation of the succeeding edition.

Bergey's Manual of Determinative Bacteriology occupies a unique position in bacteriological literature and misconceptions about its status and acceptability outweigh recognition of its fallibility and its lack of official status. For the early editions Bergey was encouraged by the Society of American Bacteriologists (SAB) but *Bergey's Manual* was never officially recognized by the SAB. In 1936 a self-perpetuating, non-profit-making Trust was formed, the Trustees acting as an Editorial Board, and an increasing number of bacteriologists helped in the preparation of succeeding editions. Many of the earlier Trustees were men who had been founder members of the International Committee on Bacteriological Nomenclature, and it was often incorrectly assumed that *Bergey's Manual* had the blessing of the International Committee, and that the system of classification and the nomenclature used in it were official, or at least approved by the Committee. This misconception seems to have been especially prevalent in US, where editors of journals often insisted on use of the nomenclature in the current edition of *Bergey's Manual*, so that to Americans its nomenclature was almost sacrosanct. Unfortunately such ideas were not confined to the US, for editors of some journals in the UK accorded to *Bergey's Manual* a much more authoritative status then the Manual's Trustees ever claimed for it.

Contributions to later editions have been written by many experienced bacteriologists (not all of them taxonomists) who had ideas of their own, but the overall hierarchical scheme was the work of the editors, individualists rather than creators. In general the classificatory schemes adopted by the editors were developments of the arrangement of families and genera of bacteria recommended by an SAB-sponsored Committee (Winslow *et al.*, 1917, *J. Bact.* **2**, 505; 1920, *J. Bact.* **5**, 191), but successive editions have necessarily shown changes dictated by advances in knowledge and a better understanding of the physiology and metabolism of bacteria. Changes in classification were often accompanied by changes in nomenclature, and it was on nomenclature that critics outside the US focused their attention. It was unfortunate that they criticized this aspect of the Manual because before 1947 the nomenclature of bacteria was undisciplined as few bacteriologists knew or cared for the Botanical Code, the old (self-appointed) custodian of bacterial nomenclature.

The uncritical acceptance by biochemists of names used in *Bergey's Manual* sometimes had unfortunate consequences and resulted in the frequent appearance of incorrect names in scientific papers, and by their repetition they became familiar and therefore more readily accepted. A name popularized and accepted by inculcation is *Serratia*; this name is now accepted for a bacterium but when first used it was probably applied to a yeast (see **starting date for nomenclature**).

The length of this article will be justified if the following five points are recognized and accepted: (1) the classification used in *Bergey's Manual* is as subjective as any other classification, and consequently is no more permanent than any other scheme. (2) The nomenclature used is that which, *at the time of publication* seems, to the authors and editors, best to fit the requirements of BAC., but again has no greater permanency than the classification on which it is based. (3) *Bergey's Manual* is not divinely inspired; it is the effort of men and women whose work is fallible. (4) *Bergey's Manual* is not sponsored and approved by any international organization, and has no connexion with IAMS or its Committee on the Nomenclature of Bacteria (now the Committee on Systematic Bacteriology). (5) Any recognition given to the Manual by editors is not sought by the Trustees; it is merely an indication of editors' liking of standardization and their (correct) acceptance that *Bergey's Manual* is the most comprehensive book on systematic bacteriology.

better-known name. In comparing the claims of two or more names, the popularity of a name or evidence that it is well known or much used should not be considered in deciding which of the rivals is the **legitimate** or **correct name**. A legitimate name may not be rejected or replaced because another is better known.[1] But when a case is presented to one of the Commissions for the **conservation** of a name, a well-known name will be at an advantage against a name that is little known or has seldom been used since its first application.[2] Although it is often quoted, frequency of usage cannot be estimated by the number of times a name appears in an abstract journal; indeed, ZOO.[2] does not accept appearance of a name in an abstract journal as usage.

[1, BAC.55(4); BOT.62. 2, ZOO.79(b).]

bibliographic(al) error. An error in the citation of a paper or book, such as the wrong year, volume or page number; not a reference to the wrong paper which would be a misquotation.

BOT.[1] seems to encourage bibliographic laxity in the extraordinary statement that *Bibliographic errors of citation do not invalidate the publication of a new combination.* [We always considered that taxonomists in general and, with Buchanan in mind, nomenclators in particular, were punctilious and pedantic. How wrong we were!]

[1, BOT.33N2.]

bibliography. This grandiloquent title is seldom applicable to the often inadequate list of references appended to a scientific paper. It might be justified if the list contained all the references on the subject, but few lists are so exhaustive. It is better to be more truthful and to describe the appended list as the 'Literature cited' or simply 'References'.

binary. In numerical taxonomy means two state, e.g. a binary character is one that be expressed in one of two ways, as positive (+) or negative (−), or as present or not present.

binary combination. In bact.[1] nomenclature a binary combination consists of two words, the generic name and the specific epithet, as *Staphylococcus aureus*. The term binary combination is not used in zoo.[2] or bot.,[3] in which the equivalents are **binomen** and **binomial** respectively.
[1, BAC.12a. 2, ZOO.5. 3, BOT.23]

binary fission. Division of the microbial cell transversally into two usually equal parts. Fission is preceded by division of the nuclear material into two parts, indentation of the cell membrane, followed by indentation of the cell wall. Appendages such as flagella and fimbriae are represented in the daughter cells.

binary system. A system of naming biological units in which binomial nomenclature is used for species.

binom. The unit of **biosystematy** for a taxon which, for lack of knowledge, is ill-defined in respect of 'biosystematic' data but has been given a binomial. It can be clearly defined morphologically and equals the 'morphospecies' or basic taxonomic (as opposed to biosystematic) unit according to Grant and other biosystematists.

binomen. A name made up of two words. In zoo.[1] binomen is the combination of the name of the genus and the specific name, and it is the equivalent of **binary combination** and also **specific name** in bact. and bot.

The name of the genus is written with a capital letter but the second word associated with the species is written entirely in lower case letters, even when formed from the name of a person or place.[2]
[1, ZOO.5. 2, ZOO.28.]

binomial. From the Latin, *binomius*; used for a name of two words. In biology the words are Latin or latinized and are usually printed in italic. Binominal, derived from *binominis* (having two names) would be more strictly correct for this biological usage, but on a popularity vote binomial would win.

The first of the two words is printed with an initial capital and is the name of the genus or generic name; the second word is printed with an initial lower case letter; in bact.[1] and bot.[2] the second word is the specific epithet, in zoo. it is the specific name. In all cases the binomial is the name of the species. The lower case initial letter of a specific epithet or name applies even when the word is derived from the name of a person or place, but in bot. (and hence in mycology) authors who wish may use a capital for a **patronymic** or for a former generic name used as an epithet.[2]
[1, BAC.59. 2, BOT.R73F.]

binomial system. A system of biological nomenclature in which a species is named by two words. The system is generally attributed to Linnaeus (though it had been used earlier by others), a Swedish naturalist who developed classifications of plants and animals to the different species to which he gave two names, one generic and the other specific; together they formed the binomial. Binomials replaced the older descriptive phrases in Latin (polynominals) used in names by pre-Linnaean authors.

binominal is from the Latin *binominis*, meaning having two names. With the same meaning the word **binomial** has been in use for about 200 years and does not show signs of losing it popularity in favour of the etymologically more correct binominal.

biochemical profile. The description of a micro-organism based mainly on the results of various biochemical tests which detect the end-products of metabolism but do not give much (or any) information about the enzymes used or the metabolic pathway followed. The information may be suplemented by electrophoretic studies intended to give details of the enzymes and proteins which go to make up the molecular structure of the organism.

bioluminescent, *adj.* Applicable to photobacteria which generate and emit visible light in aerated cultures, i.e. as an end product of an oxidative process.

biomass. Volume of organisms. This volume is influenced by the nutritional substances in the medium and is not necessarily proportional to the number of microbial cells.

biometrics, biometry. Measurement or mathematical expression of living things; application of statistical methods of biology, particularly to variability, similarity, and differences. Statistical methods formed the basis of **exact systematics** of Smirnov in the 1920s, and continue in numerical taxonomy as developed by Sneath and others.

bionomics. A branch of biology dealing with the economy of living organisms, their relations to one another and to their surroundings. Synonymous with ecology.

bionumeric code. A numerical code in which different groups of digits represent different categories (ranks) of a hierarchy. Before being coded an organism must be characterized and allotted to its appropriate genus and species. The bionumeric code is a code of names; synonyms have different numbers in such a code.

Several such bionumeric codes have been proposed, and, as they are suitable for use with computers, some central authority is needed to designate the numerical assignment.

bioserotype, *abbr.* **bioser.** A biochemical variant (or variety) of a serotype, e.g. *Salmonella paratyphi-B* var. *java* is a bioserotype (differing mainly in that it attacks D-tartrate) of *S. paratyphi-B*.

biospecies. Biological species concept in which a species consists of interbreeding forms that are reproductively isolated from all other forms.

biosystematics. The part of systematics that concerns itself (and biosystematists) with variation, adaptation, and the influence of nuclear transfer which, briefly, is the study of the evolution of new forms.

biosystematy. An attempt by Camp & Gilly to get away from classical taxonomy, in which they tried to define natural biotic units and to label them in an informative manner. Founded in California in 1942, biosystematics has come to mean 'experimental' as opposed to 'classical' taxonomy.

biota. The biological population; all the living things (plant, animal, and microbial) of an area.

biotaxonomy. Classification of living things based on samples of the population. As it is impossible to examine all the specimens (individuals) that make up a population it is necessary to assume that the sampling procedure follows statistical principles in such things as numbers and randomness of sampling, and that statistical methods are applicable to the results of the sampling. Biological classification has been described by DuPraw as a *probabilistic or statistical, procedure*.

biotest reference strain. An adaptation of the nomenclatural type method to biochemical tests used in microbial characterization. The idea is based on the well-established principle of having positive and negative controls, and biotest reference strains are those selected to control the sensitivity of biochemical tests by running the tests simultaneously on the knowns and unknowns.

biotype. 1. In higher plants a group of individuals with the same genetic constitution, but in microbiology a subdivision of a taxon, usually of specific rank, but also applied to serotypes. The subdivisions are made on differences in biochemical reactions, or in the ability to produce acid from various carbohydrates and alcohols. Kauffmann avoids the word biotype but uses the more involved **sero-fermentative phage-type**, which apparently means a biotype of a serotype and/or phage type.

2. BAC.[1] recommends **biovar** for infrasubspecific forms based on physiological (biochemical) characters.

3. Mycologists use biotype for the subdivisions of physiological races, and like them the labels attached (which may be numbers) are not subject to BOT. (Ainsworth, 1973).

[1, BAC.Ap10B.]

biovar. Recommended by BAC.[1] as a substitute for biotype, and to avoid the suffix -type.

[1, BAC.Ap10B.]

biplicate, -ity. Terms used to describe the state of a bacterial cell that has flagella of two different wavelengths.

BOT. Abbreviation used in this dictionary for Botanical Code, or International Code of Botanical Nomenclature. References to BOT. without qualification are to the Seattle (1969) code, published 1972; BOT. (1975) refers to changes proposed (Stafleu & Voss, 1975, *Taxon*, **24**, 201–54) and accepted at the Leningrad Congress, 1975 (1976, *Taxon*, **25**, 169–74).

Bot. (with or without a capital B) is used for botany; in botany; and botanical.

BOT. regulates the nomenclature of recent and fossil plants, algae, fungi, and lichens, but *not* bacteria.[1]

[1, BOT.(1975) 13.]

brackets. 1. In nomenclature *round brackets* (**parentheses**) are used to enclose (1) the name of a subgenus[1] which is placed between the name of the genus and specific epithet (name); (2) the name of the author of an epithet originally proposed for the species as a member of another genus, the enclosed name appears immediately after the specific epithet.[2]

2. *Square brackets* are used (1) to indicate words or phrases inserted to clarify a quotation taken out of context; (2) in salmonella formulae to enclose antigenic factors not present in every strain of the serotype. (3) In zoo.[3] an author's name in square brackets indicates that the name was published anonymously, but the author's name became known later either directly or indirectly. (4) Also in zoo. square brackets may enclose the name of a superspecies placed between the

generic name and the **allospecies** epithet. (5) In synonymies a name in square brackets indicates a misidentification.
[1, BAC.10c; BAC.R21A; ZOO.6. 2, BAC.34b; BOT.49; ZOO.51(d). 3, ZOO.22A(3); R51A.]

branch. **1.** A category inserted by some zoologists between subkingdom and phylum.
2. An arm of an evolutionary tree, dendrogram, or phylogram.

Breed, Robert Stanley, 1877–1956. Dairy bacteriologist who was associated with Bergey in the production of the original *Bergey's Manual of Determinative Bacteriology*; after Bergey died in 1937 Breed became chief editor and was responsible for subsequent editions up to the 7th, which he got ready for the press.

Breed was not only widely read in bacteriological literature but he had a good working acquaintance with the bacteria about which he wrote. With St John-Brooks he was Joint Secretary of the International Committee on Bacteriological Nomenclature formed in 1930 and, with Buchanan and St John-Brooks, was a joint author of the first BAC.

Buchanan, Robert Earle, 1883–1973. American bacteriologist who devoted his life to bacterial taxonomy and to nomenclature in particular. He was a member of the Society of American Bacteriologists (SAB) committee on the classification of bacteria which, under the chairmanship of C.-E. A. Winslow, published the two reports (1917, *J. Bact.* **2**, 505; 1920, *J. Bact.* **5**, 191) that formed the foundation of the present-day classification of bacteria. For many years Buchanan was Professor of Bacteriology at Iowa State College, and he became Dean of Graduate Studies and Director of the Agriculture Experiment Station at Ames.

An original member of the International Committee on Bacteriological Nomenclature set up in 1930, Buchanan became the first Chairman of its Judicial Commission in 1939, and was largely responsible for producing the draft Bacteriological Code of Nomenclature discussed and approved at Copenhagen in 1947. For the 1953 revision of the code (published in 1958) he wrote the very useful annotations which, although unofficial, helped to clarify the pseudolegalistic jargon and made the rules intelligible to the average bacteriologist.

When the *International Bulletin of Bacteriological Nomenclature and Taxonomy* was authorized in 1950 Buchanan became its first editor. On the death of R. S. Breed, Buchanan became Chairman of the Trustees of *Bergey's Manual*, and was responsible for seeing the 7th edition through the press.

Buchanan believed that every action should be in accordance with the precepts of orthodox taxonomy and, when it dealt with nomenclature, should be in strict accordance with the rules of nomenclature laid down in the Code. Essentially he was a typologist as shown by his definition of species (see **Buchananism**). His compilation of *Index Bergeyana* was the result of his life-long collection of names used (up to about 1960) for different ranks of bacteria.

In 1966 Buchanan resigned the chairmanship of the Judicial Commission; his pioneer work in bacteriological nomenclature and his paternity of BAC. were recognized by his election to Life Membership of the Nomenclature Committee. He played an active part in the preparatory work leading up to the 8th edition of *Bergey's Manual* but his age and health prevented him from fulfilling his wish to supervise all the nomenclature used in that edition, and he died before the final manuscript went to press.

Buchananism. A term used to describe succinctly the essentially typological concept of species held by R. E. Buchanan. In 1955 he defined a bacterial species as *the type culture together with other cultures or strains of bacteria as are accepted by bacteriologists as sufficiently closely related.*

buoyant density of DNA. One of the parameters of DNA from which base composition can be estimated, hence the interest for taxonomists. Under ultracentrifugation conditions, heavy metals such as cesium will, despite being in solution, concentrate towards the bottom of the tube, setting up a gradient of concentration from top to bottom. If the original density of a cesium chloride solution is made approximately equal to that of DNA then when the gradient is equilibrated DNA in the solution will sink down from the less dense upper layers and float up from the more dense layers, and so band (concentrate) somewhere in the middle. The density of that position is called the buoyant density. The precise buoyant density of a DNA sample will depend on its **base composition**; the greater the GC content, the greater the buoyant density. LRH.

C

c. In determining the homonymy of zoo. species-group names[1] the letter *c* is regarded as the same as the letter *k*. BOT.[2] shows *Skytanthus* and *Scytanthus* as examples of orthographic variants that should be treated as homonyms.
[1, ZOO.58(3). 2, BOT.75.]

calculated mean organism (CMO). The nomenclatural type of a species may be chosen arbitrarily by the author of a name and the describer of a species, and he is not required to choose a strain with characters typical of the species. In an attempt to break away from this unpractical situation, Liston, Wiebe & Colwell (1963, *J. Bact.* **85**, 1061) devised a unit based on a simple assessment of the frequencies of characters (both positive and negative), and used the mean values (means) to produce a description of characters shared by the majority of strains of the taxon. This unit, which is called the Calculated Mean Organism (CMO), is a practical one and being based on the means is representative of the population that makes up the taxon.

The method of working out the CMO is shown in simplified form in Table C. 1, in which five characters of only five strains of this imaginary taxon are shown; in practice many more characters of many more strains would be determined and tabulated.

The CMO appeals to taxonomists on several counts; (1) the simplicity of the arithmetic to characterize the unit; (2) the ability to use it for characters (such as size or metabolite production) that can be expressed quantitatively; (3) a strain does not need to have all the positive characters of the CMO to be included in the taxon; and (4) none of the characters is **pathognomonic** or essential for inclusion in the taxon, which is **polythetic**.

Table C. 1. *Illustration of method for working out CMO*

Character	Strain 1	2	3	4	5	Number positive	% positive	CMO organism (read down)
A	+	+	−	+	+	4	80	+
B	+	+	+	−	−	3	60	+
C	−	−	−	+	−	1	20	−
D	−	+	+	+	+	4	80	+
E	+	−	+	+	+	4	80	+

Tsukamura's **hypothetical mean organism** (HMO) is a derivative of the CMO, but it has the disadvantage that only positive characters contribute to the HMO (leaving negative characters), and it cannot be used for characters expressed quantitatively.

Compare with **centrotype**.

Candida. Generic name of a yeast conserved against *Syringospora*, *Parendomyces*, *Parasaccharomyces*, and *Pseudomonilia*. Common name is Monilia.

capitalization of names and epithets. 1. Established custom requires that names of the rank of subgenus and above shall have initial capitals, and that specific epithets (specific names in zoo.) and names of lower ranks shall *not* be capitalized, even when the epithet (name) is derived from the proper name of a person or from a geographical name (see **decapitalize**). The only authorities that can be quoted are BAC.[1] (high rank to species), BOT.[2] (subgenus to infraspecific taxa), and ZOO.[3] (family-group to species-group), which leaves the botanist (and mycologist) without official guidance on the capitalization of names of genera and higher ranks.

2. Proper nouns used as adjectives generally have capitals in English and lower case letters in US, as Gram, Petri, and gram, petri. A few English journals follow American practice which, in the absence of a rule, is subject to **editorial whim**. [1, BAC.7; 10a; 59. 2, BOT.21; R73F. 3, ZOO.28.]

capneic. Term used to describe organisms that grow or grow better under increased CO_2 tension.

capsule, capsulation. The surface of many micro-organisms consists of a layer (often mucopolysaccharide but occasionally protein) which can be shown by certain staining methods or can be inferred from a halo-like appearance in so-called negative staining (e.g. india ink) preparations. An immunological method of showing the capsule is applicable to certain bacteria (e.g. pneumococci) in which the capsule appears to swell when the homologous antiserum is added (Quellung reaction).

Capsulation is often associated with the organism in its most natural form, and among pathogens is the form in which bacteria (that have capsules) appear in the animal body. Thus the capsulation of pneumococci is associated with virulence or pathogenicity, and acapsulate forms are avirulent and non-pathogenic. Loss of capsule by pneumococci, when the organism loses its surface (capsular) antigen, is part of the S $\rightarrow$ R change.

carboxyphilic. Liking or favoured by carbon dioxide. Describes an organism whose growth is made possible or is improved by an increase in the concentration of CO_2 in the atmosphere.

catalase. An enzyme that breaks down hydrogen peroxide to water and oxygen. Catalase is produced by some (but not all) bacteria grown on heated blood- or haematin-containing media. Its action is inhibited by azide and/or cyanide; it is not sensitive to acid (cf. **pseudocatalase**).

Because catalase is not produced by all bacteria it has distinguishing value and occupies an important place in some identification schemes, but its value as an *overriding classification feature* has been questioned.

catalogue. In biology a list of genera and species in a collection or museum, together with literature references to original descriptions and such other information as is pertinent or available, including synonyms, type species of genera, habitat.

In BAC.,[1] publication of a name in a culture collection catalogue does not constitute valid publication. LRH.
[1, BAC.25b(2).]

catalogue information. A term used for the sort of information one would expect to find in a culture collection catalogue. This would include details of the isolation and identification of a micro-organism, the accession number in different culture collections, and the names given to it by the author and perhaps a different name under which it is kept in the culture collection.

catalogue number. In connexion with a culture deposited in a **culture collection**, a serial number given to each culture accepted; this number is permanently attached to the culture and does not change with a change in name or taxonomic status of the organism. A strain deposited in more than one culture collection will, of course, have more than one catalogue number; usually distinguished by the initials of the collection, e.g. *Bacillus subtilis* Marburg strain, NCTC 3610; ATCC 6051 (see **abbreviation 5**).

Catalogue number is commoner in bact. than in mycology in which the synonym, **accession number**, is more often used.

catalyst. A substance which, while not itself being used in a chemical reaction, speeds up the reaction.

catchword. The heading or word printed in boldface to lead the reader to a definition or article in a dictionary or to an entry in an index. When the lead into a dictionary consists of several words it may be known as a headphrase.

category. **1.** The equivalent of taxonomic rank in a hierarchical system. The genus or species is a category, but the word should be distinguished from **taxon**, which is a taxonomic group. Thus *Bacillus subtilis* is a taxon in the category of species. If category is to be distinguished from **rank** (as some may wish), rank can be considered as a level or shelf, and the category as *all* the units (taxa) of that level or on that shelf.

2. A common but undesirable usage is as the equivalent of kind, or even of taxon Mayr approves only usage **1** above, but confesses that formerly he misused the word; Cowan *et al.* misused it in sense **2**.

CBS. Initials (and so may precede accession numbers) of the **Centraalbureau voor Schimmelcultures**.

cell. In biology the cell is a structural unit made up of a wall surrounding cytoplasm which contains nuclear material. Fungi are multicellular but bacteria and yeasts are normally regarded as unicellular, though Bisset describes the cells of some bacterial species as multicellular.

Because of its many different meanings the word cell was for many years anathema to the editors of the *Journal of General Microbiology*, but they fought a losing battle and general usage among biologists won the day.

cell wall. The bacterial cell is surrounded by a rigid wall which imparts a shape to the organism; when the cell wall is destroyed the cell assumes a spherical shape (**sphaeroplast**). The cell wall plays some, but not the whole, part in the **Gram reaction**. The wall of many Gram-positive bacteria can be dissolved by **lysozyme**.

The chemical nature of the cell wall has been thoroughly investigated and the presence or absence of different amino sugars and diaminopimelic acid has considerable taxonomic value.

Centraalbureau voor Schimmelcultures, *abbr.* **CBS.** A collection of fungi and yeasts organized in two divisions: (1) Fungi at Centraalbureau voor Schimmelcultures, Oosterstraat 1, Baarn, Netherlands; and (2) Yeasts at Centraalbureau voor Schimmelcultures, Delft, Netherlands.

CBS was founded in 1906 by the Association Internationale des Botanistes, under the care of Professor F. A. F. C. Went of Utrecht. In the following year Professor Johanna Westerdijk of Baarn became Director; the yeast division was removed to Delft in 1922.

Centre de Collection de Types Microbiens, *abbr.* **CCTM.** An information centre about cultures available from collections and specialist laboratories throughout the world. Founded by Paul Hauduroy of Paris, with the backing of Académie Suisse des Sciences Médicales, at Lausanne, the Centre had an ambitious programme to compile and keep up to date a catalogue containing exact and complete descriptions of microbes available for distribution, and to make it easier for scientists to obtain cultures from other laboratories. Supported by Ralph St John-Brooks, at that time Secretary General of the International Association of Microbiologists (later IAMS), the Centre seemed to get off to a good start, but when Brooks left Lausanne the activities of the Centre began to languish. (Meanwhile BCCM-O had been established in 1948 and soon produced a series of directories of collections within the British Commonwealth, and lists of the cultures maintained by many of them. Hauduroy wished to translate these lists and directories into French but permission was refused.) CCTM remained something of a white elephant until, after Hauduroy's death, its activities ceased for a time.

centrotype. In numerical taxonomy, that **OTU** closest to or at the geometrical centre of its cluster. LRH.

change of name. All the codes of nomenclature agree that the name of a taxon should not be changed merely because it is inappropriate; BAC.[1] and BOT.[2] go further and state that a legitimate name must not be changed because it is disagreeable, or because another is preferable or better known, or because the name has lost any meaning it may have had. However, there are reasons why some names should be changed and these are discussed under **rejection of names**. ZOO. says little about changes of names, but a new name or the adoption of a synonym to replace a homonym is called a **replacement name.**[3]

BAC.[4] and BOT.[5] deal with the problems that arise when taxa are transferred or changed in rank, or when taxa are remodelled, divided, or united; these are the only occasions when a change of name is justifiable.

Most argument in microbiology centres around the choice, from among many synonyms, of the name to be used in everyday practice. One cause of the argument is that many old names were coined and used before the taxa to which they were applied had been adequately defined; thus there is often a doubt about the rightness or wrongness (as distinct from the nomenclatural validity or legality) of associating some names with the taxa to which they are now attached. Where there is doubt there will be different usage and, judging from the writings of different authors, changes of name will seem more frequent than they are in fact.

To avoid and reduce name changes different solutions have been proposed: the first is **conservation** of a name and designation of a **neotype culture** of the micro-organism, which is then deposited in a culture collection and is available for study and comparison with cultures of the same and different taxa. A second is an extension of this, namely, the **registration** of new names of taxa by the appropriate nomenclature committee or commision, together with a detailed **protologue**. Official registration would avoid the publication of duplicates (homonyms) of names already in use, and insistence on a fully documented protologue would avoid the creation of ill-defined taxa. Although only suggested for new names, registration could be applied to older names and, to some extent, this will be seen in the **Approved Lists of Bacterial Names** which become effective in 1980.

[1, BAC.55. 2, BOT.62. 3, ZOO.60. 4, BAC.37–50. 5, BOT.51–7.]

character. An object is made up of many characters, and each contributes something to the whole. Among the characters of a micro-organism some are morphological (shape and size; presence or absence of flagella, spore, or capsule), others are physiological (require or abhor oxygen; nutritional requirements; production of enzymes), chemical or serological (chemistry of cell wall or antigenic formula), or pathological (virulence for different animals or pathogenicity for plants).

In the description of an organism as many characters as possible should be listed, for it is only by making extensive comparisons that realistic taxonomic groups (taxa) can be recognized. The importance to be attached to characters is the subject of much argument; there are at least two schools of thought, and many shades of opinion. What is called the **Adansonian** school believes that all characters used should be given equal weight in constructing a classification; the anti-Adansonians disagree and believe that selective weighting is essential to good classification. Among bacteriologists Kauffmann is a leader of those who give great weight to one set of characters (antigens) and little to others (those based on biochemical tests). Cowan is Adansonian in classification but applies selective weighting for identification, and believes that by so doing he gets the best of both schools.

The elegant variation beloved of writers of English has produced many synonyms of the word character, and taxonomists run to attribute, feature, property, quality, and test without making any great distinction between them. In addition the word may be qualified as in taxonomic character, and may itself qualify other words, and these will be dealt with below.

A **'good' character** is defined by Heywood as one that is not subject to wide variation in the sample being studied, is not easily modified by environmental factors, and has such a genetic base that it is unlikely to change readily. Ruth Gordon would say that these are the characteristics of stock cultures of bacteria. I would add an essentially subjective point to this definition, namely that a good character should be easily determined, for however good in theory a character may be, it is of little value if it cannot be seen or measured.

A **diagnostic character** is one that is (almost) peculiar to a taxon, so that its presence clinches the identification; a **pathognomonic character**.

A *confirmatory* or **auxiliary character** is one that is not diagnostic in itself but, with others of a similar nature, corroborates an identification.

character coding. See **characterizing code**.

character cost. The difficulty or complexity of character determination. The **base composition** of bacteria determination is a procedure justified in research and in descriptions for taxonomic work but its complexity does not justify its cost in identification work unless this is of a very special nature.

character couplet. Paired statements, conflicting or contrasting, used in the preparation of dichotomous **keys**.

character list. Check list of characters, usually made for a particular purpose, e.g. a diagnostic table for which the list would form the contents of the character column.

character nesting. See **linked character**.

character set. The characters used in a table, punched card set, or other identification device.

character set minimization. Term used by numerical taxonomists for the equivalent of characters for a **minidefinition**; the fewest characters essential for making an identification.

character space. Numerical taxonomists regard each organism as occupying a point in space, and the greater the affinity between two organisms the less is the space between them. This **space** has as many dimensions as the number of characters that are being compared.

character state. The condition of a character at the time of observation. A character may have only two states (e.g. motility of a bacterium), positive or negative. It may be multistate in which the organism shows one of several possible states, qualities, or abilities, e.g. in gelatin medium a bacterium may show **stratiform**, **saccate**, or other type of liquefaction, or none at all.

characteristic. 1, *n.* A distinguishing feature.

2, *adj.* Having qualities typical of the subject. Because of the special implications of the word type in nomenclature the word characteristic is often preferable to typical.

characteristic root. In numerical taxonomy, a technical term for a particular number computed during **principal component analysis** (q.v.) and similar methods. The characteristic root indicates the proportion of the total variance accounted for by the corresponding principal component. Also called eigenvalue, or latent root. LRH.

characterization. The determination and listing of the characters of an organism to form a **description**. Because of the absence of standardized methods, the tests and techniques used in different laboratories do not always give comparable results, or within the same laboratory reproducible results, so that the description of the same organism carried out in different places or at different times is not constant.

Although it is obvious that the characterization should be made with a pure culture of the organism, not enough attention is always given to making sure that the culture used as inoculum is pure. The commonest cause of impure cultures is the assumption that a single plating is sufficient to purify a culture, and that a discrete colony arises from one microbial cell; this seldom is so, and on inhibitory or selective medium should never be assumed.

Characterization of a micro-organism involves a description of the characters that can be seen (morphology) or tested (enzyme systems). Various methods are used to show the potential and actual enzymic make-up, ranging from simple tests in growing cultures to **micromethods**, and the use of paper disks impregnated with substrate. Furthermore, tests of the nutritional requirements of an organism may, taxonomically, be most satisfying and informative.

characterizing code. A code, usually of numbers but may consist of letters, letters and numbers, or signs (+, −, etc.) by which the characters of an organism can be described. One well known to bacteriologists is the IMViC code made up by Parr, which records indole production, the methyl red reaction (red = +, yellow = −), the Voges–Proskauer reaction, and ability to grow in a medium containing citrate as sole carbon source, thus expressing in four signs + + − − what has taken about two lines of type.

More elaborate kinds of characterizing code, for which the name **Chester code** is suggested, have longer numbers. Grouping the characters (suggested by Gage & Phelps in 1903) had advantages by reducing the number of digits required to express the characters. The history of the SAB (Society of American Bacteriologists) characterizing or descriptive codes was given in *Manual of Microbiological Methods* (1957) and the subject was reviewed in Cowan (1965*a*).

chemolithotroph, -ic bacteria are able to use CO_2 as their only source of carbon and obtain energy from inorganic compounds. Most chemolithotrophs are unable to use organic compounds and are, therefore, obligate; hydrogen oxidizers are exceptions in that none is obligate.

chemolithotrophy. The nutritional requirements in respect of mineral salts and gases, which may be determined by methods developed by den Dooren de Jong. Stanier found that studies of this kind have taxonomic value, e.g. in distinguishing strains of *Pseudomonas* spp. from those that require H_2 (*Hydrogenomonas*).

chemo(-)organotroph, -ic, *n.* and *adj.* Micro-organisms that obtain their energy by the oxidation of organic material, which may serve as the main source of carbon. These are newer and more descriptive words for **heterotroph, -ic**.

chemostat. Apparatus in which bacteria can be grown continuously, or until such time as the culture becomes contaminated, an all-too-frequent occurrence.

Additions of medium are made automatically to the liquid culture at a rate previously found to maintain the supply of nutrients; at the same time some of the culture fluid is withdrawn so that (1) the bacterial population does not build up to the limiting concentration, and (2) the volume is kept constant. The culture medium may be aerated and different gases may be bubbled through it.

chemotoxonomy. The application of taxonomy of information obtained by the techniques of the chemist; the chemical nature of the structure and function of organisms applied to their taxonomy. Use of the word seems to be limited to the application of chemical knowledge of plants and animals to their classification; microbiologists do not use the term but refer simply to biochemistry or, more fashionably, to molecular biology.

Without using the word chemotaxonomy, botanists have used chemical characterization from time immemorial in the use of scent and colour of flowers as taxonomic characters, and Bate-Smith is confident that olfactory characters will find a place in chemotaxonomy. Bacteriologists, too, know that some bacteria (*Staphylococcus aureus, Proteus* spp., Ballerup–Bethesda group of *Citrobacter*) produce characteristic odours but so far have failed to include these subjective features in characterizations. H_2S is a product of growth of many clostridia and this makes for the unpopularity of these anaerobes as research tools.

Bacteriologists rely little on their olfactory sense and chemotaxonomy to them means biochemical study of bacterial nutrition, biosynthetic pathways, the nature of metabolites, and the chemical nature of cell walls and other anatomical features of the bacterium.

chemotype. Term used to describe a taxon in which the lipopolysaccharides of the O antigens are made up of the same (sugar) constituents.

Chester code. Name suggested for a numerical descriptive code of the kind proposed by F. D. Chester when chairman of the SAB Committee on Methods for the Identification of Bacterial Species which published its report containing the code in 1905. Bergey & Bates in 1906 used the term numerical classification for this numerical descriptive code, but to resurrect it now would cause confusion with the numerical taxonomy of today.

Chester's kind of code needs a name to distinguish it from **bionumeric** codes which record only the placement of an organism in a hierarchical system.

It was on the Chester code that the obsolete SAB Descriptive Card was based. By it *B. coli* (*Escherichia coli*) was coded as 212.11110, and *B. enteritidis* (*Salmonella enteritidis*) as 212.13310.

chromogenic. Able to produce colour or pigment under suitable conditions. The suitable conditions include temperature, light, supply of certain nutritional substances (e.g. mannitol), and presence or absence of oxygen.

chronistics. Time relations.

cilium, *pl.* **-a.** Hair-like processes on the surface of cells. Found in protozoa, for the motility of which they are probably responsible, but not found in bacteria or fungi.

circular reasoning. Reasoning in which one point is dependent on another which, in turn, is based on the first point. In taxonomy this term may relate to so-called phylogenetic classifications, the correlations of which are said, *a posteriori*, to be phylogenetic relationships.

circumscribe, *v.* To define the boundaries, indicate the extent and the limits of a taxon.

circumscription, *n.* Statement defining the limits of a taxon and, by implication, showing how it differs from similar taxa.

citation. Indication of the source of or authority for a name or statement. In taxonomy used in two main senses: **1**, **author citation** in which the first author to publish a name is indicated, and **2**, a literature reference to an earlier paper or description.

With the introduction of **Approved Lists** in bact. a new form of citation, the citation of a name, will be used after 1980, and these forms are described below.

citation of an author, citation of authority for a name. See **author citation.**

citation of a name. 1. A new species-group name should be *cited in full*,[1] which means that the generic name with which it is combined must not be abbreviated. **2.** BAC. uses the phrase *citation of a name* in connexion with the procedure to be followed after 1 January 1980 in quoting the authority for a name on an **Approved List.**[2] Three methods of citation are indicated: (1) the name of the original author, the date of publication, followed by 'Approved List no. . . . , 1980' in parentheses, as in *Bacillus cereus* Frankland & Frankland 1888 (Approved List no. 1, 1980); *Bacillus subtilis* (Ehrenberg 1835) Cohn 1872 (Approved List no. 1, 1980). (2) Without citation of the original author the citation will be 'Approved List no. . . . , 1980' in parentheses, as *Bacillus cereus* (Approved List no. 1, 1980). (3) Without specification of the list in which the name occurs 'nom. approb.' (*nomen approbatum*) is added as in *Bacillus subtilis* nom. approb.
[1, ZOO.ApE8. 2, BAC.Prov.A1.]

citation of a reused name. In bact. names not in an **Approved List** may be reused after 1 January 1980; these are to be regarded as unpublished and treated as new names.[1] Author citation is to follow the usual procedure and the author proposing the reuse of the name (after 1980) is to be cited as the author of the name.[2] When an author wants to show that the reused name is applied to a different taxon the normal author citation will be followed by 'non' (or not), the name of the original author, and the year of publication (pre-1980) of the name.[3]
[1, BAC.24aN1. 2, BAC.Prov.B1. 3, BAC.Prov.B.4.]

citation of a revived name. 1. When a bact. name not on an Approved List is to be used (revived) for what is believed to be the same taxon this may be shown by adding 'nom. rev.' (*nomen revictum*) to the indicator of rank, as in *Bacillus palustris* sp. nov. nom. rev.[1] [The absence of a comma between the two indicators is probably an orthographic error in BAC.]
2. When the author of a revived name attributes the name to the original author he does so by adding to the name the word ex, the name of the original author and the date of publication in parentheses, and follows these by nom. rev., as in *Bacillus palustris* (ex Sickles & Shaw 1934) nom. rev.[2]
[1, BAC.Prov.B2. 2, BAC.Prov.B3.]

citation of date. The date of publication follows an author's name without punctuation (bact. and bot.). In zoo., where author citation is optional,[1] a comma should be inserted between the name of the author and the date of publication.[2]
[1, ZOO.51(a). 2, ZOO.22.]

citation of the name of a transferred species. 1. When a species is transferred from one genus to another the name of the author of the specific epithet (name) is enclosed in parentheses and followed by the name of the author who made the transfer; this is the form usually taken by the author citation, but BAC.[1] requires for *full citation* that the year of publication and the page number in the main text on which the name or combination was proposed.

2. A species whose name is in a bact. **Approved List** may be transferred to another genus, and the name will be changed to form a new combination. The proposal to make the transfer is indicated by adding to the new combination the abbreviation comb. nov. followed in parentheses by the name given in the Approved List, and the citation to the appropriate list. Subsequent authors using the new combination would give the authority as the name of the new combination (name of original author and date) name of author who made the transfer (name in Approved List, dash, Approved List no. . . . , 1980).[2] Hypothetical examples should help to clarify this description of a complex procedure.

Assume that *Bacillus chinensis* Smith 1812 is in Approved List no. 1066; Jones in 1999 wants to transfer the species to the genus *Clostridium*; he makes his proposal and cites the new combination as *Clostridium chinensis* (Smith 1812) comb. nov. (*Bacillus chinensis* – Approved List no. 1066, 1980). Later authors would cite this as *Clostridium chinensis* (Smith 1812) Jones 1999 (*Bacillus chinensis* – Approved List no. 1066, 1980).

There is a note to the rule which indicates that the correct form of citation will be found in the Approved List itself.

[1, BAC.33bN.2. 2, BAC.Prov.A2.]

clade. A phyletic branch; an evolutionary branch or line.

cladism. Mayr's fourth basic theory of taxonomy, in which rank depends on the position of branching points on the phylogenetic tree. It is not applicable to micro-organisms.

cladistic indicates the degree of relatedness as shown by the pathways or phyletic lines by which taxa are linked. As we do not know anything about phyletic lines, or convergence, among microbes the term is inapplicable to microbial taxonomy.

cladogenesis. Term used by zoologists for the multiplication of species at any one time; a form of **speciation**.

cladogram. Dendrogram showing the (most often speculative) pathways of evolution. Generally, cladograms should have a time-scale, which may be real time (e.g. millions of years or epochs) or relative time (e.g. numbers of generations). LRH.

class (and **subclass).** Ranks in bact. and bot. between **division** and **order**; names of these ranks are not regulated by ZOO., but class is mentioned in Appendix E6.

The names given to the taxa of class and subclass are formed in different ways; BOT.[1] gives the following endings: for algae, *-phyceae* (class) and *-phycidae* (subclass); for fungi, *-mycetes* and *-mycetidae*; for cormophyta, *-opsida* and *-idae*. BAC.[2] states that names of taxa above the rank of order (i.e. class and subclass) should be based on a combination of characters of the taxon, or on a single character of outstanding importance.

[1, BOT.R16A. 2, BAC.8; R6.]

classification may be defined as the orderly arrangement of individuals into units composed of 'likes', each unit being homogeneous but different from every other unit.

For some obscure reason classification is often confused with identification. Van Niel thinks that while bacterial classification has developed mainly along determinative lines (i.e. with identification in mind), an unfounded phylogenetic implication has been assumed, and he would like to see determinative keys kept separate from classifications. Mayr points out that many new classifications are really new schemes for identification in which the key characters are changed. He also stresses that a classification based on a few key characters is bound to break down; in his view a classification is a scientific theory and he would deny the artistry of the classifier.

Every taxonomist has his own ideas on which organisms are alike and which are different, on what qualities make differences and what can be ignored; in brief, classification is subjective. Most microbial classifications use morphological characters for the primary breakdown. Orla-Jensen was against this, and preferred to use characters that could be expressed in biochemical terms. Pribram tried to correlate pathogenicity with taxonomic position, in other words to use pathogenicity for the initial subdivision. Ravin divides the microbial universe into populations between which gene flow is impossible; within each population the individuals are potentially capable of interbreeding by genetic recombination when the base compositions are similar but cannot do so when they are very different. Ravin's views are widely accepted and must form the basis for any modern classification of bacteria. Wharton expressed a simpler view, namely that classification was arranging animals in hierarchies.

For any group of objects there are several ways in which they can be classified, and there is not one ideal classification to be aimed at, but many different classifications, each with a different objective. The first step to be taken is to define the reason for making the classification; is it to be a general classification or a specialized one? General classifications are often falsely assumed to show a truer picture of the relations between the classes than do those made to satisfy a particular purpose or need. General classifications are likely to be **polythetic**, specialized ones **monothetic**. A general classification should avoid giving undue weight to particular characters and should be based on what have become known as **Adansonian** principles. In making a classification the taxonomist must avoid preconceived ideas based on previous experience or even on an earlier classification, and he should try to approach the problem objectively.

So-called natural classifications may be based on quite different concepts and have different objectives, so that it is common for authors to write 'natural (in Gilmour's sense)' when they refer to a classification based on resemblances and overall similarities. Other workers think of a natural classification as one based on phylogeny; since most phylogeny is unproven, most of these classifications are really made on other grounds and the phylogeny is an *a posteriori* assumption. Sneath describes a natural classification as **idealistic typology**.

classify, *v.* The act of arranging objects into groups of similar objects, each of which can be taken as a unit to which other objects can be referred. Sometimes loosely (and wrongly) used to mean identify.

cline. A gradation in a measurable character or characters in a population. A technical term introduced by Julian Huxley to replace the more general character-gradient; intended to draw attention to variation within groups.

clone. The descendants of a single microbial cell; in higher plants the progeny of the single individual as a result of *asexual* reproduction. BAC.[1] describes it as a population of all the cells derived from a single parent cell.
[1, BAC.Ap10A.]

clumper. Descriptive term (sometimes **lumper**) for a taxonomist whose philosophy demands as few divisions as possible, and whose actions lead to the fusion or clumping of taxa of the same rank. A good example of clumping is the combination of a large number of species of *Pseudomonas* into a smaller number of well-described species proposed by Stainer and his colleagues in 1966. The opposite of (c)lumper is **splitter**.

cluster. 1. Group of similar strains produced when characterizations are expressed graphically or better, in models (theoretically in multidimensional space); the clusters are usually spherical (hyperspherical).
2. Term used for a group of similar organisms not allocated to a taxon of a particular category or rank; it was introduced to avoid the non-specific use of the word group, to which exception was taken in earlier editions of BAC. A cluster then became an operational unit. Although originally intended as a temporary designation for a unit that would eventually be placed on a shelf in a hierarchy, it found more permanent favour in the eyes of those who did not wish to use categories in a hierarchical system, and who preferred to treat clusters as **OTU**s of more or less equal status.
3, *v.* To gather into groups. A verb to be deplored.

cluster analysis. Mathematical treatment of character comparisons used in identification of taxonomic groups from **similarity values** (q.v.). The comparisons are complex because when several strains are under examination the individual characters of each are compared with those of all the other strains (**multivariate**). Cluster analysis is essentially associated with complex comparisons that require mechanical aid.

CMI. Commonwealth Mycological Institute, Ferry Lane, Kew, Surrey TW9 3AF, which houses a good collection of fungi, excluding those that are pathogenic to man and animals.

CMO. See **calculated mean organism.**

coccus, *pl.* **-i.** A spherical-shaped bacterium.

code. 1. In nomenclature, a set of rules of laws that govern nomenclatural practice.
2. The expression of characters in a standardized form so that the information can be stored, analysed, and retrieved when wanted. There are various methods of coding and many different reasons for making codes. Characters may be coded (qualitatively by +, −, or by a series of numbers to give quantitative expression to them) for numerical taxonomy. Characters can also be expressed sequentially to characterize an organism for classification, or as an alternative to a system of nomenclature. In one such system *Escherichia coli* is coded as 212.11110 and *Salmonella enteritidis* as 212.13310. An appropriate name for such a code, and to distinguish it from other numerical codes (see **3** below) would be **Chester code,**

after F. D. Chester, chairman of the SAB Committee which pioneered an elaborate descriptive code (1905, *Science* **21**, 485). Codes based on groups of characters have been developed; these are even more complex but one code of this kind was fully described by Cowan (1965*a*).

3. Numerical expression of the position of an organism in a hierarchical classification is a **bionumeric code**; this is a code of names, not of characters.

4. Numerical taxonomy uses (1) two-symbol codes in which characters can be expressed in one of two states, as + (or 1) or − (0), but a third symbol **NC** is also used, and (2) multiple-symbol codes in which several symbols are available, and these are mutually exclusive.

5. Automatic coding systems ('languages') for computers have been developed, e.g. Fortran and Algol; these are written in what have been called macro-assembly language.

coefficient of alienation. Loss of information when data are transferred from a statistical matrix to a dendrogram. Where there is no loss the coefficient is 1, a lesser figure indicates a loss.

coefficient of association. The computation of resemblance or similarity by finding the proportion of characters that are shared by two specimens. Many coefficients of association have been proposed, and they can be divided into two main groups which differ in the use or omission of **negative matches**, i.e. when the character is scored as negative for both organisms of the comparison. Most coefficients exclude characters recorded as **NC**, but this is illogical, for in any comparison of two strains the chances are equal that the NC character state will be the same.

coefficient of variability, *abbr.* **V.** A measure of relative variability, calculated by dividing the standard deviation by the mean; sometimes multiplied by 100 and expressed as a percentage.

coenospecies. Species or group of species potentially capable of hybridization.

cognomen. Nickname or surname. Microbiologists often use a form of laboratory shorthand to label by simple names or numbers strains with which they work every day. Such a cognomen may be the name of a patient, a collection accession number, or an apparently meaningless combination of letters and figures (e.g. PW8 which means Park Williams 8). This form of shorthand should not be used without amplification in papers intended for publication. Also called **tag**.

Cohn, Ferdinand, 1828–1898. German botanist who did much to establish bacteriology as a science and, in particular, to arrange bacteria into genera and species. His genera (or form genera) were based on morphological differences and he regarded them as natural; his species, based on smaller differences, he thought were much less well founded and he regarded them as provisional. Unlike his predecessors, Cohn thought that bacteria should be considered to be plants rather than animals.

colicin(e). Antibacterial substance produced by some bacteria (usually coliforms) active against other bacteria; each colicin has its distinctive characteristics in the antibacterial spectrum it presents, so that they can be put to taxonomic purpose in two ways: (1) by the colicins produced, and (2) by the sensitivity of an organism to a set of colicins. Unlike **bacteriophages** the colicins do not reproduce in susceptible cells.

The word colicine is of French origin, hence the terminal *-e*; the English deletion of the *-e* brings the spelling into line with **bacteriocin**.

colicinogenic, *adj*. Able to produce colicin.

coliform. A general term for fermentative Gram-negative rods that inhabit the intestinal tract of man and other animals. The term has not been strictly defined; some would include in it all the enteric bacteria, others would limit its use to lactose-fermenting enteric bacteria, and exclude the lactose non-fermenters. It should not be applied to **pseudomonads** isolated from the intestinal tract.

Coliform bacteria have a special importance to water bacteriologists who have developed a classification intended to pick out those of sanitary significance. The characters used in this classification are few and are expressed by the acronym IMViC.

collective group. A zoo. term for a collection of animals which can be divided into identifiable species but whose generic position is uncertain. ZOO.[1] does not require a type species for such a group, but treats the name of the whole group as if it were a generic name.
[1, ZOO.42(c); Gl.]

collective species (*species collectiva*). A bot. term for a collection of species that are individually recognizable but are closely similar. The term, and the group, do not have any nomenclatural standing, and are equivalent to **species aggregate.**

colony. A word that, in microbiology, should have only one meaning, namely the growth arising on the surface or in a solidified medium from the implantation of one or a small group of organisms. Unfortunately it has been used (mainly by biochemists) for the growth developing from an inoculum of any size spread out to form a giant colony (macrocolony) of confluent growth. In the first (and bacteriological) sense a discrete colony is assumed to have grown from a single microbial cell or, when cells stick to each other as do streptococci or staphylococci, from a single chain or clump of cells. But caution must be exercised in making this assumption, and replating is necessary to be reasonably sure of obtaining a pure culture.

colony form is a character, or series of characters used in the description of a bacterial species. In the early days of bacteriology colony form was regarded as distinctive of a species, but with the multiplication of recognizable species we now know that it has little distinguishing value. Colony form is influenced by the culture medium on which the organism is grown, to a lesser degree by the temperature of incubation, and by the moisture content of the medium and the atmosphere in the incubator or jar.

In describing colony form many terms are used that are common in non-technical speech (butyrous consistency, convex elevation, etc.), and most of the other terms (fimbriate, entire, crenated, rhizoid), although technical jargon, will be found in most non-technical dictionaries.

Colony form is one of the relics of nineteenth-century bacteriology.

combination. 1. In bact. and bot. nomenclature a combination is a sequence of words that forms the name of a taxon of rank below genus, a binary combination for species and subgenus, and a ternary combination for a subspecies. The zoo. equivalents are **binomen** and **trinomen**.[1]

Names of species are **binary combinations**[2] of the generic name and specific epithet. The name of a subgenus is a combination of a generic name and a single subgeneric epithet (bot.[3]) or name (bact., zoo.); the subgeneric name (epithet) is printed with an initial capital.[3,4,5] When used in combination with both the generic name and specific epithet (name) the subgeneric name is placed between them and in parentheses,[4,6,7] and is not counted as a word in the binominal name of a species or the trinominal name of a subspecies.[7]

The name of a subspecies is a **ternary combination**[8] made up of the name of the genus, the specific epithet, and the subspecific epithet. In zoo. the trinomen is the generic name+specific name+subspecific name.[1]

Examples of binary combinations: *Bacillus subtilis*; *Bacillus* (*Bacillus*) *subtilis*. Ternary combinations: *Bacillus subtilis* subsp. *niger*; *Bacillus* (*Bacillus*) *subtilis* subsp. *niger*.

2. Requirements for valid publication of combinations are that an author must show clearly that the generic name and epithet are to be used in that particular combination,[9,10] and indicate the **basionym** with complete reference to the author and publication.[9,11] BOT.[12] and BAC.[8] require that infraspecific epithets be preceded by a term denoting their rank, and BOT.[3] also requires a similar term before a subgeneric epithet.

In deciding the correctness and priority of the different elements of binary and ternary combinations each part must be considered separately.[13]

[1, ZOO.5. 2, BAC.12a; BOT.23. 3, BOT.21. 4, BAC.10c. 5, ZOO.28. 6, BOT.R21A. 7, ZOO.6. 8, BAC.13a. 9, BOT.33. 10, BAC.30(2). 11, BAC.34N1. 12, BOT.24. 13, BAC.23aN1,2.]

combinatio nova, *abbr.* **comb. nov.** New combination. The abbreviation is used by an author who moves a species from one genus to another and so proposes a new combination of generic name and specific epithet (specific name in zoo.). The name of the author and year of publication[1] of the epithet are put in parentheses, and the abbreviation comb. nov. (in roman) immediately follows and other forms of punctuation are not needed. Example: *Bacillus diphtheriae* Kruse in Flügge 1886 was moved to a new genus by Lehmann & Neumann to become *Corynebacterium diphtheriae* (Kruse 1886) comb. nov. Subsequent authors who use the combination replace the abbreviation comb. nov. by the names of the authors who made the change, as *Corynebacterium diphtheriae* (Kruse) Lehmann & Newmann.

Two other points about new combinations need mention. (1) Incorrect author citation or errors in references do not invalidate the publication of the new combination,[2] and (2) an illegitimate epithet may be adopted by a later author to make a new combination for the same taxon.[3]

[1, BAC.34bN1Ex. 2, BOT.33N2. 3, BOT.68N.]

commensal. An organism that can colonize the surface of healthy tissue without attacking the tissue or itself being damaged by any tissue response. Commensalism is a system of mutual convenience without apparent benefit or ill-effect to either parasite or host (cf. **symbiosis**).

Commission. Permanent group set up to consider, adjudicate on, and deal with problems of nomenclature, particularly those that arise between congresses, or that necessitate revision or change in a rule. The bact. reference body is named the **Judicial Commission**, and the zoo. equalivalent is the **International Commission on Zoological Nomenclature**, usually shortened to 'the Commission'. The bot. equivalent is the General Committee.

The results of the deliberations of the Commissions are expressed as Opinions;[1,2] if a temporary amendment of the Code is needed the zoo. Commission may issue a **Declaration**;[3] decisions completing earlier rulings are issued as **Directions**.[4]

[1, BAC.23aN4(ii). 2, ZOO.78(b). 3, ZOO.78(a). 4, ZOO.78(d).]

common name. English is a language that flows better when it is not interrupted by words or phrases in Latin or latinized words; consequently common or vernacular names are often preferable to scientific names when writing about relatively large groups of micro-organisms. Scientific names should be used mainly when specificity is necessary and always, of course, when a particular taxon is being defined, or in the interest of clarity. In many other cases common names serve our purpose better when writing in English and 'Coliforms are found in the intestinal tract' is preferred to 'Members of the Enterobacteriaceae are found...' Sir Graham Wilson (1965) gives sound advice: *Remember that you are writing in English, not in Latin.*

It is only fair to say that botanists are horrified by this and similar recommendations made in this dictionary. Specific epithets should not be made into common names and it is loose editing that allows an author to say 'strains of coli (or *coli*) ferment lactose'.

When using common names the plural should be in the same language except where the result would not be euphoneous, thus in English salmonellas is preferable to salmonellae, but staphylococci to staphylococcuses, which I have never seen used. In American English the Latin plural of most generic names is preferred, but the *CBE Manual* admits that for certain genera the English plural is always used, e.g. pseudomonads.

Commonwealth Collections of Micro-organisms. An organization set up in 1948 on the recommendation of a Specialist Conference on Culture Collections of Micro-organisms held in London in August, 1947, (1) to foster the maintenance and expansion of collections existing at that time; (2) to encourage the establishment of new collections in different parts of the Commonwealth; (3) to issue Directories of Collections and Lists of the Species maintained in them, and (4) generally to make cultures easier to locate and more readily available.

Commonwealth Mycological Institute, *abbr.* **CMI.** At Ferry Lane, Kew, Surrey TW9 3AF the CMI is the home of a large collection of fungus cultures.

comparison method. Any method that involves the simultaneous determination (in contrast to the step-by-step method) of all the characters of the unknown to be compared with those of taxa considered as possibly identical. The methods can be used with large diagnostic tables intended to make an identification in one step; punched cards and computer systems can also be used for making the comparisons; for the results can be calculated a similarity coefficient, index, or value. Also known as matching methods.

competence, -ent. The ability to incorporate DNA from another bacterium in a transformation experiment. Competence indicates a relatedness between the donor and the recipient.

competitor DNA. DNA from the test organism thought to be related to another, is denatured and in that form is used in so-called in-vitro hybridization experiments, in which it competes with homologous DNA.

complex. Term sometimes used in taxonomic literature for a set or organisms among which there is evidence for more than one species, but the definitions of these and their arrangement in a formal taxonomic hierarchy remains to be clarified. An example from bact. taxonomy is the 'Rhodochrous complex'; uncertainty exists as to the generic position of what was once thought to be a good species; additionally, it is now thought that there may be more than one species involved. LRH.

compounding form. Link between stem and ending used in the formation of a compound name or epithet. Usually is a connecting vowel, but may be a hyphen (see *Taxon*, 1975, **24**, 227).

compound word. A word made by joining the stems of two or more words; only the last part has a case ending; e.g. the generic name *Leptospira* is made from two Greek words, *leptus*, thin or small, and *spira*, a spiral.

Sometimes compound words are made from the stem of words in different languages, a procedure that offends the purist; an example is in the specific epithet *pseudomallei* (cf. **pseudocompound**).

computer. An instrument for making calculations, usually of an advanced or involved nature. An electronic computer is one using electronic (as distinct from mechanical) devices to make the calculations.

In microbial taxonomy computers may be used to compare similarities (S value) and are therefore useful both in classification and in identification. Some attribute more skill to computers than they possess, forgetting (or not knowing) that the value of a computer depends on (1) the program used to instruct it, and therefore on the skill of the programmer, and (2) the quality of the information fed to it (input data). LRH. adds an adage relating to computers which reads 'garbage in, garbage out'.

CO_2 mutant. Term used by Charles & Broadbent (1964, *Nature*, **201**, 1004) for mutants that can grow in minimal medium without their specific growth factors provided that the gas phase is air plus 20–30% CO_2.

In contrast, other mutants may be inhibited by CO_2 (Roberts & Charles, 1970, *J. gen. Microbiol.* **63**, 21); such mutants have been isolated from *Escherichia coli* and from *Salmonella typhimurium*.

con-. Prefix derived from the Latin *com-* (together with, in combination or union); the *con-* form is used before consonants except *h*, *r*, and *l*.

concordances. Little used term in numerical taxonomy for characters for which the two **OTU**s being compared have the same responses. LRH.

conditional name. ZOO. has a rule that a new name proposed conditionally after 1960 shall not be **available**,[1] but conditional names proposed before 1961 are to remain available.[2] As ZOO. does not deal with the problems of provisional names it seems reasonable to assume that conditional names are equivalent to **provisional names** in bact. and bot.

[1, ZOO.15. 2, ZOO.17(8).]

confluent. Running together, joining, fusing. Term used in describing colony morphology on a plate. Confluent growth may be due to (1) too heavy an inoculum, or (2) a characteristic spreading (sometimes mucoid) type of growth. Antonym is discrete.

congener. Members of the same kind; but *SOED* says *rarely 'of the same genus'*, for which use **congeneric**.

congeneric, *adj*. Or the same kind or race; of the same genus.

congruence. Degree of similarity between different classifications of the same **OTUs** (see **cophenetic correlations and values**). LRH.

conjugation is the joining together of sexual forms of bacteria, which is followed by transfer of the chromosome genetic material from one cell to another. Mostly studied in *Escherichia coli*.

connecting vowel. The vowel between the two parts of a compound name or epithet. Both BAC.[1] and BOT.[2] have notes on the correct vowel to be used (generally *-o-* for Greek words and *-i-* for Latin), and each code allows a wrong connecting vowel to be treated as an **orthographic variant** or error which may be corrected.[3] ZOO.,[4] on the other hand, takes the view that the use of the incorrect vowel is not inadvertent, and that it is an intended spelling, and the code does not sanction its correction.

Different connecting vowels in specific epithets (species-group names[5]) based on different types do not, in themselves, avoid the homonymy of the epithets. [1, BAC.Ap9AT. 2, BOT.(1975)R73G. 3, BAC.61; BOT.73. 4, ZOO.32(a)(ii). 5, ZOO.58(8).]

conservation. The legalization of a name that is not an available name or is not **legitimate** in a particular taxonomic position. It is the procedure adopted to protect a name that is in general use but is not the **correct name** for that taxon. The name generally used (a later synonym) is conserved against the correct (valid) name that has fallen into disuse and against all other earlier names that come to light. The act of conservation is carried out by a judicial commission or committee on the recommendation, and at the behest of workers who are familiar with the organism named; the commission does not initiate action, and it is up to the proposers to state the case for conservation.

Conservation is a procedure that finds favour with bacteriologists who are willing to conserve both generic names and specific epithets, but botanists do not permit the conservation of specific epithets. In bact. and zoo. official **Opinions** are issued between congresses so that conserved (accepted in zoo.) names are made legitimate (available) with the least possible delay. The action taken in bot. is different but the end result is the same. BAC. and BOT. contain appendixes which list conserved names, and usually give the names of synonyms against which they are conserved; in zoo. there are Official Lists of accepted names, different lists for family-group, generic, and specific names. As conserved names are conserved against all earlier synonyms, these may be declared **rejected names** (suppressed names in zoo.).

conserved name. A name conserved against earlier **synonyms**[1,2] and **homonyms**[3] by the official body empowered to do so by the international congresses of the discipline. Lists of conserved generic names (and the names against which they are conserved) of bacteria and fungi are contained in BAC. and BOT. respectively.

A conserved name is *conserved against all other names in the same rank* based on the same type (nomenclatural synonyms, which must be rejected) whether they are, or are not included in the list of rejected names; they are also conserved against names based on different types (taxonomic synonyms) that are included in that list.[2]

In zoo. the term **accepted name** is the equivalent to conserved name, and names are entered in Lists of Accepted Names.[4]

[1, BAC.56bN1. 2, BOT.14N3. 3, BOT.14N5. 4, ZOO.77(5).]

consonant. In nomenclature the letter *y*, which does not occur in classical Latin, is sometimes treated as a consonant and sometimes as a vowel; it occurs most often in names formed from the names of people, as in Murray, where it is treated as a vowel.

All the codes make recommendations on the formation of generic and specific names from surnames; these are detailed and are described more fully under **personal name** and in Tables P. 1–P. 3, pp. 198–9.

1. In bact.[1] and bot.[2] a personal name ending in *-er* should be made into a *generic name* by adding *-a*; the name *Zinssera* is correctly formed but more often bact. generic names have been formed by adding *-ia* as in *Listeria* and *Neisseria*; both these names have been conserved and can be used legitimately. To names ending in any constant other than in *-er* the letters *-ia* are added, as in *Pasteuria, Kurthia, Nocardia.*

In forming *specific epithets* from personal names ending in a consonant distinction is again made between those names that end in *-er* and in any other consonant, including *-r* (i.e. not *-er*).[1,3] When a man's name ends in *-er* an epithet can be formed by adding the letter *-i*, as in *flexneri*, and when a name ends in another consonant the letters *-ii* are added to the name of a man (as *pasteurii, barrii, morganii*), or *-ae* to the name of a woman (as *gordonae*). Note that a feminine ending (*-ae*) is given to the name of a woman ending in *-er*, as *klienebergerae* and *hookerae.*[1,3]

2. ZOO. does not distinguish between personal names ending in *-er* and any other consonant; it recommends that modern patronymics ending in a consonant should add the appropriate suffix (*-ius, ia, -ium*) to form a *generic name*,[4] and gives different guidance from the other codes on the formation of *specific names*; the letter *-i* is added to the name of a man,[5] and the letters *-ae* to the name of a woman.[6]

[1, BAC.Ap9B. 2, BOT.R73B. 3, BOT.R73C. 4, ZOO.ApD.37. 5, ZOO.ApD16. 6, ZOO.R31A.]

consortium. Defined by BAC.[1] as an *aggregate or association of two or more organisms.* The name of a consortium is not governed by the rules of BAC.

[1, BAC.31b.]

conspecies, *pl. n.* Two or more species of the same genus.

conspecific. Belonging to the same species.

constituent element of a taxon is the type of the taxon; for a species or subspecies of bacteria it is the type strain or culture.[1] In bot. it is that specimen or part of a specimen to which the name of a taxon is permanently attached; the type is *not necessarily the most typical or representative element of a taxon.*[2]

[1, BAC.(1966)Pr11N. 2, BOT.7N1.]

constitutive enzyme. An enzyme that is produced by a micro-organism during growth, the production of the enzyme being independent of the presence in the medium of the substance acted upon. Thus *Proteus* species produce urease irrespective of the presence or absence of urea in the growth medium (cf. **inducible enzyme**). Note that the terms induced and constitutive are relative; many constitutive enzymes may be formed by an organism in greater amount when the inducer is present.

contaminant, contamination. An impurity; something that spoils the purity of something else. In microbiology the contamination of a culture is the introduction of another species (the contaminant) to the medium in which both organisms grow together.

A contaminant may also be a substance (an impurity) in a chemical used for solutions, stains, or media.

In food microbiology, contaminant is the name applied to the extraneous organism that has got into the food and multiplied there; when ingested the food may be toxic (staphylococcal food-poisoning) or infective (salmonella food-poisoning).

continuous classification and identification. Taxonomy does not stand still and new knowledge is continually being added to the information stored on punched cards or in computers. This new information may upset an existing classification so that it must be revised; in turn, a revision may dictate changes in identification procedures or require the change of name of an organism – and that is when trouble really begins.

This can be summarized in the statement that taxonomy, in all its branches, is a continuing subject.

continuous culture. Culture in a **chemostat**, in contrast to a **batch culture** in a tube or flask. In a continuous culture, medium is both added and withdrawn simultaneously so that the volume is maintained and, since exhaustion of nutrients is avoided, growth proceeds logarithmically. Organisms are removed with the medium so that the population should not become overcrowded. Growth continues until (1) the end of the experiment, or (2) contamination occurs.

conventional classification. A term employed by users of recent methods of classification to refer to older methods. It is a dangerous practice as what is modern today will, if successful, be conventional tomorrow. LRH.

conventions in the use and presentation of words can be considered as both editorial and typographical.

1. *Editorial.* Each journal has its own style, and this must be followed; therefore, before writing a paper an author should decide on the journal, study recent issues of that journal, and look for any instructions (sometimes labelled advice) to contributors, usually contained in the end papers or pages. On taxonomic subjects some editors are much more expert than others (many are woefully ignorant); those interested in the subject will have ideas on the presentation of the material and the deductions made; the less interested are more likely to let the author get away with anything.

Editorial fads and fancies relate chiefly to the use of different kinds of printing type, to the use and misuse of capitals in words such as type and group, and to methods of citing references in the text. In scientific papers a capital is needed for the name of a taxon above genus and the word should be printed in roman type; generic names as nouns should each have a capital initial letter and be printed in italic type; when they are used adjectivally lower case initial letters are used and the names are printed in roman; thus, *Staphylococcus aureus*, *Brucella abortus*, but a brucella infection. The specific epithet (species-group name), even when formed from the name of a person or place should not have an initial capital.

2. *Printing* practice does not vary as much as editorial whim. In most countries scientific names of lower ranks are printed in italic, but in Germany the old style was to print in roman type with the letters spaced out as B a c i l l u s s u b t i l i s. In some countries there is a free use of italic type (Scandinavian journals sometimes use italic type for authors' names whenever these appear in the text), but it may be difficult to distinguish between editorial and printing (house) conventions.

convergence. Two organisms with many characters in common but which are descended from widely separated stocks are said to show convergence. It is doubtful whether we ever have enough knowledge of the ancestry of micro-organisms to justify use of this word in microbiology. It is difficult to distinguish from parallelism due to intermediate conditions.

converting phage. A lysogenic bacteriophage that alters (or converts) the characters of a bacterium. It is seen best in salmonellas in which the presence of such a phage is shown by a change in the antigenic structure, e.g. phage ϵ_{15} changes *Salmonella anatum* (3, 10: e, h: 1, 6) to *S. newington* (3, *15*: e, h: 1, 6). In describing the new antigenic formula, the newly introduced factor is underlined in typescript and printed in italic (see **antigen, antigenic formula**).

Sometimes incorrectly described as phage conversion, which means a conversion of the phage and not, as happens, conversion by the phage.

cooperative studies. In bact. taxonomy, recent years have seen the emergence of working groups or parties, sometimes international in character, which study one and the same set of **OTU**s. This is particularly useful as it permits measures of between laboratory discrepancies to be made. See also **permissive philosophy**. LRH.

co-ordinate, *adj*. In taxonomy, names or ranks of equal nomenclatural status.[1] Used for names that may compete for priority, such as names of genera and subgenera, and names subject to the same rules and recommendations of a code.[2]

Specific and subspecific epithets (names) are co-ordinate, and within a genus only those based on the same type may be the same.[3] Thus if *Bacillus subtilis* (which is typified by the Marburg strain) is divided into subspecies, the subspecies that contains the Marburg strain is the one to be named *B. subtilis* subsp. *subtilis*.

[1, ZOO.Gl. 2, BAC.10b. 3, BAC.12b; 40cN.]

cophenetic correlations and **values.** In numerical taxonomy, a method of estimating the similarity of two or more dendrograms to each other. More than one dendrogram may be obtained for the same **OTU**s as a result of using different characters, similarity coefficients, or clustering techniques. Sokal & Rohlf (*Taxon*, 1962, **11**, 33–40) proposed a method whereby the separate similarity scales are divided into classes and the cophenetic value between a pair of OTUs is the class number in which the two OTUs are linked in the dendrogram.

The overall similarity between dendrograms is calculated as the correlation coefficient between pairs, triplicates, etc., of cophenetic values derived from the two, three, etc. dendrograms. LRH.

core-storage. Capacity of the memory of a computer.

correction of errors. The codes of nomenclature authorize the correction of unintentional errors, which are often euphemistically called typographic or orthographic errors, supposedly made by typists and compositors. In fact, the errors are as likely to be attributable to the carelessness of their authors (as *Neisseria mucocus* instead of *Neisseria mucosa*, when *Diplococcus mucosus* was moved from *Diplococcus* to *Neisseria*); Buchanan rightly deprecated such errors and castigated their perpetrators as bacteriological illiterates.

Although the correction of errors is approved, the liberty of doing so should not be lightly taken,[1] and we should correct our own mistakes before condemning others (the example quoted is one of my solecisms).

[1, BOT.73N2.]

correct name. The earliest name for a taxon (with a given circumscription, position, and rank) that is in accordance with the rules of nomenclature applicable to the discipline;[1] it is, therefore, the name that must be adopted under those rules.[1,2] BAC.[1] applies the principle of priority to names of orders and lower ranks; BOT.[3] limits it to names of family and ranks below. Equivalent to **valid name** in zoo. A taxon may have more than one legitimate (available) name but it can have only one correct name. In combinations the rules should be applied to each element separately.[4]

The correct name must (1) not be a junior homonym (a name exactly repeating a name attached to another taxon of the same rank); (2) be accompanied by an adequate description of the organisms forming the taxon; (3) be accompanied by the designation of a type specimen (and in bact. cultures of the type should be deposited in one or more culture collection, which should be named); the name and description should be published in a journal or book that is on sale to scientists and libraries throughout the world (see **valid publication**); after 1975 BAC. requires that new names of bacteria should be published in the *International Journal of Systematic Bacteriology.*

[1, BAC.Pr8; 23a; BOT.PrIV. 2, BOT.6. 3, BOT.11. 4, BAC.23aN1.]

correctus, *abbr.* **corr.** Correction; corrected by. Usually the correction of errors is made discreetly and anonymously, but when attention is to be drawn to it the abbreviation corr. is appended to the **author citation** and is followed by the name of the correcting author.

corrigendum, *abbr.* **corrig.** When an author corrects an orthographic error of another author he may add the abbreviation corrig. if he wants to draw attention to the correction, but he is not obliged to do so. Such a correction does not affect the validity of the original name or its date of publication.[1]

[1, BAC.61.]

coryneform. A general term applied to Gram-positive, non-sporeforming rods, some but not all of which are club-like in shape; all may show clump formation rather than chains. Like coliform and actinomycete, coryneform is a designation for a large, ill-defined group of bacteria, and the term covers more than the genus *Corynebacterium*.

cotype. A strain (specimen) of a nominal (named) species used by author who failed to designate a holotype. It is equivalent to **syntype**; ZOO.[1] recommends avoidance of the term cotype.
[1, ZOO.R73E.]

country. The phrase 'in this country' should be avoided when writing a scientific paper; instead the name of the country or the part of the world should be indicated clearly.

crateriform liquefaction. Appearance of growth of a gelatin-liquefying aerobic organism, in which there is a filiform growth along the stab, and at the air surface a saucer-shaped zone of liquefaction.

critique. A branch of logic which, as Blackwelder (1964) puts it, examines statements made by scientists and not the statements of science. Without critique inaccurate scientific statements, faulty conclusions, and even illogical ideas would abound, for *it is not only human to err but to use other ideas out of context.*

cryobiology. The study of life and living things at low temperatures. The so-called spectrum of cryobiology ranges from the formation of ice crystals and nuclei to the effect of freezing on the whole living organism. Freeze-drying, which is important to microbial taxonomists as the best method for preserving and maintaining micro-organisms in their original state, is included in the subject.

cryophilic. Synonym of **psychrophilic** (q.v.).

crypticity. The state of cells towards a substrate; particularly the inability of the cells to metabolize the substrate in spite of the presence within the cells of the appropriate enzyme. Apparently the effect depends on the permeability of the cell wall to the substrate.

cryptogram. Descriptive code proposed by Gibbs *et al.* to form the second (and characterizing) part of the name of a virus. As knowledge of the virus increases so the cryptograms may change; it was suggested that bold type should be used for properties determined by experiment, and non-bold type when the property had been assumed by analogy from a supposedly related virus, but examples given by Wildy (1971) do not make this distinction in typeface.

While the complete cryptogram would provide a *unique specification* of the virus, an abbreviated form of four terms would be used in the virus code. The four pairs of symbols of each cryptogram represent:

(1) type of nucleic acid/strandedness of nucleic acid
(2) molecular weight of nucleic acid (in millions)/percentage of nucleic acid in infective particles.
(3) outline of particle/outline of 'nucleocapsid'
(4) kinds of host infected/mode of transmission/kinds of vector

Example: Poxvirus group D/2: 130–240/5–7.5: X/*: I, V/O, R, Ve/Ac, Di, Si. Decoded, this means DNA/double strand: 130–240 × 10^6 daltons/5–7.5% nucleic acid per infective particle: X = complex/*property unknown: invertebrate, vertebrate hosts/contact, respiratory tract, invertebrate vector modes of transmission/Acarina (mites and ticks), Diptera (fly, mosquito), Siphonaptera (fleas) vectors.

See Fenner (1976), where symbols used are listed.

CSU. Carbon source utilization used as characterization tests for taxonomic work.

The Koser test for citrate utilization is one of the oldest of these tests but the range has been extended; there is about 5% variability in the results of individual tests, but in some species, nobably *Pseudomonas mallei*, a greater variability is found. If this variability is repeatable and characteristic, *Fascinating problems... would be raised* (Snell & Lapage, 1973).

cultivar. **1.** A cultivated strain of plant. In mycology the term is probably *only correctly used... for trade varieties of mushrooms* (Ainsworth).
2. BAC.[1] defines it as a *cultivated strain with special properties*; could not this be a mutant?
[1, BAC.Ap10B.]

culto-type. The term type culture does not apply to fungi and BOT.[1] specifically excludes living plants or cultures as type specimens. However, the dried material that forms a mycological holotype may remain viable for years, and a living culture derived from the plate or specimen dried (not freeze-dried) is a culto-type. Neither the term culto-type nor the culture has any status in nomenclature.
[1, BOT.9.]

cultural character, -istic. Appearances of a culture grown on solid or liquid medium. Includes details such as size, shape, surface, edge, consistency, and emulsifiability of colonies, haemolysis in medium around colonies on blood agar, appearances of growth in stab cultures (fir-tree, inverted fir-tree), liquefaction of gelatin, presence or absence of surface growth or of deposit in broth cultures. Although essential for a complete description of a micro-organism, these characters introduce a number of descriptive terms of limited value and use, and they are better defined by drawings of the colony shape, elevation, and edge. These characteristics find little place in day-to-day diagnostic bacteriology and in their *Manual* Cowan & Steel describe them as *relics of nineteenth-century bacteriology.*

culture. **1,** *v.* To attempt to grow living organisms from a **specimen 2** by inoculating a small portion on a suitable nutrient medium.
2, *n.* Successful growth after incubation of nutrient medium inoculated with a specimen or with another culture. The growth consists of millions of micro-organisms; if from a specimen it may be a mixture of several different kinds of organism, and much work may have to be done to obtain each different organism in what is termed a **pure culture**, i.e. only one kind (species, serotype, or other low-rank taxon) present in each of the subcultures.
3. BAC.[1] defines a culture as *a population of bacterial cells in a given place at a given time.*
See also **batch culture**, **chemostat**, and **continuous culture**.
[1, BAC.Ap10.]

culture collection. The microbiological equivalent of botanical and zoological gardens (though generally not open to the public!) with some of the functions of the botanical herbarium and zoological museum. Culture collections are repositories of living cultures of bacteria, fungi, yeasts, algae, and protozoa; a few of the cultures are 'type' specimens (**nomenclatural types**), others are typical (or characteristic) strains of the taxa they represent, and yet others are kept because they have unusual properties or characteristics of genetic or economic importance, among which are micro-organisms capable of producing antibiotic substances.

The living cultures may be maintained by serial subculture at appropriate intervals or better, by freeze-drying (lyophilization) and storage *in vacuo*, in which state many micro-organisms will survive for years. During storage the metabolism of the organism appears to be suspended but when nutrient medium is added to the dried powder the organisms soon resume their characteristic shapes and start to multiply. The great advantage of freeze-drying over serial subculture is that the mutation that may occur in subculture does not take place while the organism is desiccated, and there is no evidence of any selection in the freeze-drying process. Consequently, once the organisms start to grow normally they show the characters of their ancestors which were dried many years before. Some culture collections also use storage in or above liquid nitrogen, especially for seed stocks.

A directory of almost 350 culture collections has been compiled for the World Federation for Culture Collections: Martin, S. M. & Skerman, V. B. D. (1972). *World Directory of Collections of Cultures of Microorganisms*. Wiley-Interscience, New York. Addresses of well-known collections will be found in this Dictionary under **ATCC**, **Centraalbureau voor Schimmelcultures**, **CMI**, **NCDO**, **NCIB**, **NCMB**, **NCPPB**, **NCTC** and **NCYC**, the initials of some of the larger collections.

culture medium. Solid or liquid substrate containing nutritional substances suitable for the growth of micro-organisms. Since microbes may be fastidious, many different kinds of media are needed for the successful cultivation of organisms from different sources.

cultype. Culture of a fungus given the status of a nomenclatural type. This is not allowed under BOT. which requires type material to be a dead specimen, or a description, or an illustration[1] (cf. **culto-type**).
[1, BOT.9.]

D

Darwinism. From Charles Darwin. The theory of evolution by natural selection.

data, *pl.* of **datum.** Facts, figures, information, and observations. These should be available before the experiment, but as used by most authors the data are the findings or results of experiment. It is a word often used in the adjectival form, essential to statistically-minded taxonomists.

data bank. Information stored in a form that can be read by a machine. Data banks containing descriptions of bacteria are already established and these should be useful in preparing manuals for the identification and characterization of known and recognized taxa.

data chart. A **diagnostic table**.

data column. **1.** Column (or line) in a **diagnostic table** in which the characters are listed (= character list).

2. Space or box on a working sheet for recording the characters of a micro-organism.

data matrix, *pl.* **matrixes.** A table of character states plotted against taxa (see

diagnostic table). The left-hand column is the character list, the headings are the names (or numbers) of the taxa. The character state is best shown as the percentage positive of the strains tested, when this figure is based on figures too small to be meaningful, the state is shown by symbols such as +, d, and −.

Synonyms given by Morse *et al.* (1975): comparision chart, data chart, data table, taxonomic data matrix.

date and author citation. ZOO. requires that the author of a name (of genus or lower rank) and the date should be cited *at least once in each publication*[1] and that they should be separated by a comma.[2] BAC.[3] and BOT.[4] also require the date of a previously published name of a taxon or new combination to follow the name of the author, but without the interposition of punctuation. The date of the name is that of its valid publication.

[1, ZOO.ApE10. 2, ZOO.22. 3, BAC.33b. 4, BOT.46.]

date of a name or epithet. The date of its **valid publication.**[1,2] In the absence of other information the date on which the publisher delivers the printed matter to the carrier should be accepted.[3] The correction of an error in the original spelling of a name does not affect the date of that name.[2]

In zoo.[4] a name becomes **available** and is dated from the time it satisfies the requirements of Art. 11, or from the time when an infrasubspecific name is raised to species-group rank. Names published after 1930 must be accompanied by a description or reference to a description, and genus-group names must be accompanied by the designation of a type-species.[5] See also **interrupted publication of a name**.

[1, BAC.23b. 2, BOT.45. 3, BOT.R30A. 4, ZOO.10. 5, ZOO.13.]

date of publication. In nomenclature this means the date on which the publication containing the name was distributed to the general public. Weekly journals print the exact date of issue on the copy; journals published less frequently often give the date of publication of the previous issue, or, in the list of contents, of each issue of the volume. In the absence of such exact information the official date or the date printed on the cover (sometimes months before the actual date of publication) must be regarded as correct.[1] Librarians are enjoined not to remove covers that bear information on the date of publication if this information is not contained in the material printed between the covers.[2]

When the date of publication is appended to the author citation a comma should be interposed only in zoo.[3] names.

The date of acceptance of a paper for publication is without significance in determining the date of effective publication, or in the establishment of priority.

The dating of a report of a congress or meeting may present difficulty to an author when the year of publication does not coincide with that of the congress, e.g. the *Report of Proceedings of the Fourth International Congress for Microbiology* (held in 1947) was not published until 1949. The date of publication should be given correctly for it is on this date that priority is determined, but it may be possible in the text to indicate the data of the actual meeting.

The date of the first Bacteriological Code is 1948, when it was published in the *Journal of Bacteriology*, a year before publication of the Proceedings of the Congress at which the Code was approved. The Code approved in 1973 was published in 1975, but on the title-page it is stated to be the 1976 revision.

From 1 January 1976 the date of publication of a new name or combination in bact. is the date of publication in the *International Journal of Systematic Bacteriology* (*IJSB*).[4]

After 1 January 1980 a bact. name not in an **Approved List** may be reused and published as a new name; this will be regarded as an entirely new name and, to make it valid, it must be published (with a description of the taxon to which it is applied) in *IJSB*.[5] The author should claim authority for the name and he should not attribute it to another author who used it before 1980; to indicate that it is an earlier author's name that has been reused he can append the English 'not' or the Latin *non*, followed by the name of the original author and the date of publication before 1980.

[1, BAC.26a; BOT.30; ZOO.21(a). 2, ZOO.R21C. 3, ZOO.22. 4, BAC.27N. 5, BAC.28aN2.]

date of receipt. Receipt of a paper at an editiorial office does not establish the **priority** of a name, and is without significance in nomenclature.

date(s) of the start of nomenclature. Starting dates are laid down in the codes of nomenclature; bot.[1] has different dates for different groups, 1753 for algae and either 1801 or 1821 for fungi. Zoo. nomenclature[2] starts in 1758. Under the older versions of BAC. the nomenclature of bacteria started in 1753 but a new starting date, 1 January 1980 has been agreed.[3] Some names published before 1980 will be put on **Approved Lists** (q.v.) and these names will be assumed to have been published on 1 January 1980; all other names will be available for reuse. An official date for the start of virus nomenclature has not been fixed.

[1, BOT.13. 2, ZOO.3. 3, BAC.24a.]

date opinions become effective. In bact. a request for an Opinion is published in ***IJSB*** and six monthes later a vote of the Judicial Commission is taken. If it is approved by ten or more Commissioners the Opinion is published in *IJSB* and reported to ICSB; it becomes final unless rescinded by a majority of those voting in that Committee.[1] The Statutes do not say when or how (by mail or at a meeting) the Opinion is considered by ICSB, but presumably it becomes effective unless rescinded at the conclusion of the next meeting of the full Committee; it does not appear to need confirmation by the next Congress.

In zoo. an Opinion becomes effective immediately the ruling of the Zoological Commission is published, but it must be reported to the next Congress.[2]

[1, BAC.S8a, c. 2 ZOO.78(c).]

datum, *pl.* **a.** A thing that is known or assumed to be a fact, and the premise from which inferences may be drawn. Generally used in the plural, and in scientific writing often incorrectly used for the results of an experiment.

de-. This prefix may be attached to a verb to give the sense of reversing the action specified by the verb, but it should be used only in the absence of a simple verb for the action. Hydrate is made into dehydrate, and fuse into defuse, but an example of an unnecessary *de-* form is destain as a substitute for decolorize or bleach.

decapitalize. To replace the initial upper case (capital) letter of a personal or geographical name by a lower case letter when the name is used to form a specific epithet[1,2] or specific name.[3] However BOT.[2] allows authors to use a capital for an epithet formed from a personal name, from a vernacular (non-Latin) name, or from a name formerly used as a generic name.

Stafleu & Voss (1975, *Taxon*, **24**, 230) advise caution in using capitals; those who do so should be sure that the epithet is entitled to a capital, and they cite *Lobelia Cardinalis* and *Mimulus cardinalis*.

[1, BAC.59. 2, BOT.R73F. 3, ZOO.28.]

decimal point. The decimal point is printed in English and US journals as a full stop (period) which is on the line (low stop). In European (continental) publications it is usually printed as a comma on the line. Because of this continental usage English authors should avoid using commas as spacers in long numbers. Some continental writers use stops as spacers in numbers, a practice that is equally undesirable.

declaration. Provisional modification of ZOO. made by the Zoological Commission between congresses; each is considered to be a part of ZOO. until the next congress ratifies, modifies, or rejects it. Afterwards the Declaration is deemed to be repealed.[1]

[1, ZOO.78(a).]

deep. 1. Medium tubed in sufficient depth (> 50 mm) to be inoculated by stabbing with a straight wire; i.e. make a stab culture.

2. A shake culture (US).

define, *v.* **1.** To state the characters of a taxon, especially those characters (***differentiae***) by which it differs from other taxa (*cf.* **circumscribe**).

2. To state that a particular name applies to a species of which the type specimen is representative. According to Chiselin it is nonsense to say that we define species; we discover them, describe them, and name them.

defined, past participle used as an *adj.* Stated, known. Better than **synthetic** when applied to a culture medium made up of ingredients of known composition.

definition, *n.* **1.** A list of characters of a bact. or bot. taxonomic unit, particularly of those characters by which the unit can be distinguished from other (often closely similar) units.

2. The distinguishing characters of a zoo. taxon;[1] the equivalent of **diagnosis** in bact.[2] and bot.[3]

[1, ZOO.Gl. 2, BAC.R30b. 3, BOT.32F0.]

definitive, *adj.* Final, perfect. Taxonomy is never perfect or perfected, and the results of taxonomic thought are never final.

degenerate peritrichous flagellation. A term to describe the flagella insertion seen in bacteria in which flagellation appears to be modified by the conditions of growth, especially by the medium. In what is considered the normal state the peritrichous flagella are inserted fairly uniformly over the surface of the bacterial body, but under certain conditions (possibly less favourable for multiplication) the flagella are inserted near (but not at) the pole, a subpolar position.

dehydrogenation. One of the mechanisms by which living cells obtain energy; they do this by removing hydrogen from substrates and transferring it to hydrogen acceptors; when oxygen is the acceptor the process is described as **oxidation**, when the hydrogen acceptor is organic it is described as **fermentation**. The two kinds of metabolism have taxonomic significance and seem to distinguish between two of the major groups of bacteria.

Delft collection. A collection of yeast cultures started by Professor A. J. Kluyver. The address is Centraalbureau voor Schimmelcultures (Yeast Division), Delft, Netherlands.

deme, -deme. 1. The deme system of category-terms was introduced by Gilmour & Gregor (*Nature*, 1939, **144**, 333) and extended by Gilmour & Heslop-Harrison (*Genetica*, 1954, **27**, 147–61); it was intended for special-purpose categories, and to be completely divorced from the categories of hierarchical and orthodox taxonomy. The category-terms are formed by adding the suffix *-deme*, denoting a group of individuals of a specified taxon, to the appropriate root, such as topodeme for a group of plants occurring in a particular geographical area. The suffix should not be used alone.

2. Deme is used by Simpson for small units of population of evolutionary significance (he quotes species and subspecies) that are in close contact with each other, thereby misunderstanding the deme terminology of Gilmour, Gregor and Heslop-Harrison. He advises against naming the demes because they are evanescent. The deme in Simpson's sense is equivalent to a gamodeme.

denaturation temperature of DNA is the temperature at which, under a given set of conditions, double-strand DNA changes to single-strand DNA. This usually occurs over a range of temperature, and the denaturation temperature (also called melting temperature, or Tm) is defined as the midpoint of the transition. One frequently used method determining the midpoint is from analysis of the curve showing the increase in the ultraviolet absorption (old usage: optical density) that accompanies the transition. Under standard conditions, the DNA base composition can be estimated from the denaturation temperature, for the greater the GC content, the higher the Tm. LRH.

dendrogram. Tree-like figure used to represent a hierarchy or a figure to show different levels of similarity coefficient. Mayr introduced the term to show lines of descent obtained from evidence of existing form, in contrast to 'phylogenetic trees' which are based on fossil evidence. Others, e.g. Sneath, use the term in ways that rouse Mayr's ire; the figures may be supposed to express phylogenies (**phylogram**), but the fact that they are constructed from the results obtained by numerical methods does not, in Mayr's view, make them any more realistic or expressive of evolutionary trends.

denitrification. Microbial breakdown of nitrates and nitrites and the liberation of free nitrogen.

dependent characters. See **linked characters**.

depersonalization. Virologists do not intend to perpetuate the names of those who first name a taxon whatever its rank, and the name of the virus should stand alone without author citation. The decision to do this is in tune with the opinion among virologists that it is the identification, characterization, and classification of viruses that matter; that names attached to them are of minor importance, and the people who give names to them are of no importance whatever.

Virologists have chosen the ugly word depersonalization for the term 'without author citation'; in the later Rules of Nomenclature (VR.) rule 8 says cryptically *No person's name shall be used*; this refers both to author citations and to the use of a personal name in the formation of the name of a virus (H. G. Pereira, personal communication).

deposition of culture. Authors of names of new taxa are recommended to deposit the **type strain** in a culture collection.[1] It is helpful to workers if the accession or catalogue number in the culture collection is quoted in the paper in which the new taxon is described.
[1, BAC.AdC.]

derive, derivative, derivation, *v.* and *n.* In grammar these terms deal with the origin of a word. In classical biological nomenclature a name may be derived from any source whatever, but it is generally accepted that Latin words (or latinized forms of Greek words) make the most satisfactory names. Whatever its origin, a name is treated as a Latin word and given an ending of the kind used when Latin was a living language.

Names can be coined quite arbitrarily but the codes[1] recommend that the derivation of the name should be given and, if necessary, explained.
[1, BAC.R6(5); BOT.R73I; ZOO.ApE16.]

derived character. A character that has changed from the form in which it appeared in the ancestral form of the organism. A term preferred by Mayr to 'advanced' or to **plesiomorph** for use in zoo. See also **uniquely derived character**.

derived data. A term used by Quadling & Martin for conclusions drawn from **laboratory data** which have been used to make estimates of the probability that a strain is a member of a certain taxon.

describe, *v.* The act of listing the various characters of an organism. An essential task of any taxonomist but one that is unfortunately often done in a slip-shod manner.

description. A list of characters of an organism or group by which it is hoped that subsequent workers will be able to recognize similar organisms or groups; its most important use is that it can form the basis of a diagnosis. In the older literature descriptions are often vague and contain insufficient information on which to base a diagnosis. Present-day workers should try to produce worthwhile descriptions founded on thorough investigations of the strains they describe; this is particularly important in the description of new species, and a conscious effort should be made to show how the new differs from the older, well-established species.

Unlike descriptions in bot. and zoo. in which the characters described are selected (see Davis & Heywood (1973), pp. 114–15, 303–6), descriptions of micro-organisms should be as complete as knowledge allows. Different organisms (especially those of interest to specialists in the subdisciplines of microbiology, whose techniques differ considerably) have been studied by such a variety of technical methods that standardization of descriptions cannot be attempted, but BAC.[1] recommends the introduction of certain **minimal standards** by which it is hoped to improve the usefulness of descriptions to future workers.

It is usual to start with morphology, go on to cultural characteristics, physiological reactions, nutritional requirements, metabolic products, antigenic structure, pathogenicity, and then special features. Two features useful in bot. and zoo. have no place in microbial descriptions; these are the geographical distribution (most microbes have a world-wide distribution though this may only be discovered when it is looked for), and the source of the culture, which is useful in the limited field of the pathogens.

Descriptions must be based on observations made on pure cultures and every effort should be made to maintain cultures in that state.
[1, BAC.R30b.]

descriptive chart. Name given to a sheet of heavy paper or card on which are printed numerous cultural characteristics; the user underscores characters that apply or enters an appropriate sign (+, −) in a box or draws a diagram of colony shape. These charts served a useful purpose but have been superseded by punched cards, some of which have space for descriptive information in the centre of the card.

descriptive name. A name formed from a descriptive character; recommended in bot. for the name of a **division.**[1]
[1, BOT.(1975)R16A.]

descriptor. Word or term that indicates the contents of a paper in a journal or other publication. Descriptors are becoming increasingly useful as computers take over the task of information retrieval.

desiccate, *v.* and *n.* The removal of moisture from an organism (*v.*) and the end product (*n.*). When desiccation is carried out slowly the organism will almost certainly die (unless it has formed a spore) but if it is done quickly, especially from the frozen state without going through the liquid phase, the organism will survive and if kept in a vacuum and protected from light will remain viable for years (see **freeze-drying**).

designate, *v.* To state categorically the nomenclatural type of a taxon (see **typification**). The use of designate is much stronger (and better) than the word 'suggest' when proposing a type (strain) for a new taxon. Firmness in this matter does not leave room for doubt and so lessens the chance of nomenclatural dispute. Type strains of species may be designated (1) by the original author by a deliberate act, (2) by **monotypy**, (3) as a **lectotype**, or (4) as a **neotype**.[1] In bot. type specimens are designated in a similar manner and, as in bact., the designation of a **holotype** by the author takes precedence over all other forms;[2] for zoo. usage see **designation 3**.
[1, BAC.18a–d. 2, BOT.7.]

designation. 1. A title or descriptive name.
2. The result of selecting and appointing; indicating positively. BAC. uses the phrase *Designation by international agreement* in connexion with the selection of a type species for a genus that did not have an identifiable type species, and cites the genus *Vibrio* for which a type species was designated in an Opinion issued by the Judicial Commission.[1]
3. For the designation of the type species of a genus ZOO.[2] lists (1) type by original designation, (2) type by indication (see **indication 2**) and (3) type by subsequent designation. Among the last occurs one of those absurdities for which codes of nomenclature and taxonomists are renowned, viz. *A nominal species is not rendered ineligible for designation as a type-species by reason of being the type-species of another genus.*[3]
[1, BAC.20e; Ap5. Op.31. 2, ZOO.68. 3, ZOO.69(a)(v).]

destain. A word for decolorize, and one to be deplored.

determination. The identification of a biological taxon, usually of a species or subspecies.

determinative, *adj*. Finding the identity and perhaps the name of a taxon. In *Bergey's Manual of Determinative Bacteriology* the title shows that the book is intended for use in identifying an unknown bacterium and, whenever possible, finding a name for it; the editors believe (a belief based on a subjective judgement) that the names used are nomenclaturally correct within the classification used in *Bergey's Manual*. Users may look to *Bergey's Manual* for an inspired classification, but that is making an unwarranted extension of the word determinative, and one not encouraged by the editors.

determinator. 1. One who makes an identification (determination).
2. Device to aid identification when using **diagnostic tables**.

determine, *v*. **1.** A common laboratory use of this verb, to find out, is not given in dictionaries but is popular in scientific writing; for example, a worker determines by experiment whether an organism produces or does not produce a particular end product.
2. To identify, name, and place an unknown organism in its rightful place in a **hierarchy**. Among microbiologists it is used more by mycologists than by bacteriologists. Bentham & Hooker used the verb in a more restrictive way, as *ascertaining the name* of a plant, whether for *ulterior study* or *intellectual exercise*.

devalid, devalidate. Not valid because the name was proposed before the official starting date of the nomenclature of the discipline. Holmes's names of viruses were devalidated when it was agreed by the Virus Subcommittee of the International Committee on Bacteriological Nomenclature (at the time responsible for virus nomenclature) that the starting date of virus nomenclature should be some time in the future, and that the principle of priority should not apply to names that had been proposed in the past. The new virus nomenclature committee has not yet proposed a starting date for virus nomenclature.

See also **revalidate**.

deviant index. A measure, proposed by Goodall (*Nature*, 1966, **210**, 216), of the way in which an individual differs from the norm of a population; it can be applied to characters that are expressed qualitatively or quantitively.

diacritic signs. Marks used in printing to indicate a different sound. They are not used in scientific (i.e. Latin) names or epithets.

In names formed from words with diacritic signs the German *ä*, *ö*, and *ü* drop the umlaut mark and add *-e* to become *ae*, *oe*, and *ue*;[1,2,3] French *é*, *è*, and *ê* become *e*; Spanish *ñ* becomes *n*; and the Scandinavian *ø* becomes *oe*, æ becomes *ae*, and å becomes *aa* in bact.[1] and *ao* in bot.[2]

[1, BAC.64. 2, BOT.73. 3, ZOO.32(c)(i).]

diaeresis. Mark of two dots over the second of two adjacent vowels to show that they are to be pronounced separately, e.g. microörganism. This archaic device is permitted by BOT.[1] for names of taxa, but is prohibited by ZOO.[2] It is seldom used by modern writers in English but is occasionally seen in US; *CBE Manual* recommends that when a foreign name has an English form without a diacritic sign, the English form should be used, e.g. Cologne, not Köln.

The diaeresis mark is not now mentioned in BAC. but it was allowed by earlier editions.

[1, BOT.73. 2, ZOO.27; 32(c)(i).]

diagnosis. A brief statement of characters that are valuable in distinguishing one taxon from another;[1] a listing of the ***differentiae*** and a comparison with similar organisms.

BOT.[2] requires that the proposal of a new name should be accompanied by a diagnosis in Latin; this applies to names of algae proposed after 1957. The diagnosis should not be based solely on the type but on as many specimens as possible; it is recommended that it should be accompanied by a description, also in Latin.

BAC. and ZOO. do not require diagnoses or descriptions to be in Latin.

[1, BOT.32F0. 2, BOT.36; R36A.]

diagnostic character. A character that has a particular value in distinguishing one kind of something from another; in biology a distinguishing character of this kind is known as a ***differentia***, from which is formed the incorrect term, **differential character**.

diagnostician. Used in the medical sense as one who makes an identification by comparing the characteristics of an unknown with those of the known, and not, as a biologist might expect, one who writes a diagnosis (i.e. a list of *differentiae*).

diagnostics, *n.* General term, seldom used by the native English, for the methods used in the investigation of the cause of diseases. In microbiology it can be applied to the identification of micro-organisms.

diagnostic table, *syn.* taxonomic data matrix. A table of characters for the identification of taxa that are based on the sharing of characters. To simplify the identification more than one set of tables may be used and the result is obtained in a series of steps (the step-by-step method of identification). Because they contain more information, diagnostic tables look more complicated than **dichotomous keys** but, except with the simplest of keys, identifications are easier to make with the help of tables. When several characters are variable the table is more successful as a determinative aid than a key in which there are too many routes leading to the same identification.

Table D. 1. *Diagnostic table made to show polythetic groups*

Character	Polythetic group				
	1	2	3	4	5
Motility	+	+	+	–	–
Catalase	+	+	+	+	–
Gas from glucose	+	+	–	–	–
Lactose fermented	+	–	–	–	–

These characters could be those of: 1, *Escherichia coli*; 2, most salmonellas; 3, *Salmonella typhi*; 4, *Shigella flexneri*; and 5, *Shigella dysenteriae* serotype 1 (Shiga's bacillus).

From the characters shown in this table a dichotomous key can be made, see Table D. 2.

Some diagnostic tables (e.g. those of Cowan & Steel) contain so many characters that the use of all of them is not essential to make an identification; this gives some freedom of choice to the user of the tables, a freedom appreciated particularly by workers in medical diagnostic laboratories.

dichotomous, dichotomy. Cutting in two. Used in dichotomous **keys**, which were invented by Lamarck for making an identification by asking a series of questions that can be answered directly and without qualification, e.g. motile?, rod-shaped?, spore produced?, all of which can be answered by yes or no.

The characters used are known as key characters (character couplets are the questions or contrasting statements). Characters common to all the individuals to be distinguished are useless as key characters, which must be of the kind that are positive in some individuals and negative in others. It may happen that within one of the smaller units (e.g. species) a character may seem to be variable and an ambivalent answer given to the leading question (lead). In such a case the character will be keyed out twice; where several characters are variable the keying can become very complicated, and several branches of the tree may end up with the same taxon.

An example of a simple key is shown in Table D. 2; a more complicated one designed to identify a bacterium down to the level of genus will be found in the 8th edition of *Bergey's Manual*, pp. 1098–1146; this was compiled by V. B. D. Skerman from the information contained in the Manual.

Table D. 2. *Simple dichotomous key made from the characters shown in Table D. 1*

Character	Organism
(*a*) Motile	
(*b*) Gas produced from glucose	
(*c*) Lactose fermented	1. *Escherichia coli*
(*cc*) Lactose not fermented	2. Most salmonellas
(*bb*) Gas not produced from glucose	3. *Salmonella typhi*
(*aa*) Not motile	
(*b*) Catalase produced	4. *Shigella flexneri*
(*bb*) Catalase not produced	5. *Shigella dysenteriae* (serotype 1)

difference. A difference between two organisms exists when the same character is recorded in different ways; e.g. indole production, + for one, − for the other; colony consistency, butyrous and friable; gelatin liquefaction, napiform and crateriform, and so on.

differentia, *pl.* **-ae.** A character that has distinguishing value and consequently is useful in compiling a **diagnosis** or making an **identification**. Sometimes but inadvisedly called a taxonomic character; all features of a taxon are taxonomic characters but they need not be *differentiae*, diagnostic, or distinguishing characters.

differential character. Derived from ***differentia***, this is a term to be avoided; a better one is **distinguishing character**, which means just what it says.

differential hosts. Term used in plant pathology for the species or varieties of plants whose reactions determine the physiological race of the micro-organism.

differential media. Media containing indicator and substances that give distinctive reactions (often colour changes) when certain organisms are grown on them. Examples are Russell's double sugar medium, Gillies's tubes, Knox's plate.

differentially-shaded similarity matrix. See **Sneath diagram**.

digraph. Two consonants or two vowels which in speech form a single sound, as *ph* (pronounced f), and *ui* (in fruit).

diminutive endings are often used to form generic names from the names of people; in bact. -*(i)ella* seems to be the most popular of the diminutives and is seen in *Edwardsiella, Klebsiella, Loefflerella, Shigella*. The suffix *-illus* was used by Heller to name 23 of the 25 genera into which she split *Clostridium*; these names included such prize specimens as *Ermengemillus*, *Metchnikovillus*, and *Omelianskillus*. The *-illium* ending is found in *Penicillium*.

dimorphic. Having two shapes or morphological states.

diphasic variation. F. W. Andrewes showed that many salmonellas had flagella in which different antigens predominated at different times; he called the phenomenon diphasic variation, but it has been shown since that three or four phase variations may occur, and it is better to use the term **phasic variation**.

diphtheroid. Term used in medical bacteriology for Gram-positive rods that resemble and may be confused with the diphtheria bacillus; usually a species of the genus *Corynebacterium*. Corresponds (in part) to the non-medical **coryneform**.

diphthong. Two vowels printed in appostion to indicate one vowel sound. Strictly, diphthong should be applied only to a compound vowel sound, such as *ou* in loud, made by combining two simple vowels. A single letter can be a diphthong, as the *ī* in *idle*, and the *ū* in *duty*. Pairs of vowels pronounced with a simple vowel sound (as *ou* in soup) are monophthongs or **digraphs** (Fowler, 1965).

The nomenclatural codes all use the word diphthong for **ligature**; in scientific names the letters should be separated,[1] but BOT.[2] allows both ligatures and the **diaeresis**.

[1, ZOO.ApE3. 2, BOT.73Fo.]

diplococcus, *pl.* **-i. 1.** A pair of spherical bacteria.

2. Formerly used as a generic name for many different paired cocci, e.g. *Diplococcus mucosus, Diplococcus intracellularis, Diplococcus pneumoniae.*

diploid. A term that describes the state of a cell that has a double set of chromosomes.

Direction. Term for (1) a decision that completes an earlier ruling of the Zoological Commission; (2) a formal instrument required under automatic provisions of ZOO. A Direction has the same status as an Opinion.[1]

[1, ZOO.78(d).]

disagreeable name. A legitimate name may not be changed or rejected on grounds of disagreeableness.[1]

[1, BAC.55(2); BOT.62.]

discontinuous variation makes classification possible. If variation were continuous it would not be possible to see breaks in an ever-changing spectrum of features. Biologists assume, without much justification, that during the course of evolution there are pauses during which lines of organisms are temporarily stabilized. At such stages it is possible to recognize groups such a species, and to define them by their phenotypic characters.

discordant elements. A term formerly used in BOT.[1] to describe a taxon in which all the units were not of the same kind, although when the taxon was originally described it was believed to be homogeneous. The term applied not only to plant specimens, but also to descriptions and illustrations.

In bact. a mixed culture is made up of discordant elements.

[1, BOT.70 – repealed 1975.]

discover, -y, -er. In microbial taxonomy the verb refers to the act of the individual who first found (and perhaps isolated) a micro-organism; the discovery itself does not play any part in taxonomy unless at the time it is published a name is attached to the organism discovered. If he does not publish his discovery legitimately (in terms of the appropriate nomenclatural code) and name the organism, the discoverer's name will not be given in the **author citation** when a later worker names the organism. Many microbiologists feel that the discoverer is more important than the man who names an organism, and they object strongly to the present rules. An example of the difficulty and confusion that may arise was Pfeiffer's discovery of a bacterium which, believing it to be the cause of influenza, he referred to as Influenzabacillus. This was in 1892, but he did not anticipate the Bacteriological Code (first published in 1948) or realize that he should have given it a two-part name in Latin and not a monominal in German. The common name of the organism is still Pfeiffer's bacillus (which is sensible as we now know that it does not cause influenza), but the scientific name is *Haemophilus influenzae* (Lehmann & Neumann) Winslow *et al.*, and it is probably true that none of the people named in the author citation, even when the *et al.* is spelt out, had at any time contributed anything to our knowledge of the organism.

discrete. Separate, not joining. A term used in describing colony form or morphology. A discrete colony is seldom an attribute of the colony but a demonstration of the technical ability of the person who spread the plate. Antonym: confluent.

discriminate. Distinguish between objects or organisms. A discriminating character is one that is present in every strain of the taxon, and may be unique to that taxon. A discriminating test is one that reveals a characteristic and constant feature of a taxon.

dissimilarity. See **index of dissimilarity**.

distance, to a numerical taxonomist, is a measure of similarity or difference. Similarity is regarded as zero distance, other distances represent the extent of the difference up to $\sqrt{n}$ when n = number of characters being compared. See **Euclidean distance, Manhattan distance**.

distinguishing character. Characteristic of an organism; commonly used in the wide sense of any one of many characters that distinguish the organism from others; occasionally limited to a character unique to the taxon, for which a better adjective would be exclusive or unique.

divide, *v.* To split into two or more parts. When a taxon is divided into nomenclaturally co-ordinate subdivisions (e.g. genus into two or more subgenera), one of the parts must retain the name of the original taxon and is the nominate subtaxon; this must be the part that includes the **type** of the taxon that was divided.[1] [1, BAC.39; 40; BOT.52.]

division. 1. In bot. the rank next below kingdom. The name of a division is uninominal, and BOT.[1] recommends that it should be taken from the distinctive characters of the division, or from the name of an included genus; it should end in *-phyta*, except where it designates a division of fungi when it should end in *-mycota*. A subdivisional name should end in *-phytina* (fungi, *-mycotina*).

BAC. is silent on the naming of these ranks, and bacteriologists make little or no use of taxa above **class**.

2. In zoo. the terms division or section have been used for primary subdivisions of a genus; the name (uninominal) proposed for such a division or section has the nomenclatural status of a subgeneric name provided that it fulfils the criteria of availability[2] (see **available**).
[1, BOT.(1975)R16A. 2, ZOO.42(d).]

divisive strategy. See **agglomerative strategies**.

DNA. Deoxyribonucleic acid, and acid derived from nuclear material, the specificity of which is determined by the four bases guanine (G), cytosine (C), thymine (T), and adenine (A). See **base composition** for a discussion of taxonomic implications. DNA occurs as a double-stranded molecule, termed 'native' when stranded together by specific interaction between complementary nucleotide pairs; A and T will pair and so will G and C, and these are called complementary base pairs. The two strands are also referred to as complementary strands; when separated they are said to be dissociated (or denatured); under suitable conditions complementary strands will reassociate to form a double-stranded helix similar to native DNA.

DNA homology is the degree (or percentage) of hybridizing capability between the DNA of micro-organisms.

DNA-hybridization. Common term used to describe the reassociation of single-strand DNA to form double-strand DNA, when one strand originates from one organism and the other strand from another organism. The double-strand DNA so formed is, therefore, a molecular hybrid. Because of the time-honoured use in biology of the word 'hybridization', there is some objection to using it in this context (see **duplex formation**), but in any case should be qualified to 'DNA molecular hybridization *in vitro*'.

The reassociation can take place when the **base compositions** (q.v.) of the two DNA samples are similar *and* when the base sequences are the same. The base sequences need not be similar over the whole length of the macromolecules, and quantitative methods have been developed to measure the extent of the homology in base sequences present. LRH.

DNase. Enzyme that breaks down DNA.

double citation. The **author citation** of the species name in parentheses, followed by the name of the author who placed that species in the genus. A double citation shows that the species named has been moved from one genus to another or raised in rank from a subspecies to species.

doublet analysis. Estimation of the frequency of occurrance of each of the 16 possible dinucleotides (GpG, GpC., GpA, GpT; CpG, CpC. . . etc.) in DNA. If the bases are distributed randomly, the expected frequencies of occurrence of each dinucleotide can be easily calculated. Actual doublet frequencies differ from random and it is of taxonomic interest to study systematic differences from random. LRH.

doubt. A name, although accepted by its author, that is followed by a question mark or other indication of **taxonomic doubt** is treated differently under the rules of BAC. and BOT. In bact.[1] such a name is not regarded as validly published, but in bot.[2] such a name is excluded from the provision that requires an unaccepted name to be treated as not validly published (i.e. it is regarded as validly published). The proliferation of negatives in BOT. obscures the

meaning and makes interpretation difficult.
[1, BAC.28b(1). 2, BOT.34N1.]

duplex formation. Term introduced by Walker & McLaren (1965, *Nature*, **208**, 1175) as a replacement for terms such as DNA hybridization. They strongly object to the use of the word hybridization for in-vitro molecular DNA/DNA or DNA/RNA duplex formation, since hybridization has a long-established meaning in biology. LRH.

D_T, expression used in numerical taxonomy for the total difference between two strains, taking into account differences in growth rate, temperature of incubation, and duration of incubation; described by Sneath (1968, *J. gen. Microbiol.* **54**, 1) as the sum of the Vigour Difference and the Pattern Difference.

D_T is the complement of the Simple Matching Coefficient and equals $1 - S_{SM}$.

E

-e. 1. Suffix used in bact. and bot.[1] to form a specific epithet from a personal name ending in *-a*, as *shigae* from Shiga and *balansae* from Balansa.

2. A personal name ending in *-e* can be used as the stem of a generic name or for a specific name or epithet. (1) To form a *generic name* BAC.[1] and BOT.[2] add *-a*, as *Beneckea* from Benecke; ZOO.[3] adds -us, -a, -um, as *Milneum* from Milne. (2) *Specific epithets*[1] (*names*[3]) are formed by adding *-i* to the name of a man or *-ae* to the name of a woman, as *sonnei* from Sonne (♂) and *josephineae* (or *josephinae*[4]) from Josephine.
[1, BAC.Ap9B; BOT.R73C. 2, BOT.R73B. 3, ZOO.ApD37. 4, ZOO.ApD18.]

-ea. 1. Suffix added to a personal name ending in *-a* to form a new generic name,[1,2] e.g. *Rochalimaea*.

2. When a personal name ends in *-ea* and it is to be used to form a generic name a suffix is not needed.[2]
[1, BAC.Ap9B. 2, BOT.R73B(a).]

-eae. Suffix attached to the stem of a generic name to form the name of a tribe in bact.[1] and bot.[2] The tribe that contains the type genus of the family must have its name based on the name of the type genus with its ending modified to *-eae*. Other tribal names used must be based on a legitimate name of an included genus.

The ending is not used in zoo.
[1, BAC.9; 21a. 2, BOT.19.]

ecads. Phenotypic modifications, or a range of variations from one genotype which are developed in response to differences in environment. The phenomen is also known as *phenotypic plasticity*.

ecological characters. Spatial relations of a parasite to its host which have been used to distinguish genera and species, e.g. *Phoma* occurs on the stems and *Phyllosticta* on the leaves of host plants.

ecology. Study of the relations between organisms and their surroundings, biological and physical.

ecotype, ecological type. A race genetically adapted to the environment.

edition. All the copies printed from one typesetting, or from a set of plates reproduced by offset. A book may be reprinted using the same type except for the correction of typographical errors, and this is generally termed a new impression. A second and subsequent edition generally implies a major revision or correction of the text with consequent resetting of the type.

The date of publication of each edition is normally printed either at the foot of the title page or, as in this dictionary, on the back (verso) of the title page; dates of reprintings (with or without revision) are also printed on the verso.

editorial whim. Editors are human, often overworked, and have their likes and dislikes. Young authors will find them kind and helpful people who will overlook (and correct) errors in journal style or convention. More experienced authors are expected to prepare papers in the style used by the journal of their choice. Editors vary in their preference for (or abhorrence of) italic type, use of capital letters, and so on, but they usually have good reason for their choice of style. Authors who disagree should send their papers elsewhere.

editors and publication. The codes of nomenclature make certain demands of editors (over whom they have no control) by instructing authors what they must do in their publications. BOT.[1] requires authors to *indicate precisely the dates of publication of their works*, and when a paper appears in more than one part to *indicate the precise date* of publication of the parts, and the number of pages and plates in each part. Separates are required to be dated with the year, month, and day of publication of the journal.

These demands may not be compatible with the **style** of the chosen journal, and while editors are open to suggestions and agree to reasonable requests, authors should approach them in a tactful and undemanding manner. They should bear in mind that an editor is not under any obligation to publish the author's *magnum opus*. However, some senior microbiologists think that editors have a duty to make sure that authors fulfil the requirements of the codes. [1, BOT.R45B.]

EDP. Electronic data processing.

effective publication. A name is effectively published in bact. and bot. at the time of its first use in accordance with the rules of BAC.[1] and BOT.;[2] the rules of the two codes are very similar. Publication is effective only (1) in printed form, and (2) when made available by sale or gift to the public and to scientific institutions. BAC.[3] states categorically that *No other kind of publication... is accepted as effective*. In contrast BOT.[2] will accept indelible **autographs** before 1953; this includes lithography and photographic reproduction of handwritten material. As so much printing today is by lithography, xerography, or other form of photographic reproduction, it is becoming increasingly difficult to interpret the rules concerning effective publication.

Publication is not effected by reading a paper at a scientific meeting or in printed abstracts of a paper read at a meeting, or in the minutes of a meeting; by the appearance of a name on a demonstration, on a label of a specimen (culture), or in a catalogue of a culture collection, museum, or seedsman, or by the issue of microfilm, microcards, or reports in ephemeral publications or non-scientific periodicals; or by the appearance of a name of a new taxon in a patent application or the issued patent.

elective culture. Method for the isolation of nutritionally exacting organisms by incubating the inoculum in a medium containing the essential substrate. The medium may contain substances inhibitory to other (unwanted) bacteria, or have a pH value unsuitable for the growth of many unwanted organisms. See also **enrichment media**.

electrophoregram, electrophoretogram. The migration pattern produced by proteins and enzymes in gel electrophoresis; the bands may be stained by the addition of appropriate substrates. Many factors affect the gel pattern and identical patterns obtained with different genera do not necessarily indicate that the proteins are identical. With closely related species similar patterns are more likely to be due to homologous proteins.

Many terms are given to the patterns; in addition to the above (electrophoregram is preferred by Strickland) others include electrophorogram, electrophoreogram, electrochromatogram, electrophoretic pattern, phoregram, pherogram, proteinogram, and proteinographic chart.

electrophoresis. Movement of charged particles towards one or other pole when an electric current is passed through a liquid or colloidal conducting system. Substances of different electrical charge migrate at different speeds and are said to have different mobilities; when the system is a colloid such as a layer of starch gel, different substances will be separated and, suitably stained, are seen as separate bands. The chemical nature of the bands can be determined by comparison with controls of known substances.

In taxonomy the technique is used to detect (and identify) proteins and enzymes in cell-free extracts of micro-organisms. Boulter & Thurman (1968) think that reliance on electrophoretic analysis to establish taxonomic relationships may mislead unless the information is supplemented by other studies on the properties of the proteins.

Electrophoresis may be used to separate proteins before they are identified by serological means (immuno-electrophoresis).

element. Part, specimen, strain (sometimes).

1. Term used in typification to describe the type specimen (strain) to which a name is permanently attached; it is not necessarily the most typical or representative part of the taxon.[1,2]

2. BAC. uses the term to mean strain or category in the statement *A taxon consists of one or more elements.*[2]

[1, BOT.7. 2, BAC.15.]

elevate. Used occasionally in taxonomy in the sense of raising a taxon from the level of one rank to another of higher rank. In the 1966 version of BAC. the term was used incorrectly when it was stated *A . . . name may be elevated*[1] for, as Skerman pointed out in his comments on the revision of the code, only a taxon can be raised in rank; a name cannot be elevated. In BAC. as revised, the phrase has become *designation may be elevated,*[2] which does not overcome Skerman's objection.

[1, BAC.(1966)8. 2, BAC.14b.]

elimination systems of identification. Methods in which one or a group of characters of an unknown organism is compared with the characters of known taxa to eliminate those that do not resemble the unknown; the procedure is continued with comparisons of other characters or groups of characters. Essentially this is identification by elimination, and is the basic principle of the step-by-step (progressive) method of Cowan & Steel, in which a series of **diagnostic tables** is used in sequence. It is also used in **punched-card systems** and in dichotomous keys in which one or two characters are considered at a time.

elision. The omission or suppression of a letter; common in speech, as *didn't*, but not in printed matter.

In nomenclature seen in the name *Haemophilus paraphrophilus* Zinnemann *et al.* (1968) the authors wished to indicate a similarly to *H. aphrophilus*, and obviously thought of para-aphrophilus for the specific epithet. However, before a vowel the prefix *para-* becomes *par-*, hence *paraphrophilus*.

In forming a zoo. species-group name from the name of a woman, a final *-a* or *-e* may be elided to make it euphoneous, e.g. *josephineae* or *josephinae* from Josephine.[1]

[1, ZOO.ApD18.]

-ella, -iella. Diminutives that may be appended to a personal name to form a generic name; the *-iella* form is used after names ending in *-s* or *-x*, as in *Edwardsiella* from Edwards, and *Coxiella* from Cox.

ellipsis, *pl.* **-es.** Omission of words. Apart from the grammatical uses of ellipsis (for which see *Fowler*), the omission of words from a direct (exact) quotation is shown by three periods (full stops).

emend, *v.* To correct an error (cf. **amend**).

emendation. 1. In zoo., an intentional change in the spelling of a name.[1] Two kinds of emendation are recognized: (1) *justified*, in correcting a mis-spelling; this retains its original author citation and date; and (2) *unjustified*, any other emendation; its nomenclatural status is determined by the new author and date; it becomes a later synonym of the original name. Both kinds of emendation become available names.[2]

2. In bact. and bot., emendation refers to revision of the definition of a taxon, or an alteration in its **circumscription**. When the alteration is considerable the word **emendavit** or the abbreviation emend., together with the name of the author who revised the definition, should follow the name.[3] Prudence is needed because nearly every author alters to some extent the definition or circumscription of a named taxon. In bot. little use is made of this convention unless the alteration is a major one affecting, for example, typification. Bacteriologists, on the other hand, have been liberal in the misuse of the abbreviation emend. for quite trivial alterations, which justifies the use in the codes of the word considerable to qualify the alteration.

[1, ZOO.33. 2, ZOO.19. 3, BAC.35; BOT.47; R47A.]

emendavit, *abbr.* **emend.** Appended to the name of a taxon of bacteria or plants when the diagnostic characters of the taxon have been considerably altered; see **emendation 2**; it is followed by the name of the author who makes the change.

In bact. and bot. the qualification *emendavit* and the qualifying phrase *mutatis characteribus* refer only to changes in definition; in zoo. the emend. refers to the spelling of the name.

empiricism. The third of Mayr's basic theories of taxonomy, by which taxa were based on all the known characters of organisms; this is essentially **Adansonism**. Between 1965 and 1968 Mayr changed his mind on whether it was fair to Adanson to attribute to him the unbiased *a priori* approach. Mayr's chief criticism in 1968 of the empirical approach is that it does not supply a reason for what he calls natural groups.

empiricists. Those taxonomists who do not base classifications on known or assumed phylogenetic relations, but on arrangements that depend on the facts of character distribution. The maximum association of such characters gives so-called phenetic classification; groups based on one or a few characters may be deliberately constructed to give frankly artificial groups.

encode, *v.* The act of representing the characters of an organism by symbols, a symbol in a particular situation having a specific meaning in relation to one character.

ending. Botanists show a preference for the word *termination* rather than *ending* or *suffix* (*Taxon*, 1976, **25**, 170).

endings of names. 1. Above genus the ending of a name can indicate the rank of the taxon to which the name is applied (see **suffix**).

2. The formation of generic and species names from **personal names** is subject to recommendations made by all the codes; in addition diminutive endings, *-ella*, *-iella*, and *-illus* may be added to a surname to form a generic name in bact.

3. The principle of priority does not extend to the endings of specific epithets (names). Because the codes differ in their recommendations for the latinization of personal names, a common error is to form the genitive incorrectly and use *-i* when *-ii* should be used; this is commoner than the opposite error, *-ii* for *-i*; ZOO.[1] treats these as homonyms. Such errors are also regarded as **orthographic errors**[2] which subsequent authors and editors have a duty to correct, and to ignore cries of 'priority' or 'original spelling' when criticized for so doing.

4. A specific epithet (species-group name) that is an adjective in the nominative singular should agree in gender and number with the generic name;[3] when the species in translated to another genus it may be necessary to change the ending of the epithet, e.g. *Sarcina flava* moved to the genus *Micrococcus* would become *Micrococcus flavus*.

[1, ZOO.58(10). 2, BOT.73. 3, BAC.12c; BOT.23; ZOO.34.]

endospore. A spore formed within the vegetative cell of a micro-organism. Bacterial spores are much more resistant to drying, to disinfection, and to rises in temperature than are their own vegetative forms, or the spores of fungi.

enrichment media. Used to isolate a particular organism from a source that probably contains several different organisms. They are of two kinds: (1) simple media to which is added a substrate attacked specificially by the wanted organism, and (2) complex media to which an inhibitor is added to suppress the growth of unwanted organisms. See also **selective medium**.

-ensis. Suffix added to a geographical name, or to its stem, to make an adjectival specific epithet, as *somaliensis* from Somalia. ZOO.[1] regards geographical names ending in *-ensis* and *-iensis* as homonyms.

[1, ZOO.58(11).]

enterobacteria. Common name for members of the family Enterobacteriaceae; used particularly by those who would reserve **coliform** for the lactose-fermenting members. Nowadays less importance is attached to the fermentation of lactose by members of the group, and both lactose fermenters and non-fermenters may be included in the same species. There is no correlation between pathogenicity and non-fermentation of lactose.

enzyme. A term for a **catalyst** of biological origin. Enzymes are proteins of high molecular weight; each is specific and catalyses only one kind of reaction. The substances acted upon are **substrates**.

The nomenclature of enzymology is as complicated as that of microbiology, but the microbial systematist can survive if he knows that in most cases the suffix *-ase* is added to the name of the substrate, e.g. urease, or to the reaction produced (dehydrogenase). The specificity of enzymes and the large number of substrates that can be attacked by bacteria make them useful taxonomic tools. See also **constitutive enzyme** and **inducible enzyme**.

enzyme inducer. A substance that induces the synthesis of an enzyme. Often the inducer is the substrate acted upon by the enzyme, but the terms inducer and substrate are not synonymous as some substrates for induced enzymes are not themselves inducers.

ephemeral publication. Brochures, leaflets, magazines, catalogues, society notices, lists of papers, and demonstrations are examples of ephemeral publications; such 'publications' are usually of limited interest and are unlikely to be preserved or stored in a library. They are not acceptable for the valid publication of scientific names.[1]

[1, BAC.25b(4); BOT.R29A.]

episome. A genetic element which is not essential to the reproduction of the organism since it may be absent. When present it may be part of the genome or of the cytoplasm; consequently the phenotype may be in one of two states, integrated or cytoplasmic. Maintenance of episomes such as colicin factors in a culture needs care, and single-colony isolations, which may not contain the factor, are to be avoided.

epistemology. The study of knowledge. Gilmour describes scientific epistemology as *the philosophy of science*.

epithet. 1. In bact. the second word of a binomial and peculiar to a *species* is called the specific epithet, as coli in the **species name** *Escherichia coli*. Bacteriological and mycological practice follow the same conventions but protozoologists follow ZOO. which does not now use the word epithet; instead it uses **specific name**, and the combination of the generic and specific names is called the **binomen**.

A specific epithet may be treated in one of three ways:[1] (1) as an adjective agreeing in gender with the generic name, (2) as a substantive in **apposition** in the nominative case, or (3) as a substantive in the genitive case. Epithets may be taken from any source or even be composed arbitrarily; if they consist of two or more words (expressing a single concept) they must be joined.[2] In bot. the specific epithet must not repeat exactly the generic name (**tautonomy**).[3] The formation of epithets from the names of people is regulated by the codes (see **personal name**).

2. In bact. and bot. the third element of the name of a *subspecies* (or variety)

is an epithet. Within the same species (or even within the same genus) the same subspecific epithet may not be borne by more than one subspecies;[4] this sensible requirement eliminates possible confusion should two or more subspecies (or species) be combined, or raised to the rank of species at some future time.
3. In bot. the word denoting a ***subgenus*** is an epithet[5] (in bact. it is a name[6]); it is either the same as the generic name or is a substantive.[7] It is printed with an intial capital letter and enclosed in parentheses between the generic name and the specific epithet.[8] It is always part of a combination of generic name+sub-generic+specific epithets, and is never used as the first part of a binomial. It is not counted as one of the words of a binominal name of a species or of the trinominal name of a subspecies.[9]
[1, BAC.12c. 2, BAC.12a. 3, BOT.23. 4, BAC.12b; 13c; BOT.24N; 64. 5, BOT.21. 6, BAC.10a. 7, BOT.R21B. 8, BOT.R21A. 9, ZOO.6.]

epitheta hybrida. Specific epithets made up of parts of words from two different languages.

epitheta specifica conservanda. Conserved specific epithets.

epitheta specifica rejicienda. Rejected specific epithets.

eponym. **1.** A person from whose name another name (often that of a place or institution) or word is derived; a name-giver. Examples in microbiology are Pasteur (Institut Pasteur, Pasteurian), Lister (Lister Institute, Listerian).
2. In transferred use means the person whose name has become a synonym, e.g. Bunsen (burner).

-er. (Bact., bot.) A **personal name** that ends in *-er* may be used to form a new generic name or specific epithet but is treated differently from a name ending in any other consonant, and even from one ending in *-r* not immediately preceded by the letter *e* (as Pasteur). To a name ending in *-er* add *-a* to form a generic name, and *-i* to form a specific epithet from the name of a man or *-ae* to the name of a woman.[1]

The generic names *Listeria* and *Neisseria* were not formed in accordance with the rules and recommendations of BAC. but were placed in the list of conserved generic names in 1954 (Opinion 12, 13) and their validity thus assured.

Zoologists do not treat personal names ending in *-er* differently from any other names ending in a consonant.
[1, BAC.Ap9B; BOT.73B; BOT.(1975)R73C.]

erect, *v.* To separate the component members, define the boundaries, select a type, and create a taxon. To complete the task the taxon must be named, and a description published in accordance with the appropriate code of nomenclature.

erose. Bot. term for denticulate edge; also used to describe the edge of a bacterial colony with fine, pointed projections. Alternative, dentate.

errors in names. Unintentional errors may be corrected by later authors who use a name, but the privilege should be exercised with caution, lest the finger of scorn be turned upon the corrector. ZOO.[1] does not allow the correction of such errors as incorrect latinization or transliteration, or the use of an inappropriate connecting vowel; on the other hand, BAC.[2] and BOT.[3] regard the correction of this kind of illiterate or **orthographic error** to be the duty of subsequent authors and the editors who publish their papers.

In zoo. an incorrectly spelt name has no standing in nomenclature and the name is not available;[4] inadvertent spelling errors (*lapsus calami*, printer's errors) may be corrected and the corrected name then becomes available but with its own date and author.[5]

[1, ZOO.32(a)(ii). 2, BAC.61. 3, BOT.73. 4, ZOO.19. 5, ZOO.33(a)(ii).]

errors in references should not be allowed to occur in print, and an author should check and double check *all* his references, especially those to papers of which he is author; memories are fallible and he may not have said what he now thinks (or wishes) he said.

What appears to be an extraordinary licence for laxity in BOT.[1] in the statement that *Bibliographic errors of citation do not invalidate the publication of a new combination* should not be taken at its face value for the 'errors' referred to may be the subjective interpretation (or misinterpretation) of the rules of the code, or of faulty attribution of names to their authors. BAC.[2] also turns a blind eye on incorrect citation, with the feeble excuse that the error can be corrected by a later author.

Unless an author is sure of his citations he should omit them. There can be no excuse for giving incorrect information in a scientific paper.

[1, BOT.33N2. 2, BAC.55(6).]

essentialism. One of Mayr's five basic theories of taxonomy (the other four are **nominalism**, **empiricism**, **cladism**, and **evolutionary classification**); it is the pursuit of knowledge to find the nature or essence of things. Applied to biological subjects, all members of a taxon would be expected to have the same essential nature, and to conform to the same type.

establish, *v.* A name becomes established when its publication satisfies the rules (articles) of the appropriate code of nomenclature. This is equivalent to being made **legitimate** (bact. and bot.) or **available** (zoo.).

established custom. This should be followed *In the absence of a relevant rule.*[1] Unfortunately established custom is not defined, may not be easy to determine, and it will vary from one country to another (cf. **usage**).

[1, BOT. Pre.]

established neotype. A **neotype** that has survived without objection for two years since its proposal was published in *IJSB.*[1]

[1, BAC.18e.]

et alii, et aliorum, *abbr.* **et al.** And others; used in literature citations in the text at the second and subsequent mention of a reference to a paper by three or more authors; at the first mention and in the list of references (**literature cited**), the surnames of all the authors are printed in full.

BAC. and BOT. recommend that *et al.* should follow the name of the first author when three or more authors are to be cited as authority for a name.[1]

[1, BAC.AdB(1); BOT.R46B.]

ethics. ZOO.[1] has a *Code of Ethics* to be observed by zoologists who publish new names; observance of the principles cannot be enforced and the Zoological Commission cannot consider complaints of their breach. A similar code is not found in BAC. or BOT.

[1, ZOO.ApA.]

etymology. The source or derivation of a word or name. The nomenclatural codes make different demands on their followers: BAC.[1] and BOT.[2] recommend that the etymology of names and epithets should be given when the meaning is not obvious; ZOO.[3] asks only for the etymology (and gender) of new genus-group names.

The usefulness of the information was illustrated in an annotation to a provisional code of nomenclature for viruses, which explained the epithet *nove-boracensis* as derived from the latinized form of New York, which in turn came from the Latin name for York in England.

[1, BAC.R6(5). 2, BOT.R73I. 3, ZOO.ApE16.]

Eu-. Prefix found in the older literature to indicate a subgenus that included the type of the genus, e.g. *Eu-Bacillus.* This usage is now obsolete.

eucaryote, -ic, *n.* and *adj.* Cell (or relating to a cell) in which the nuclear material is contained in a nucleus with a nuclear membrane; the genetic make-up of the organism is associated with the chromosomes of the nucleus, and is largely made up of DNA. Eucaryotic cells form the structure of all organisms except bacteria, blue-green algae, and viruses.

Euclidean distance. In numerical taxonomy, the distance between **OTUs** calculated by extension of Pythagoras' Theorem. OTUs can be considered as points in an *n*-dimensional space, one dimension for each character. Similar OTUs will be close to each other in this space and dissimilar OTUs widely spaced. If the space is defined as all dimensions being of equal length and each at right angles to all other dimensions, then the distances between OTUs can be calculated by simple extension of Pythagoras:

2 Dimensions: $h^2 = a_1^2 + a_2^2$ (a square),
3 Dimensions: $h^2 = a_1^2 + a_2^2 + a_3^2$ (a cube),
n Dimensions: $h^2 = a_1^2 + a_2^2 + a_3^2 \ldots + a_n^2$. (an *n*-dimensional hypercube),

where h = hypotenuse

and $a_{1,2,3\ldots n}$ = adjacent sides of the right angle.

LRH.

evolutionary classification. The fifth of Mayr's basic theories of taxonomy; this is Darwinism and depends on common ancestry and later divergence. In practice relationship is inferred from the similarity of *a posteriori* weighted characters. Mayr believes that these methods (of which he heartily approves) are essentially those which *the great masters of taxonomy have practised for more than a hundred years* as if such long usage endowed the resulting taxa with greater reality and an existence more credible than the man-made taxa of those who avoid any weighting of characters.

ex. From. Used in author citations to indicate the original source of a name; bact. and bot. usage differs and are dealt with separately.

1. In bot. *ex* can mean one of three things: (*a*) an author A proposed a name but did not publish it validly; later another author B published the name and attributed it to A; in this case the citation should read A *ex* B.[1] (*b*) Author C published as a synonym a manuscript (i.e. unpublished) name of author D, the word *ex* is used to connect the names of the two authors as D *ex* C pro syn.[2] (*c*) BOT.[3] has a new use for *ex* (instead of square brackets) when a publishing author E ascribes a name to author F who used the name before the starting date for nomenclature of the group concerned;[4] the author citation will be F *ex* E. Note that in all these examples from botanical usage the name of the publishing author follows the word *ex*.

2. (a) BAC.[5] deals with the use of *ex* in a manner contrary to the bot. usage shown above in **1** (*a*) and (*b*). The advice given is that the name of the publishing author precedes that of the original author, as *Bacillus caryocyaneus* Dupaix 1930 *ex* Beijerinck. BAC.[6] also deals with the use after 1980 of *ex* (in parentheses) when a **revived name** is published by author G and the original (pre-1980) author H is to be indicated. Author G would publish the name as *Bebus suis* (*ex* H 1900) nom. rev. and subsequent authors should use the form *Bebus suis* (*ex* H 1900) G.

[1, BOT.R46C. 2, BOT.R50A. 3, BOT.R46E. 4, BOT.13. 5, BAC.AdB(3)(b). 6, BAC.Prov.B3.]

exact systematics. Name given by Smirnov to the application of statistical methods to taxonomy, claimed to be the forerunner of numerical taxonomy, from which it differed in using characters weighted by frequency, the infrequently-occurring character receiving the greater weight.

example. All the codes give examples to show how the rules work in practice. Examples are interpretative and are useful in showing the different ways in which names and citations are printed, but they are not an integral part of a code and when an example conflicts with a rule it is the rule that must be followed (*Reg. Veget.* **56**).

exclusion of type. When a taxon is divided the name is retained by that part which includes the type; the other part, from which the nomenclatural type has been excluded, must be given another name.[1]

[1, BOT.52; 53; BAC.37a.]

exclusis generibus (specibus, varietatibus), *abbr.* excl. gen. (sp., var.). Excluding genera (species, varieties). A qualifying phrase appended to an author citation to show that some genera (species, varieties) included by the original author are excluded from the taxon as redefined.[1]

[1, BOT.R47A.]

exemplar. A representative sample of a taxon, chosen at random in the belief that variation within the specimen will be less than among all organisms in the survey, and with the realization that it may be atypical in some characters.

exemption. When a biologist thinks that the application of the rules affecting the nomenclature of his branch of biology will upset existing usage or produce instability of nomenclature he may apply for the rules to be set aside and an action taken that is contrary to the rules, or he may argue that the rules have been misinterpreted by users of the code. His application may request rejection of one or more names and the conservation of another.

In bact. and zoo. the procedure is to ask for an **Opinion** of the appropriate Commission on the point at issue. ZOO.[1] states (and it speaks for both Commissions) that it is not under any obligation to look for violation or evasion of the code, to verify information supplied by the applicant, or even to initiate action, though it may do all or any of these things. In other words, the Commission does not act as a police force but rather as a judiciary; it is up to the individual

who thinks he has found a flaw in the code (or in its application) to start proceedings to right the wrong.
[1, ZOO.81.]

existing usage. Defined by ZOO.[1] as the most common usage. It is to be applied to the name of a taxon when that name or a synonym is under review by the Zoological Commission or by the General Committee responsible for botanical nomenclature.[2]
[1, ZOO.80(ii). 2, BOT.R15A.]

exobiologist. Biologist with an eye on the future, perhaps studying extraterrestrial forms of life.

exobiology. 1. The biological study of extraterrestrial matter; the study of meteorites, cosmic dust, and material from the moon; in the future possibly material from other extraterrestrial sources.
2. Biology, especially taxonomy, as it may appear in the future. A visionary pastime that supposes, among other things, that chemical methods such as analysis of DNA base sequence will be practicable for everyday use and the results will be recorded, stored, and recovered by machine. Methods such as these may allow a classification or an identification to be made without looking at the organism.

exotic organism. 1. An organism with unusual characters; one that seems likely to be a member of a hitherto unknown taxon. On further examination it may be found to be a mixed culture, perhaps growing in symbiosis. This is the biologist's conception of an exotic organism.
2. Biochemists have other ideas; to them an exotic organism is one that has one or more extreme characteristic, such as an ability to grow at a particularly high or low temperature.

exsiccata, pl. -ae, n. A set of dried specimens, particularly one distributed by a herbarium or an individual. After 1 January 1953 a name included in printed matter accompanying exsiccata is not effectively published, but if the printed matter is distributed independently of the dried material, then publication of the name may be effective.[1]
[1, BOT.31.]

exsiccate. A specimen of a fungus dried by slow evaporation of water. Such a specimen (*exsiccatum*) could be the type of a name whereas a **desiccate** (a viable product of freeze-drying) could not be.[1] The type is a specimen of a taxon; the taxon is of a certain rank, e.g. species, and the type is then the type of the species name. Although exsiccates are not dried with the object of maintaining their viability, Ainsworth (1962, *Nature*, **195**, 1120) has reported survival for 52 years, and many other specimens have remained viable for a great many years.
[1, BOT.9.]

exsiccatum, pl. -a. Dried specimen of a fungus, usually permanently mounted. May be designated as a type specimen.

exsiccatus, -a, -um, adj. Dried. Used in mycology to describe a dried specimen (*fungus exsiccatus*) but not a freeze-dried culture; a dried specimen can also be described as *specimen exsiccatum*, and a dried plant as *planta exsiccata*.

F

f or ph? SOED describes *ph* as a consonantal digraph; the combination used by the Romans to represent the Greek letter ϕ, phi. In mediaeval Latin, the Romanic languages, and popular usage, *f* was often substituted for *ph*, hence fancy, fantastic.

In scientific literature *ph* is preferred by writers of English, *f* by French, Spanish, Italian, and US writers. A WHO expert committee on pharmaceutical preparations advised that *f* should be used in preference to *ph*.

In the scientific names of micro-organisms the spelling used should be that of the original author, even to the extent that an English writer should use the spelling *Desulfovibrio*. ZOO.[1] regards species-group names differing in spelling only by *f* or *ph* as homonyms.

[1, ZOO.58(7).]

factor analysis. See **principal component, coordinate analysis.**

familia nova, *abbr.* **fam.nov.** New family. Qualifying phrase used by the author who proposes and names a new family and it appears only in the original publication in which the family is named and described. In any later publication, even by the original author, it is replaced by the name of the original author.

family (and **subfamily**). **1.** Ranks between **order** and **genus**. Names used for the ranks in bact. and bot. will be considered together; they are plural adjectives used as substantives made by adding the suffixes *-aceae* (for family)[1,2] and *-oideae* (for subfamily)[1,3] to the stem of the name of an included genus. For the subfamily that includes the type genus of the family, the suffix is added to the stem of the name of the type genus.[3] In bot. the merits of long usage are recognized and alternatives are permitted, e.g. Labiatae (Lamiaceae), Compositae (Asteraceae), Cruciferae (Brassicaceae); bact. has an exception in Enterobacteriaceae in which the type is *Escherichia* and not *Enterobacter*.

In zoo. the name must be based on a valid name of a contained genus and must be a noun in the nominative plural;[4] the suffixes added to the stem of the type genus name are *-idae* (for family) and *-inae* (for subfamily).[5]

2. An individual taxon of the rank family (or subfamily).

[1, BAC.9T. 2, BOT.18. 32, BOT.19. 4, ZOO.11(e). 5, ZOO.29.]

family-group. Designation used in ZOO.[1] to include the ranks superfamily, family, subfamily, and tribe. Suffixes added to the stem of the name of the type genus to form names of these taxa are: superfamily, *-oidea*[2] (note the similarity to the suffix *-oideae* used in bact. and bot. for subfamily names); family, *-idae*[3] (note the identity with the suffix for subclass of Cormophyta); subfamily, *-inae*[3] (identical with the suffix for subtribe in bact. and bot.); tribe, *-ini*.[2] A family-group name based on an incorrectly formed stem should not be amended if it was proposed before 1961 and is in general use, but it must be corrected if it was formed after 1960.[3]

Each taxon of the family-group is defined by reference to its type-genus.[4] All the categories are nomenclaturally co-ordinate, and identical family-group names based on different type-genera are homonyms, even if they are of different rank (and consequently have different endings).[5]

[1, ZOO.35(a). 2, ZOO.R29A. 3, ZOO.29. 4, ZOO.35(b). 5, ZOO.55.]

family tree. A tree-like figure intended to represent a classification; the verticals show a change in the number of similar features, not a change in time and in this differ from a genealogical tree.

See **cladogram**.

fancy name. A term used for a cultivar (cultivated plant) that is neither the botanically correct name nor the common name. Not used in microbiology.

feature. **1.** A synonym for **attribute** or **character** (q.v.)

2. In numerical taxonomy each state of a character is a feature. In a two-state (+, −) character each is a feature; when a character may exist in several states (as colony form) several symbols are used and each represents a feature.

ferment, fermentation. **1.** A general term for the reaction between the enzymes of micro-organisms and a chemical substance.

2. Used in relation to the anaerobic attack on carbohydrates and alcohols, in which acid production is detected by the change in colour of an indicator. Fermentation is distinguished from the aerobic attack on carbohydrates, which is known as **oxidation**.

fermentative type. **1.** Used by Kauffmann for subdivisions of a serotype in which the fermentation reactions differ; at one time he regarded these as subspecific entities, and the serotypes as species. Later he modified the term to sero-fermentative type.

2. Used in defining the method by which a carbohydrate (usually glucose) is broken down. A simple test to distinguish between fermentative and oxidative types of bacteria was introduced by Hugh & Leifson (1953, *J. Bact.* **66**, 24), and this has proved to be a test of considerable importance in taxonomic work.

fide. When a reference has not been seen in the original (or photocopy) and is quoted second hand, this fact should be stated either by the words 'cited by' or '*fide*' as Smith (1900), cited by Jones (1950) or as Smith (1900) *fide* Jones (1950).

field strains. In bact. frequently used for strains of recent isolation which need to be distinguished in a study from older, usually culture collection deposited and well-documented, strains. LRH.

filament. Descriptive term for morphological features, some of which, like axial filament, may be important taxonomically.

Many rod-shaped bacteria produced elongated or filamentous forms.

filiform. Growth along the line of inoculation in a stab culture, usually in a nutrient gelatin medium incubated at a temperature below the melting point of the medium (i.e. < 25 °C). Some bacteria later begin to hydrolyse the gelatin, which liquefies the medium, sometimes in a characteristic manner (**stratiform**, **infundibuliform**, or **saccate**).

filter. To explain the principles and the working of the codes of nomenclature Jeffrey (1973) described a *nomenclatural filter* in the form of a dendrogram, and a modified version referring particularly to BOT. was drawn up by Hawksworth (1974). The filters deal successively with the publication of names (unpublished names are rejected in the divergent branch), the formation of names, their legitimacy; subsequent layers in the filter deal with typification and priority.

fimbria, *pl.* **-ae.** Hair-like processes on the surface of bacteria they occur on both motile and nonmotile forms and are not associated with the motility of the bacterium. They can be seen in electron micrographs and their presence assumed by certain adsorption phenomena. They appear to play a part in the conjugation of bacterial cells, and some at least have earned the name 'sex fimbriae' (**sex hair**).

The derivation of the word fimbria is discussed by Duguid (1966, *J. Path. Bact.* **92**, 137) and by Duguid & Anderson (1967, *Nature*, **215**, 89). A later synonym used by most US workers and geneticists is *pilus*, pl. *-i*. It might be helpful to use fimbria for the morphological entity and pilus for the genetic sex form.

first reviser. When more than one name for a taxon, or the same name for different taxa are published simultaneously, priority is determined by the actions of the first reviser. The term first reviser is to be *rigidly construed*,[1] see **reviser 2**. [1, ZOO.24(a)(i).]

fir-tree growth. Term used to describe the appearance produced by anaerobic non-liquefying bacteria (e.g. *Clostridium tetani*) in gelatin stab culture. To see this appearance it is essential to incubate the gelatin stab below 25 °C.

The opposite appearance, the inverted fir-tree, is seen in growth of aerobic non-liquefying bacteria such as *Bacillus anthracis*; again cultures must be incubated at a temperature below the melting point of the medium.

flagella. Plural of **flagellum**; unlike candelabra, flagella is always the plural form.

flagellar index. The percentage of cells in polarly flagellated bacteria that have more than one flagellum per pole. The index was first used by Lautrop & Jessen in an attempt to define precisely the terms **monotrichous** and **multitrichous**. A low index indicates that only a few cells have more than one flagellum at a pole (monotrichous), and a high index that many cells have several polar flagella. Among pseudomonads, *P. aeruginosa* had an index of 3, and *P. fluorescens* one of about 50.

flagellin. The protein that makes up a bacterial flagellum, and is characteristic of the strain or species. In the monomeric state the protein flagellin exists as isodiametric particules, and these units are polymerized into long, multistranded threads with a regular helicoid form. Flagellar wavelength and thickness are constant properties of a particular strain, and possibly are characteristic of a species. Differences between bacterial flagella are presumably due to differences in the primary structures of their flagellins, which in turn affect the mode of polymerization.

flagellum, *pl.* **-a.** A thread-like process found on the surface of many bacteria. In wet preparations flagella show lashing movements and are probably responsible for the **motility** of most bacteria; in this they differ from **fimbriae** which do not contribute to motility. Sometimes flagella occur in non-motile bacteria and they are then said to be paralysed. A bacterium may have a single flagellum which will generally be at one pole of a rod, or it may have many flagella collected at one or both ends (in which case they are described as *polar*) or they may arise from any or all parts of the bacterial body, when they are said to be *peritrichous*. Sometimes peritrichous flagella become twisted together and bind themselves over the body of the bacterium, to become free like a whip (or cat-o'-nine-tails) at one pole, thus simulating polar flagella. Even in an electron micrograph it may not be possible to decide whether the flagella are polar or peritrichate (see Hayward & Hodgkiss, 1961, *J. gen. Microbiol.* **26**, 133).

The site of attachment of flagella has been used as a taxonomic character and in some classifications great weight has been placed on it. The difficulty of determining the site of attachment with certainty, and the observations that it may change with conditions of culture have lessened the importance now given to the site character, but has not decreased the value of the presence or absence of flagella, or of the motility of the organism. Leifson regards the wavelength of the flagella as important and believes that classifications should pay more attention to this feature.

Antigenically the flagella form the H antigens of a bacterium.

The plural, flagella, may be erroneously treated as singular and a new plural in *-ae* has been used by those who should know better.

fluorescence, fluorescent. The property and ability to absorb light of short wavelength and emit visible rays of longer wavelength. The effect is to produce an apparent luminosity in various colours, e.g. under u.v. light many pseudomonads produce a bright green fluorescence.

fluorescent antibody staining. Bacteria and viruses stained by adding the homologous antibody labelled with a fluorescent dye; the preparation is examined under u.v. light.

fluorescent microscopy. Bacteria can be stained by fluorescent dyes (e.g. auramine, fluorescein) and will fluoresce when examined under u.v. light.

font. See **fount**.

forgotten name. See ***nomen oblitum***.

form, forma (subform, subforma). **1.** Ranks (categories) below variety and subvariety; recognized in bot.[1] but not in bact. or zoo. The names are subject to BOT., but see also ***forma specialis***.

2. An individual taxon of the rank form (or subform).

3. The word form (but not forma) is sometimes used to mean shape, as in coccal form, or other morphological state, as in L form.

[1, BOT.4.]

-form. Suffix to indicate a bacterial infrasubspecific taxon, e.g. chemoform. An alternative is **-var.**[1]

[1, BAC.Ap10BT.]

forma specialis, *pl.* ***formae speciales.*** Botanical rank used for a parasitic strain (especially of fungi) adapted to a special host, whose name the parasite often takes. The names of *formae speciales* are not subject to BOT.[1] In bact. used only occasionally; although not subject to BAC., these *formae speciales* are mentioned in it.[2]

[1, BOT.4N. 2, BAC.Ap10BT.]

formation of new names. All the codes[1,2,3] give advice on what to avoid in the formation of new names; much of this advice overlaps and, as little of it conflicts, a summary of the combined recommendations may be helpful. Each code gives different and sometimes conflicting advice on the formation of names from **geographical names** and **personal names** (q.v.).

(1) Let a new name have an agreeable form; avoid names and epithets that are long or difficult to pronounce when latinized.

(2) Avoid combining words from different languages (hybrid words).

(3) Avoid small differences in spelling between the new name and other

older names. Within one genus avoid specific epithets that are very similar in spelling.

(4) Do not adopt an unpublished name found in notes, letters, etc. unless the original author has approved publication, then see **ex 1**.

(5) Do not name an organism after a person unconnected with the discipline, or at least with biology. When a personal name is used it should recall someone connected with the organism, e.g. its discoverer.

(6) Avoid forming a new generic name that duplicates one in use in another biological discipline.

(7) Avoid adjectives used as nouns for generic names; within one genus avoid substantival and adjectival forms of the same name for different species.

(8) Do not make a generic name from a specific epithet (within the genus) or vice versa, or form an epithet with the same meaning as the generic name (**pleonasm**).

(9) Avoid hyphenated words; if they must be used, delete the hyphen and join the parts.

(10) Avoid ordinal numerals as epithets.

(11) Give the etymology of the new name if this will be helpful.

(12) Make sure that the new name (epithet) is formed and published according to the rules of the relevant code of nomenclature.

These twelve recommendations (not commandments), if followed, should lead to the better formation of new names.

ZOO.[3] gives the sort of advice needed by microbiologists who want to form a new scientific name for a taxon; it was a detailed appendix on the latinization of Greek words (BAC. 1958 editon has a shorter version) with tables showing how Latin and Greek words form their stems (the stem used in nomenclature is not always that shown in grammars), and gives examples of family names based on these stems. See **generic name** for recommendations on the formation of a new name for a genus.

A specific epithet may be (1) an adjective, which should agree grammatically with the generic name (*Neisseria flava*, *Staphylococus flavus*); (2) a noun in the nominative case in apposition with the generic name (*Vibrio comma*); or (3) a noun in the genitive (*Vibrio cholerae*, *Salmonella typhi*). Although specific epithets (and specific names in zoo.) are usually formed from Latin or Greek words, none of the codes prohibits the formation of names in an arbitrary manner, but regardless of their derivation the words formed must be treated as if they were Latin words. Letters such as *j*, *k*, *v*, *w*, and *y* may be used (or retained) but the vowels in ligatures (*æ*, *œ*) should be separated. The archaic **diaeresis** is better avoided (but is allowed by BOT.), and **diacritic** signs should be removed.

As an example of the possibilities in the formation of a new name or epithet we will consider the unpromising Turkish word yoghurt, variously spelt in English as yaourt, yogurt, and yoghurt (*SOED*) as a candidate for a specific epithet. *Index Bergeyana* has only one epithet formed from the word, namely, *Bacterium yoghurt* Kuntze 1908 which, to say the least, is unimaginative and unambitious but it is an allowable form as a Latin termination is not mandatory. *Bergey's Manual* (1974, 582) cites the spelling *jugurti* but let us adopt the English spelling yogurt. We could latinize this (as the stem ends in a dental sound) by adding a diminutive such as *-ulus*, *-ula*, *-ulum* which would be euphonic but pretentious. Instead we can add *-us*, *-a*, *-um* to make nouns, or *-ianus*, *-iana*, or *-ianum* to make adjectival forms. The adjectival forms could be used as *Micrococcus yogurtianus*, *Neisseria yogurtiana*, *Chromobacterium yogurtianum*; as nouns in apposition the epithets would be *M. yogurtus*, *N. yogurta*, and *C. yogurtum*, or in the genetive *M. yogurti*, *N. yogurtae*, *C. yogurti*.

When a new name (epithet) had been formed a **proposal** should be made in such a form as to obtain **valid publication**, and at the same time to designate the type to which the name is attached; this should be accompanied by a description and diagnosis (in Latin for algae and fungi).[4]

In contrast to ZOO., BAC. and BOT. concern themselves more with the name changes that occur when a taxon is moved laterally or vertically in the hierarchical scale; these movements do not involve the formation of new names but changes in the endings of epithets.

In all three codes there are rules and recommendations on the formation of new names (generic and specific) from **personal names**, and both BOT. and ZOO. deal with names formed from **geographical names**.[5] See also **compound words**, **family**, **family-group**, **hybrid words**, and **mongrel combinations**.

[1, BAC.R6; R10a; R12c. 2, BOT.R20A; R23B. 3, ZOO.ApD. 4, BOT.36. 5, BOT.R73D; ZOO.ApD22, 23.]

form genus. 1. Form-genus (with a hyphen) is a genus unassignable to a family, but it may be referable to a taxon of higher rank.[1] Used in bot. for fossil specimens.

2. In mycology form genus (without a hyphen) is a genus for imperfect states, e.g. Fungi Imperfecti. Some form genera may have perfect states referable to several different genera.

3. Used (without a hyphen) by some early bacteriologists for a genus, e.g. *Spirillum*, that was characterized mainly on morphology (shape or form). To Cohn every form (shape) that showed big differences was regarded as a genus, and small deviations from this became the species.

[1, BOT.3N1.]

form species. Term used by Cohn and other early bacteriologists for the small differences in morphology that characterized the different species of a genus. Although he accepted any small deviation from the type as a *form species*, Cohn recognized that further work might show that some were merely stages in one life history; he was also aware that form species based on morphological differences were inadequate substitutes (necessitated by the limitations of technical methods) for natural species. Within a morphologically-defined form species he postulated that varieties or races would be found, and that these would be characterized by chemical physiological differences.

fossil record of micro-organisms is virtually unknown, but claims are made that bacteria have survived for millions of years in paleozoic salt deposits, and also in the mud of lake beds. In the macrobiologies the fossil record plays an important part in working out evolutionary (phylogenetic) classifications, but according to Mayr even the large number of known fossil bird species has not contributed enough evidence to improve on previously accepted classifications.

fount (US, font). The characters (letters and figures) used by printers. The type faces differ in appearance (e.g. **bold-face**, *italic*, roman), in size (measured by the point system in which 72 points = 1 inch), and in legibility (cf. black-letter and sans serif). This paragraph is printed in Linotron Imprint 9 point, 2 point leaded (2 point space between lines).

The type face and size of type are normally determined by editorial, journal, or printer's practice, and the use of different faces for scientific names is governed more by **convention** than by rule. See **typography 2**.

freeze-drying (US, lyophilization). Drying biological material by sublimation of water from the frozen state without passing through a liquid phase. Microorganisms dried in this manner remain alive (but with metabolism suspended) for years, and the method is used to prevent changes due to mutation or variation, and so preserve the organisms in a state as close as possible to the original.

fundamentals. Basic, pertaining to the foundations on which all else is built. Taxonomic theories (for classification, nomenclature, or identification) have their own premises (which may be expounded as Principles, Heresies, or Fundamentals) which are usually simple and obvious to the extent that taxonomists run the risk of not questioning them. LRH.

Fungi Imperfecti. Fungi in which only asexual spores are taken into account. Ainsworth says that possibly a third of the species have named perfect (i.e. sexual) states but the connexion is unrecognized; a third may have unnamed perfect states; and the remaining third really do not have a sexual state.

fuzzy connexions. In numerical taxonomy, the connecting together of two branches of a dendrogram over a range of similarity values corresponding to all the pair-wise links between the two branches. This is more difficult to draw and to retain clarity, but the linkage of two branches by simple line at one given similarity level is a gross simplification. LRH.

G

gamete. A sex cell which carries only half the number of chromosomes of the non-sex or fertilized cell. Because it carries only one of each kind of chromosome it is described as haploid.

GC content of DNA. See **base composition**.

gap. See **phenetic gap**.

gelatin liquefaction. One of the oldest characterizing features in bact. is the liquefaction of gelatin, and its appearance in stab culture. To see the characteristic growth and shape of the liquefied area incubation at a temperature below the melting point of the medium is essential. The shape of the liquefied area depends to a large extent on the oxygen requirements of the organism.

The appearances of the growth are given the terms **filiform**, **fir-tree**, and inverted fir-tree; these are appearances given by non-liquefying bacteria that are facultatively anaerobic, strictly anaerobic, and strictly aerobic respectively. Liquefaction is described as **crateriform**, **infundibuliform**, **napiform**, **sac-**

cate, and **stratiform**. The differences between the appearances of the liquefied areas are not clear-cut, their importance is questionable, and the observations are seldom made in routine identification work. Their importance, if any, lies in making a description as complete as possible.

In twentieth-century bacteriology gelatin liquefaction in stab cultures incubated at 22 °C is replaced by other tests (e.g. Frazier's) for gelatin hydrolysis in which the cultures can be incubated at temperatures optimal for growth.

gender of names. 1. Names of bacterial taxa above the rank of genus are feminine in gender, and plural in number.[1]

2. The gender of a generic name is that of the Greek or Latin word from which it is derived; when it is a compound of two words the name takes the gender of the second. Diminutives ending in *-ella* and *-ina* are feminine. Arbitrarily formed generic names and adjectives used as nouns have the gender assigned to them by their authors.[2,3,4] BOT.[3] states that generic names ending in *-oides* and *-odes* are feminine; in this it differs from ZOO.[4]

BAC. generally follows BOT. but does not legislate for words ending in *-oides* or *-odes*. An official Opinion (no. 3) has fixed the gender of names ending in *-bacter* as masculine. A generic name formed from a **personal name**, whether that of a man or woman, should have a feminine form (e.g. *Neisseria*).[5]

ZOO. requires the author of a new generic name to state its gender.[6] Generic names ending in *-oides*, *-odes*, and *-ops* in zoo. names are masculine. A genus-group name that exactly copies a modern Indo-European language word has the gender of that word if the language has genders.[4]

3. Epithets may be (1) adjectives in the nominative singular and must agree in gender with the generic name[7] (*Staphylococcus aureus*, *Neisseria flava*), of (2) substantives in apposition when they need not agree in gender with the generic name, or (3) substances in the genitive (possessive) and again need not agree in gender with the generic name (*Salmonella typhi*).

An adjectival epithet formed from the name of a person takes the gender of the generic name, but when the epithet is a substantive its gender is determined by the sex of the individual after whom the taxon is named.[8] (See Tables P. 2 & P. 3, pp. 198–9.)

[1, BAC.7. 2, BAC.65. 3, BOT.R75A. 4, ZOO.30. 5, BAC.R10a; BOT.R20A. 6, ZOO.ApE16. 7, BAC.12c; BOT.23. 8, BAC.Ap9B; BOT.(1975)R73C.]

gene. The part of the genetic material (DNA) that determines the nature of a particular macromolecule in the cell; a sequence of 4 elements repeated in different permutations. It is an inherited unit that governs a characteristic, not necessarily shown by (in) the possessor of the gene. Genes are located in the chromosomes of the cell nucleus, and they determine the properties of the enzyme-protein that catalyses a particular reaction. *The rules of the genetic game are the same for all living beings.*

general classification. A classification based on as much data as practicable. Such classifications will be useful for the widest variety of purposes and correspond to Gilmour's concept of natural classification (*Nature*, 1937, **139**, 1040–2). It is in deriving general classifications that numerical taxonomy is most useful, but the use of a numerical taxonomy method will not, in itself, make the resultant classification a general one. If the characters used are all of one kind (e.g. exclusively biochemical), then the classification will be a **special classification**. LRH.

general considerations. An introductory section of BAC. that corresponds roughly with the Preamble of BOT. and ZOO. Shown as GC in the minireferences to the codes appended to some entries in this dictionary.

generic name. 1. Name given to a taxon of the rank of genus; the first letter is written with a capital and the word is in the singular number. The plural of a generic name is a common (non-Latin) word and should not be underlined in a manuscript or printed in italic (see **plural forms**).

The rules allow a new generic name to be formed in any way an author may wish, but see **formation of new names** for suggestions and recommendations from the various codes. Often the name is taken from a patronymic, i.e. the surname of an individual, usually the name of a person who did important work on the group of organisms named after him. Diminutives -(*i*)*ella* or -*illus*, may be used but can be abused as by Heller who added a diminutive suffix to the names of many distinguished bacteriologists associated with anaerobic bacteria, and thus formed a series of ridiculous names such as *Robertsonillus*, *Macintoshillus*. A modicum of responsible reflexion would have avoided such abominations.

The name of a genus is a noun in the singular but BAC. allows use of an adjective treated as a noun;[1] it is co-ordinate with (subject to the same rules as) a subgeneric name. When used alone as a noun a generic name is printed in a distinctive type (usually italic) with an initial capital, as in 'a species of *Brucella*', but when used as an adjective it has a lower case initial letter as in 'a brucella species'. It may be taken from any source whatever and may be an anagram of an existing name. When names are formed from patronymics BAC. and BOT.[2] recommend that a feminine form should be given to the names of individuals of either sex; ZOO.[3] lists suffixes to be added to **personal names** (q.v.).

2. Valid publication of a generic name requires that it is accompanied by a description, diagnosis, or reference to a previously published description.[4] In a monotypic genus this may be a combined generic and species description,[5] or exceptionally, an illustration. See **valid publication** for other requirements of BAC. and BOT.

3. When writing about a genus with several species the generic name may be repeated many times and some abbreviation is called for by editors. The accepted convention, approved in BAC.,[6] for binomials is to spell out the generic name at first mention and subsequently to abbreviate it to the initial capital. But if in the paper there are two different generic names beginning with the same letter the names should be spelt out at each mention.

[1, BAC.10a; BOT.20. 2, BAC.R10a; BOT.R20A. 3, ZOO.ApD37. 4, BAC.27; BOT.41. 5, BAC.29; BOT.42. 6, BAC.AdA.]

generitype. Type of the name of a genus, i.e. the type species. In bact. the type of the type species is a strain or culture (description or illustration when the organism cannot be grown or preserved in the living state). Thus the generitype is the type strain of the type species.

genetic code. The 64 different triplets (codons) that can be formed from the four bases (guanine, cytosine, adenine, and thymine) that occur in DNA each specify an amino acid. The correspondence of each triplet to an amino-acid is called the genetic code. It contains considerable redundancy since only 20 amino-acids are coded for by the 64 triplets (e.g. phenylalanine is specified by either of two triplets, leucine by either of six triplets, etc.). LRH.

genetic homology. The similarity of genetic make-up between two organisms. The likelihood of genetic homology may be estimated by the similarity of the overall **base composition** of the DNA, by the ability to form molecular hybrids, and other evidence of genetic recombination. It is determined by the similarity in the sequence of their nucleotides containing the bases adenine, thymine, guanine, and cytosine; these act as the coding symbols that store genetic information.

LRH. adds: given the known redundancies in the **genetic code** it is unlikely that similarity in base sequences could be due to anything other than common ancestry.

genetics. Defined by Hayes as the *study and analysis of heredity in all its aspects*, and described as the *fundamental biological science, the focal point upon which all other aspects of biology necessarily converge*.

Genetics involves observing the characters and particularly the differences between different generations; characters themselves are not inherited, but only the ability to produce the characters.

genetic symbols. Standardized symbols and abbreviations for use in characterizing genetic loci, genotypes, phenotypes, and also in genetic descriptions. Proposed by Demerec, Adelberg, Clark & Hartman (1966, *Genetics*, **54**, 61; reprinted 1968, *J. gen. Microbiol.* **50**, 1).

genome. The genetic constitution of a procaryotic micro-organism.

genome size can be expressed either as the molecular weight of DNA (in molecular weight units, daltons; oxygen = 16 daltons), as the number of nucleotide pairs per genome (DNA molecular weight divided by 663, the average molecular weight of the sodium salt of bound nucleotide pairs), or, in an approximate way, as the number of cistrons per genome (assuming about 1500 nucleotide pairs per cistron). Useful in the molecular biology approach to taxonomy as a gross genetic character, it varies from about 0.2×10^9 daltons (0.3×10^6 nucleotide pairs, approx. 200 cistrons) for *Mycoplasma gallisepticum*; 4×10^9 daltons (6×10^6 nucleotide pairs, approx. 4000 cistrons) for *Bacillus subtilis*; to 14.5×10^9 daltons (22×10^6 nucleotide pairs, approx. 14000 cistrons) for *Saccharomyces cerevisiae*. LRH.

genophore. The double-stranded DNA fibrils that form the morphologically distinct nucleoplasm of the procaryotic cell, not separated from the cytoplasm by a nuclear membrane.

genospecies. A term used by Ravin for organisms in which interbreeding is possible; this is shown by genetic transfer and **recombination**. This is probably the most practical definition of a bacterial species at present available.

genotype. 1. The hereditary potential of an organism forms the genotype; it is a theoretical unit since the physical unit that exists (the phenotype) is the result of the reaction between the genotype, the internal environment, and the medium in which the organism grows. The phenotypes arising from one genotype may be different because the medium may vary.

2. A single strain or species on which a genus is founded. This usage (by nomenclators) is to be avoided,[1] and the term type species should be used when referring to the type of a genus.
[1, ZOO.R67A.]

genus, *pl.* **-era. 1.** One of the basic ranks in the hierarchical systems used in biology; probably the highest rank of any significance in microbiology. In position between **family** and **species**, it is best considered as a collection of species with many characters in common; unfortunately no one has indicated the extent of this sharing of characters and it is based purely on personal judgement (in other words, it is a matter of the utmost nicety) as to what constitutes a genus; like **species** (q.v.), the genus is a subjective concept.
2. An individual taxon of generic rank.

genus-group. Term used in ZOO.[1] for ranks below **family-group** and above **species-group** in the hierarchy of classification; it includes only two categories, genus and subgenus and the names applied to them are **nomenclaturally co-ordinate**[2] and subject to the same rules of ZOO. Each taxon is objectively defined by reference to its type-species. A genus-group name must be a noun in the nominative singular or if it is not it must be treated as one.[3]

Names of **collective groups** (which do not need type-species) are treated as genus-group names.[4]
[1, ZOO.42. 2, ZOO.43. 3, ZOO.11(f). 4, ZOO.42(c); Gl.]

genus novum, *abbr.* **gen.nov., gen.n.** New genus. The qualifying phrase is appended to the name of a new genus only by the author at the time he proposes the new name.[1,2] As in the case of new species names it was considered bad form for an author to add his own name to the proposal of a new name, that was left to other (subsequent) authors; practice is changing and authors are advised to be less modest. The ZOO.[2] recommendation is gen.n.
[1, BAC.33aN1. 2, ZOO.ApE7.]

geographical name. 1. A new epithet formed from the name of a place is usually treated as an adjective and is formed by adding the suffixes *-ensis*, -(*a*)*nus*, *-inus*, *-ianus*, or *-icus* to the name.[1] When the name has a Latin form (e.g. Aquae Sulis = Bath) this should be used in preference to the modern name of the place; ZOO.[2] requires that a name in two parts shall be joined without the hyphen, as *aquaesulis*. The original spelling or a transliteration into Latin form (e.g. *sinensis* rather than *chinensis*[3]) are used, and Latin forms of mediaeval writers are preferred to latinized forms of modern names, e.g. *londiniensis* rather than *londonensis*.[4]

In geographical names the abbreviation St is spelt out and joined to the rest of the name in different ways as *sancti-johannis* (bot.[5]) or *sanctipauli* (zoo.[6]).
2. Geographical names are often used in modern (non-Latin) form in the designation of serotypes, especially in the Salmonella group of bacteria. BAC. does not deal with names for the infrasubspecific forms. Custom has established the use of italic type and a lower case initial letter for the latinized forms of these names (as *Salmonella arizonae*) and for those in modern form (*Salmonella derby*).
[1, BOT.R73D. 2, ZOO.26(a); 32(c)(i). 3, BOT.R73E. 4, ZOO.ApD23. 5, BOT.(1975)R73C(j). 6, ZOO.ApD22.]

g-form. A small-colony variant of a staphylococcus produced by growth on medium containing lithium chloride, antibiotic, or other agent that inhibits normal growth. When subcultured to nutrient media without inhibitor the g-form gradually reverts to the normal colony form.

GLC. Gas–liquid chromatography, which is beginning to assume an important role in bact. taxonomy being a **polyphenic** (q.v.) technique, yielding in one experiment considerable amounts of quantitative information. LRH.

'good' character. A character that is stable in the individual culture and is regularly found in different strains of the taxon. An added quality that will be welcomed by the taxonomist is an ease of observation or demonstration. These are essentially the qualities of 'good' characters of micro-organisms.

'good' species. A species readily separable from other species, usually by some pathognomonic character. Examples are *Staphylococcus aureus*, distinguished by being coagulase positive, and the pneumococcus (*Streptococcus pneumoniae*), distinguished from other streptococci by its bile solubility. In these two species only the expert would dare to assign a strain negative in these characteristics to the one for which it was more or less specific and unique. On the basis of this definition a 'good' species would seem to be a **monothetic** group.

grammar and nomenclature. There is no such thing as a grammar of nomenclature, but because nouns (from which generic names are usually formed) in Latin and Greek have genders, and adjectival epithets must agree with the generic name in number and gender, the ending of an epithet must be changed if a taxon to which it is attached is moved to a genus with a name of different gender. Thus, *X-us albus* transferred to a genus *Y-a* (feminine) would become *Y-a alba*.

Gram reaction. H. C. J. Gram was a Danish physician who described a method (not a stain) for staining bacteria in sections of tissue; briefly the section was stained by a violet dye, mordanted with an iodine solution, and decolorized by alcohol. Not all bacteria were stained as some had been decolorized by alcohol. Modifications were made by other workers, the most important being the showing (by a counterstain) of the bacteria that had been decolorized by the alcohol or other solvent, and by applying the method to smears or films of cultures.

Bacteria that retain the first (violet) stain are said to be Gram positive; those that are decolorized and take up the counterstain (usually pink) are Gram negative. Gram-positive bacteria may become Gram negative (i.e. lose the ability to retain the violet–iodine complex) in old cultures, or in cultures at unsuitable pH values. Although the reaction should generally be determined on young cultures, a few bacteria (e.g. *Agrobacterium* spp., *Acinetobacter anitratus*) are Gram negative in young cultures but after several days of incubation may become Gram positive.

group. 1. An assemblage of co-ordinate nomenclatural categories, such as family-group, genus-group, species-group, and collective group.[1] Note that apart from collective group, all these zoo. nomenclatural group names are hyphenated.

2. A non-commital name without nomenclatural significance for an assemblage of organisms of doubtful position within the hierarchy of classical taxonomy.

3. A collection of organisms from one source such as locality, kind of animal, plant, and so on.

4. A subdivision made on the basis of antigenic analysis, e.g. the Salmonella group, Arizona group; some workers regard these groups as the equivalent of genera, others as species, but most regard them as examples of a genetically unstable population and hence of uncertain taxonomic status.

Another example of a primary serological classification or subdivision concerns the streptococci, the groups of which are based on precipitin reactions of cell wall polysaccharides and teichoic acids. These groups can be broken down further into serotypes on the basis of carbohydrate and protein reactions; the organisms of group A can be further subdivided by the action of specific bacteriophages into phage types.

5. The word group was strongly disliked by some taxonomists and its use discouraged, but it was a losing battle and the word retained its favour in the eyes of most bacteriologists.

[1, ZOO.42(c); Gl.]

group number. Term applied to the characterizing code used in conjunction with the early versions of the SAB (Society of American Bacteriologists) **descriptive chart** or card. *Escherichia coli* had the group number B.222.11110. The group number was deleted from later versions of the chart. A suggested alternative is **Chester code**.

group variation. In biosystematics this term is used for variation between populations. It has no connexion with the term formerly used in describing diphasic variation of bacterial flagella antigens (see **phasic variation**).

H

halophil(e), halophilic. Preferring salt. Terms applied to organisms that tolerate and grow well in concentrations of NaCl above the physiological (about 0.85%). Marine forms require 1.5–5% NaCl and extreme halophils may need concentrations up to saturated NaCl.

handbook. According to Blackwelder, handbooks are designed for laymen and *do not form part of the real taxonomic literature*. They may be in the form of field guides or of student manuals, and *occasionally they are scholarly volumes*.

The original version from which this dictionary developed was described as a Handbook in a series University Reviews in Botany, and it is to be hoped that it fitted into the last category.

H antigen. Antigenic component of a bacterium contributed by the flagella. First described by Smith & Reagh (*Journal of Med. Res.*, 1971, **10**, 189), the flagellar antigens were named H (meaning Hauch) by Weil & Felix (*Wiener klin. Wochenschr.*, 1917, **30**, 1509). The H antigens are more labile than the O antigens and are destroyed by heating at 70 °C for 20 minutes.

haploid. A term that describes the state of the cell which contains one set of chromosomes.

Harvard system. A well-known **name–date** system.

headphrase. Phrase or expression defined, discussed, or commented upon in the entry that follows. This serves for phrases and terms what the **catchword** does for individual words, viz. acts as a lead to the information sought in this dictionary.

helix-coil conversion or **transition** of DNA. Change from the double-strand state (configuration) to the single-strand state; more frequently called denaturation for helix-to-coil and renaturation for coil-to-helix. LRH.

heretical taxonomy. An attempt by destructive criticism to make bacterial taxonomists think seriously about the place of taxonomy in general and nomenclature in particular, to get their priorities sorted out, to see the absurdity of some retrospective applications of rules of nomenclature, and to notice their effects on the discipline and indiscipline of bacteriologists.

The main arguments of heretical taxonomy (Cowan, 1965, *J. gen. Microbiol.* **39**, 143; 1970, **61**, 145; 1971, **67**, 1) can be summarized: (1) Good descriptions of organisms are more important than correctly formed Latin (or latinized) names. (2) Good taxonomy is based on good benchwork to characterize the organisms, not in assiduous library searches for old and often unused names. (3) Classifications are temporary arrangements made on available knowledge; as information increases classifications may change. Classification is dynamic and should not be tied to a static nomenclature. (4) Identification, which is the practical application of taxonomic thought to everyday events, will be helped by simple and logical assessment of characters; when characters can be organized in a systematic or coded form identification should be improved. (5) Names, which cannot have any meaning in microbiology, may be a hindrance to good work because members of associated disciplines (for whom identifications are made) may be naive enough to assume that a name means something.

The advantages of numerical codes and numerical taxonomy are more obvious to the heretical taxonomists than to the classical taxonomist steeped in pseudo-Latin, but the aims of all are laudable and not, as some would have it, antagonistic.

The first breakaway from the hidebound traditions inherited from botanists was made by virologists, but so far their initiative has been neither productive nor successful; bacteriologists, at whom heretical taxonomy was aimed, seem to have been influenced to some extent and this can be seen in the simplification of the Bacteriological Code in its latest revision, and in the Approved Lists of Bacterial Names which should do much to remove nomenclatural absurdities.

heteroecious. Describes those rust fungi that have one sporing part of their life cycle in one host and a second in an unrelated host species.

heterofermentation, heterofermentative. A type of fermentation used by leuconostocs and some lactobacilli in which the end products include CO_2 or acetic acid in addition to lactic acid (cf. **homofermentative**).

heterogeneous. 1. A mixture of two or more kinds in a sample or specimen. A taxon said to be heterogeneous would consist of more than one kind of organism; if the term were used to describe a microbial type specimen it would indicate a mixed culture. Difficult to distinguish from **discordant elements** of botany, but McVaugh *et al.* (1968, *Reg. Veget.* **56**, 15) imply that this is possible.

2. Leaving aside type specimens (strains, cultures), the descendants of a microbial colony could be described as heterogeneous *vis-à-vis* the descendants of a single-cell isolation (a clone).

heterokaryon. Mycelium of a fungus in which nuclei of different genetic stock share the same cytoplasm.

heterologous. Term used in serology to indicate the relation (or rather the lack of relation) between an antigen and the antibody to another (heterologous) antigen. Thus if antigen A is inoculated into a rabbit, the rabbit is stimulated to produce antibody to A, and this we will call antibody α; antigen B inoculated into a different rabbit stimulates it to produce antibody β. Antigen A and antibody β are heterologous, as are antigen B and antibody α. By contrast, A–α, and B–β are homologous antigen–antibody pairs.

heterotroph, -ic. An organism that needs organic material as an energy source (cf. **autotroph**).

heterotypic synonym. Bot. term for a taxonomic synonym., i.e. for names not based on the same type. See Table S. 3, p. 253.

hierarchical system. An arrangement of objects or organisms into a series of groups which are assigned to a succession of categories or **ranks** of different seniority. The nomenclatural codes treat an organism as a member of several consecutive categories of which the species is regarded as basic.

In microbiology it is unusual to found a new species on a single strain, but groups (taxa) higher than species may contain only one member of the next lower rank; e.g. when founded the genus *Listeria* had only one species, *L. monocytogenes.* In building up a hierarchical system we combine taxa of a low rank to form a taxon of the next higher rank; thus starting with the lowest unit, the individual (which is not a taxon), several of which have some characters in common may be collected into a taxon known as a **species**, and species with many similar features may be collected into a **genus**. If it is thought profitable to do so, taxa of higher ranks can be formed by the collection of two or more genera with some features in common.

In bact. the ranks of taxa most often used and accepted are (from below): variety or subspecies (these include serotypes and phage types, the most important categories to epidemiologists and clinicians), species, genus, family, and order. Provision is made for several intermediate categories (e.g. tribe, suborder) but these are less often used. The general opinion of bacteriologists seems to be that while it is useful to have the concepts of genus and species, higher categories are even more the figments of the imagination of taxonomists. *Only the individual is real*; all other categories are man-made collections.

hierarchy. A result of grading according to rank, in which each individual organism is a member of taxa of all the higher ranks of the hierarchy. Thus each species is a member of a genus, of a family, and so on up the scale.

BOT. makes provision for 22 subdivisions of the plant kingdom (Regnum Vegetabile) and further supplementary ranks may be intercalated or added.[1] Not all the ranks are used by a discipline, and the different disciplines vary in the ranks they made available (see Tables R. 1, R. 2, p. 220). The sequence of the ranks must not be varied.[2]

The names of most ranks are governed by the rules and recommendations of the codes of nomenclature. The names of bacterial groups below subspecies are not subject to the rules,[3] and may be represented by letters (as the groups of streptococci), numbers (as serotypes of pneumococci), names of places (the salmonella serotypes), or even by latinized words (as *Salmonella salinatis*), or by a series of numbers and/or letters to indicate sensitivity to particular phages.

Similarly in zoo., infrasubspecific category names are not subject to ZOO.;[4] and BOT. rules extend to the rank of subform, but exclude forma specialis.[5]

[1, BOT.4. 2, BAC.5a; BOT.5. 3, BAC.14a; Ap10B. 4, ZOO.1. 5, BOT.4N.]

history of a culture. In microbial taxonomy the history of a strain may be important, especially in subsequent comparative work. The kind of information that will probably be useful many years hence includes: (1) source of material from which the culture was isolated; (2) date of isolation, by whom, and on what medium; (3) subsequent subculture history – medium, frequency, temperature of incubation and of storage; (4) transfer to other laboratories, with dates; accession numbers in other collections; (5) dates when characters were checked, and by whom; (6) date on which the strain was freeze-dried, method and suspending fluid used; dates and results of viable counts; dates of character checks on dried product.

HMO. See **hypothetical mean organism**.

holograph. A handwritten document. A name or description in such a document reproduced by a photographic or mechanical process before 1953 was effectively published under BOT.[1]

[1, BOT.29.]

holomorph. All the character states of an organism; morphological, physiological, and so on.

holotype. The specimen (strain, isolate) designated by the original author of a name as the type of the name.[1] If the author based the taxon on one specimen (strain), that specimen is the holotype, whether designated or not. In bacteriology but not in mycology the holotype will be a living culture (in mycology a dead specimen), and every possible precaution must be taken to preserve it so that (1) it retains its characteristics, (2) it does not become contaminated, and (3) that it survives. When a holotype exists the application of the name is automatically fixed.

[1, BAC.18b; BOT.7; ZOO.73.]

homeotypical, *adj.* Describes a strain found to be of the same species as the nomenclatural type.

homofermentation, homofermentative. The type of fermentation carried out by streptococci and by some lactobacilli, in which carbohydrate is converted entirely to lactic acid (cf. **heterofermentation**).

homogeneous. Uniform, alike, consisting of similar units. In biology a specialized meaning: made up of elements similar by reason of a common origin; e.g. a single-cell culture, the progeny of a single microbial cell, is made up of cells all more-or-less alike, and certainly more alike (more homogeneous) than those descended from a single colony picked from a nutrient medium. The antonym is **heterogeneous**.

homoiology. Similarity due to parallelism.

homologous. 1. Similar characters (usually anatomical) due to common ancestry are described as homologous; the characters themselves are *homologues*. In this sense it has little or no use in microbiology.

2. In serology, the antigen that has been used to stimulate the production of antibody is homologous with that antibody; the antigen and antibody are homologous, and react together when mixed outside the animal body. The antonym is **heterologous**.

homology. 1. Having a common origin, often a matter of interpretation. Homologous features are those that can be traced back to a feature in a common ancestor; this involves circular reasoning since homology itself is invoked in identifying the features of the recent and the ancestral organism.

2. genetic homology, comparison of the similarity of the genomes of different organisms. Proof of genetic homology is seen in the **duplex formation** between single strands of denatured bacterial DNA, in **recombination**, and in **transformation**.

homology group. A taxon based on the results of in-vitro hybridization experiments, believed by many to provide the best basis for the circumscription of bacterial genera. But these experiments on duplex formation are not everyday tests in routine laboratories, and are not likely to become useful taxonomic tools until the results can be determined by a simple method. Meanwhile taxa will continue to be defined (or described) by their phenotypic characteristics.

homonym, homonymy. The same name (i.e. spelt in exactly the same way) given to two or more different organisms (taxa based on different types) are homonyms. This simple statement needs amplification; homonymy can exist only between names of taxa of the same rank, and at the species and subspecies level does not extend beyond the species of one genus. Generally the difference of a single letter is sufficient to avoid homonymy,[1] but orthographic variants (different connecting vowels, different transliterations, changes in gender and endings of specific names[2] or epithets, words with the same meaning but differing slightly in form[3]) are regarded as homonyms.[4,5]

Because of the interdependence of the nomenclature of microbes the homonymy of bacterial generic names extends to the names of algae, fungi, protozoa, and viruses.[6] Names that are later homonyms are illegitimate and must be rejected; even if the earlier name is illegitimate, the later homonym must be rejected unless it has been conserved.[4] After 1 January 1980 names of bacteria published under the rules of BAC. are not to be rejected as homonyms of names published under earlier editions of the code.[6]

The adjective *senior* may be attached to the first published name and *junior* to the same name (applied to a different organism) that is published later. The homonymy of names applies to the single word of generic names and names of ranks above genus. The epithets pyogenes in the names *Streptococcus pyogenes*, *Staphylococcus pyogenes*, and *Corynebacterium pyogenes* are not homonyms in a classification that accepts the three genera, but in a classification in which two of these genera are combined the epithets in the names of the combined genera will be homonyms. Names (epithets) subject to the same rules compete in matters of homonymy, except those species and subspecies names (epithets) that are based on the same type.[7]

In zoo. two other kinds of homonym are distinguished:[8] *primary homonyms* are identical species-group names applied to different taxa within the same genus-group when they are first published; *secondary homonyms* are those species-group names that become homonyms by combination of taxa or by transfer of a species from one genus to another.

BOT.[4] deals with the unlikely event of the same new name being published simultaneously for more than one taxon; if this should happen the first author who adopts the name for one of the taxa and rejects it for the other is to be followed.

[1, ZOO.56(a); 57(d). 2, ZOO.57(b). 3, BAC.Ap9A. 4, BOT.64. 5, ZOO.58. 6, BAC.51b. 7, BAC.40cN. 8, ZOO.Gl.]

homoplastic similarity. Similarity not due to common ancestry; characters may incorrectly be regarded as homologous.

homoplasy, homoplastic. Descriptive terms applied to organisms which show similarities or resemblances that are not due to common ancestry.

homotypic, *adj*. Used by botanists in connexion with synonyms based on the same type; also referred to as nomenclatural, objective, or obligate **synonyms.**

Hugh & Leifson test. A method for determining whether carbohydrates are broken down by anaerobic fermentation or by oxidation. Originally described by Hugh & Leifson (1953, *J. Bact.* **66**, 24), the test has been modified by several workers by using another indicator or by producing a greater degree of anaerobiosis in a jar of the McIntosh & Fildes or Brewer type.

hybrid, *n*. The progeny of a cross between two genetically dissimilar organisms. In the macrobiologies hybrids may be fertile or infertile; in microbiology the only hybrids that can be detected must be self-reproducing.

hybrid name. Compound word made up of stems from two languages, e.g. the specific epithet *pseudomallei* made from Greek and Latin roots. Looked upon with disfavour by purists.

hybridization. **1.** The formation of a hybrid, which may be fertile or infertile. **2.** Used by molecular biologists to mean the formation *in vitro* of double-stranded DNA molecules when one strand originates from one DNA source, and the other strand from a different DNA source. This usage is acceptable when it is qualified in such phrases as 'molecular hybridization' or 'in-vitro hybridization'.

On the other hand, in view of the classical usage (1 above), this usage is not recommended, and P. M. B. Walker insists on the term *duplex formation*. LRH.

hyperbaric. Relates to a condition where the ambient pressure is greater than the barometric or atmospheric pressure, and it is applied to air or gases.

Hyperbaric air or oxygen may be bactericidal to some bacteria (e.g. *Corynebacterium diphtheriae*), but cells of *Bacillus subtilis* and *Escherichia coli* that are synthesizing catalase can survive 10 atmospheres of oxygen for at least 18 hours.

hyperellipsoid. In numerical taxonomy indicates a cluster of points (**OTU**s) defined in a **hyperspace** and distributed in the form of an ellipsoid. LRH.

hyperoxia. Oxygen at elevated pressure.

hyperspace is used by numerical taxonomists to represent the *n* dimensions needed to plot the distance between *n* characters. As in space, the closer two specimen points lie, the more characters the specimens they represent have in common.

hypersphere. In numerical taxonomy a cluster of points (**OTU**s) defined in a **hyperspace** and distributed in the form of a sphere. LRH.

hypha, *pl.* **-ae.** One of the filaments of a mycelium. In bact. hyphae are found only in the actinomycetes.

hyphen. Earlier editions of BAC. and BOT.[1] made similar recommendations about the use of hyphens in specific epithets, viz. when an epithet consisted of two or more words, these were either to be united or hyphenated. If they were not joined or hyphenated when first published they were to be joined by subsequent authors. Editors, too, had a responsibility and should not have allowed publication of a specific epithet of more than one word.

Buchanan, who believed that a name should have a meaning, showed in his annotation to Rule 6 of the 1958 Code that if the meaning of the epithet was to be retained the problem was not quite as simple as the wording of the rule implied. For example, *Bacillus albus lactis*, meaning the white bacillus of milk, could not be hyphenated because in this name the white and the milk referred to two different things; if the words were combined to *B. albolactis* the meaning was altered to the bacillus of white milk.

The choice of hyphen or union was open to the author, and this still holds in bot. In bact. hyphens are not now permitted in newly formed epithets, and where one has been hyphenated in the past the parts should be joined.[2] The parts should refer to a single concept.

Only BOT.[3] mentions two or more words in a generic name; they should by joined by a hyphen.

ZOO.[4] allows only one use of a hyphen, when the first element of a species-group name is a Latin letter used to denote a character, e.g. *c-album*. In all other cases a name based on two separated words must be united without a hyphen.
[1, BOT.23. 2, BAC.12a. 3, BOT.20 4, ZOO.26(c); 27.]

hyphenated qualities. A composite adjective made up of two words is normally hyphenated, as Gram-positive, and such adjectives are often used in the descriptions of micro-organisms. But the same two words may be used as a predicate, when the hyphen is dropped; 'The Gram-positive bacillus' becomes 'The bacillus was Gram positive', but in some texts the hyphenated form is also used as in '*E. rhusiopathiae* is non-motile, non-capsulated, and non-sporing'.

hypodigm. Consists of all the specimens known to and regarded at a given time by an author as belonging to a taxon. Essentially it stands for the source material on which the definition and characterization of a taxon is made at a given time, and is compatible with the concept that a taxon is a population. The hypodigm changes with our knowledge of the taxon so that the term, coined by Simpson for zoologists, can be useful to microbiologists.

The hypodigm rather than a single 'type' specimen should be used for any subsequent comparison or for identification work. Even when a species was founded on one specimen, that specimen should not have any more weight or authority attached to it than is given to subsequent specimens that contribute to the hypodigm. However, Simpson limited contributing specimens to those that formed the basis of published works, which need not be contemporaneous.

A reviser may add or delete specimens from the hypodigm, which is never a static list of specimens, but the **name-bearer** should always be a specimen within the hypodigm.

hyponym. The name of a taxon that cannot be recognized because of the vagueness or inadequacy of the description as published by the author to whom the name is attributed.

hypothetical mean organism (HMO). Described by Tsukamura & Mizuno (*Japan. Journal Microbiol*, 1968, **12**, 371) as *an organism prepared from the mean of positive characters.* It is based on characters most frequently observed and omits those that occur less frequently than the mean number of positive features. If in the taxon examined there are n strains, and these have $x_1, x_2, x_3, \ldots x_n$ positive characters, the mean of positive characters (X) per strain will be

$$X = \frac{(x_1+x_2+x_3+\ldots x_n)}{n}.$$

The HMO is not a number but a list or sequence of positive characters; to work out the HMO the characters observed are recorded against strains as shown in Table H. 1 below; the number of times each character is positive is recorded in a right-hand column; the number of characters positive in each strain is shown on the bottom line of totals. In the simple example shown in the table, the total number of positive characters in the five strains is 14, giving a mean of 2.8. The HMO starts with the character occurring as positive most frequently (A), and is followed by characters positive in order of decreasing frequency until the mean number of positive characters is reached; if at the last frequency two or more characters are positive equally frequently, all are recorded.

Table H. 1. *Simple illustration of how the HMO is worked out and comparison with CMO of Liston* et al. (*see* **calculated mean organism**)

Character	Strain 1	2	3	4	5	No. positive	Median characteristic
A	+	+	+	−	+	4	+
B	+	+	−	+	−	3	+
C	+	−	+	−	−	2	−
D	+	−	−	+	+	3	+
E	+	−	−	+	−	2	−
Total positive characters	5	2	2	3	2	14	

The HMO for the example is ABD; the CMO is A+, B+, C−, D+, E−.

The HMO for the organism in Table H. 1 is ABD, and corresponds to the positive characters of the CMO of Liston *et al.* When more characters and strains are tabulated the HMO and the positive features of the CMO may not be identical but will not show any great discrepancy. The HMO is not as versatile as the CMO, and cannot express characters determined quantitatively.

It will be noticed that in the example none of the strains has all the characters of the HMO, but that does not exclude a strain from the taxon, which is **polythetic**.

hypothetical mean strain (HMS). The HMS is the description of a hypothetical strain based on the method of Liston *et al.* (1963) (see **calculated mean organism**) which takes into account the frequency of both positive and negative characteristics. In Table H. 1 the median characteristics are shown in the last column to compare them with the HMO of Tsukamura & Mizuno.

I

i. When, in a bot. name, there are alternative spellings with *i* or *j*, choose *i* before a consonant, and *j* before a vowel (*Jungia*, not *Iungia*).[1]
[1, BOT.(1975)73.]

-i. 1. Suffix used in zoo. to form a specific name (or species-group name) from the name of a man.[1]
2. Suffix used in bact. and bot.[3] to form the genitive of a personal name of a man ending in *-er*, *-y*, or a vowel except *-a*, and so form a substantival specific epithet. See also **-ii**.
3. A personal name ending in *-i* may be used in bact.[2] and bot.[4] to form a generic name by adding the letter *-a*, as *Petria* from Petri.
4. In zoo.[5] the terminations *-i* and *-ii* of patronymic genitives are regarded as homonyms.
[1, ZOO.R31A. 2, BAC.Ap9B. 3, BOT.(1975)R73C. 4, BOT.R73B. 5, ZOO.58(10).]

-ia. 1. In bact.[1] and bot.[2] the suffix added to a personal name ending in a consonant (except *-er*) to form a new generic name, as *Nocardia* from Nocard.
2. *Neisseria* Trevisan 1885 and *Listeria* Pirie 1940 do not conform to the rule but their authors could not have been expected to anticipate the BAC. first approved in 1947. The names were well-established and in constant use; they were made legitimate as conserved names by actions of the Judicial Commission in Opinions 12 (*Listeria*) and 13 (*Neisseria*).[3]
[1, BAC.Ap9BT. 2, BOT.R73B. 3, BAC.Ap5.]

-iae. In bot.[1] the suffix added to the name of a woman ending in a consonant (except *-er*) to form a substantival specific epithet, as in *wilsoniae* from the name of Ms Wilson.
[1, BOT.(1975)R73C.]

IAMS. International Association of Microbiological Societies, the organization responsible for the international congresses on microbiology, and ultimately for the nomenclature committee (ICSB) responsible for BAC.

-ian. A suffix attached to the name of an individual to form an adjective descriptive of the work, thought, or discovery associated with the person, e.g. Darwinian. As in the abstract-noun form (**-ism**), the use of the suffix is associated with honour to the person named, and is indicative of the esteem with which his views (work, discoveries) are usually held. Other examples are Adansonian, Pasteurian.

IAPT. International Association of Plant Taxonomy, q.v.

-iarum. Suffix added in bot. to the personal name of women to form a substantival specific epithet.[1] Presumably all the women should have the same name.
[1, BOT.(1975)R73C.]

ibidem**, abbr. **ibid. In the same place. The abbreviation was formerly used in lists of references when the journal in which the paper was published was the same as the preceding reference. This could be misleading if, for any reason, the previous reference was deleted or another, from a different journal, was inserted between them. Not now acceptable in a list of references.

-icola. Epithets ending in *-icola* are substantives and consequently do not modify the ending when attached to generic names of different gender.

ICSB. Abbreviation used in BAC. and in this dictionary for **International Committee on Systematic Bacteriology.**

ICTV. International Committee on Taxonomy of Viruses (q.v.).

-idae. 1. Suffix used in zoo.[1] and suggested for vir. to be added to the stem of a type genus name to form a family name (cf. *-aceae* in bact. and bot.).
2. BOT.[2] uses this suffix for the names of subclasses of the Cormophyta.
[1, ZOO.29. 2, BOT.R16A(b).]

idealistic typology. Sneath's colourful phrase for a 'natural' system of classification.

identical. Alike in every detail. Often misused for similar in most (or many) details or characters.

identicard. Card with one or two rows of holes punched near the edges; each hole represents a character and these are notched when the corresponding characters are positive. Used in a minimal-difference identification system for bacteria developed from the step-by-step approach to identification. One pack (US: deck) of cards deals with the characters of genera (usually one card to a genus but occasionally a card represents a group of genera); a second pack deals with the characters of species within one genus (but it might deal with a few genera). The details of notching cards, the different needling methods, and the interpretation (for bacteria of medical interest) are given in Cowan & Steel (1974).

identified. 1. Shown to be identical with another (known) object or organism. In practice, identification of micro-organisms is never absolute because organisms never behave *exactly* according to the book (not even *Bergey's Manual* or *Cowan & Steel*), and most workers have to be satisfied with a near identification.
2. Numerical taxonomists express the closeness of identification in a more-or-less quantitative manner, and Gyllenberg & Niemelä have introduced terms in an attempt to express the degree of identity; in their view the term identity can only be applied to an organism when both the absolute and relative affinities are known. See also **identity**.

identify, identification, *v.* and *n.* The act and result of making a comparison in which an unknown is shown to be similar to (identical with) a known, and in microbiology, to refer an unknown organism to its species. But because organisms are never identical many people prefer the terms determine or determination. Identification is the third part of **taxonomy** as defined in this dictionary, and is the practical application of the arts of classification and nomenclature.

In making an identification the unknown organism is examined (the detail is generally decided by the difficulty) and its characters are compared with those of characteristic organisms or descriptions of characteristic organisms (the word typical is deliberately avoided because of its nomenclatural connotation). Identification makes use of weighted characters, namely those that have been found to be significant in distinguishing one organism from another. In this it differs from classification (sometimes confused with identification by those who should know better), in which unweighted characters are preferable.

In practice the identification of a micro-organism is made in stages; both

diagnostic tables and **dichotomous keys** are helpful and will indicate the kind of character that should be tested, but neither is of use until the basic characters have been determined. In all forms of micro-organism the simple morphology is fundamental, and in some parasites may be the only information obtainable; among bacteria, fungi, and yeasts characters detected by biochemical tests are increasingly important in making accurate identification possible.

Probably the most important factor in making an identification is to have a pure culture of the unknown organism; although self-evident, not enough care is always taken about this, and curious cultural characters, or those of apparently **exotic organisms** are more often those of mixed cultures than of truly rare species or even newly-discovered organisms.

A vogue that has nothing to recommend it except 'trendiness' makes the organism the subject of the verb, as in 'strains of *Agrobacterium tumefaciens* failed to identify' when, in fact, they were not identified by the author.

In *numerical identification*, i.e. identification based on a statistical assessment of similarities, many terms are used to express identity and are defined in other parts of this dictionary (see **affinity** – absolute and relative, **intermediate**, **neighbour**, **outlier**, and **identified 2**).

identity. Exactly alike. A state seldom, if ever, found in the biological world.

-ii. In bact.[1] and bot.[2] a specific epithet can be made from the name of a man when his name ends in a consonant (but not in *-er*) by adding the letters *-ii* to the patronymic.

In zoo.[3] where the appropriate ending is *-i*, ZOO.[4] treats genitive patronymic endings in *-i* and *-ii* as homonyms. See **personal name**.

[1, BAC.Ap9B. 2, BOT.(1975)73C. 3, ZOO.R31A. 4, ZOO.58(10).]

IJSB. Abbreviation used in BAC. and in this dictionary for *International Journal of Systematic Bacteriology*, the official organ of the International Committee on Systematic Bacteriology (ICSB) and its Judicial Commission.

illegitimate name or **epithet** is one that, either in formation or publication, is contrary to the rules.[1] The term applies only to names validly published but which contravene rules other than those concerned with publication; thus a name (epithet) that is illegitimate in one taxonomic position may be legitimate in another.[2] A specific epithet is not illegitimate because it was published in combination with an illegitimate generic name,[3] and the legitimacy of generic names and specific epithets in a combination must be assessed separately.[4]

An illegitimate name is not taken into consideration when deciding on the correct name of an organism, that is, in matters of priority,[5] but it must be taken into account in questions of **homonymy**. When a validly published earlier homonym is illegitimate the later homonym must be rejected unless it has been conserved.[6]

[1, BAC.23aN5; BOT.6. 2, BAC.51a. 3, BOT.68. 4, BAC.23aN1. 5, BOT.45. 6, BOT.64; 66; 67.]

illustration. 1. Line drawing, photograph, diagram marked to indicate important points or parts. May be used as a supplement to a description of an organism, especially in descriptions of algae, fungi, and protozoa. BOT.[1] requires such an illustration or figure to be marked to show the distinctive morphological features of algae. In mycology a good line drawing can be more valuable than a page of description (Hawksworth, 1974).

2. For a bacterial species that cannot be grown in culture a drawing or photograph may be the most important part of the description, and indeed an illustration can be used as type material, sufficing only until the bacterium is successfully cultured and preserved.[2]

[1, BOT.39. 2, BAC.18a; 18h.]

immune, immunize. The immune state is the result of attempting to create a resistance to an antigen or series of antigens; this state is relative and ranges from absence of resistance to a high degree, sufficient to render the animal totally insusceptible to the effects of the antigen.

The verb to immunize is used to describe a course (of inoculations or other ways of introducing the antigen to the body) designed or intended to stimulate the production of antibodies. An animal that has received such a course is said to be immunized, even if it has failed to respond and, on infection, is found to be susceptible.

immunity. 1. A state of insusceptibility to an antigen. In practice it is a state of increased resistance (of varying degree) or sensitivity to the antigen.

2. Used by virologists to indicate the insusceptibility or resistance to infection possessed by lysogenic bacterial cells to the temperate phage (or prophage) that the cells carry.

immuno-electrophoresis, *abbr.* **IEP.** A method used to identify proteins especially when several occur together as in sera or tissue fluids. Preliminary separation of the proteins is made by gel electrophoresis, a long trough is then cut in the gel and antiserum placed in the trough to set up a gel diffusion precipitation system.

imperfect state. Mycological term for a stage in the life cycle of some fungi, the Fungi Imperfecti or Deuteromycetes; in this state asexual spores only are produced. Imperfect-state genera are often well-defined but may include some species for which the perfect state is not known. The imperfect fungi included in one imperfect-state genus may have perfect states referable to more than one perfect-state genus (Ainsworth, 1973, *Rev. Pl. Path.* **52**, 59).

Separate names are given to an organism in the two states, perfect and imperfect, but the correct name of all states of a species is the earliest legitimate name typified by the perfect state,[1] and this name has precedence over the name given to the imperfect state of the same species.

[1, BOT.59.]

impression. 1. (Printing) A number of copies printed at one time. When a book is reprinted without change in type (apart from the correction of minor errors) it is designated a new impression, and this is indicated on the back of the title page.

2. A technique (impression preparation) for the study of intracellular inclusions.

impure culture. More than one organism growing in a nutrient medium; this may be a culture of an organism (1) growing in symbiosis with another organism, or (2) contaminated by the growth of another organism.

Before the era of pure cultures many names of microbial species were based on observations made on mixed cultures; consequently the descriptions were sources of confusion and this in turn affected the status of the name applied to the organism. The name of a taxon must be accompanied by a description (or reference to a description) of the taxon, but the name will be illegitimate if the description is based on an impure or mixed culture.[1]

The commonest cause of an impure culture in modern bact. is the assumption that a well-isolated single colony arose from a single bacterium; this may be so on a non-inhibitory medium, but is seldom the case on selective or inhibitory media.

[1, BAC.56a(3).]

IMViC. An acronym in which certain basic characters of coliform bacteria can be expressed simply in the form of a code made up of + and − signs. The letters IMV and C stand for indole, methyl red reaction, Voges-Proskauer test, and citrate utilization; the lower case letter *i* is inserted to make a pronounceable word. These 'reaction combinations' were used in this order some years before Parr used the acronym IMViC.

in. The word *in* is used as an indication (or qualifying phrase) to show in an author citation that the first named author is responsible for the name which he published in a book or report written or edited by the second named person.[1] The person named before the word *in* is the *publishing author*, and if only one author's name is mentioned it should be his.[2]

[1, BAC.AdB. 2, BOT.R46D.]

inadvertent error. 1. ZOO.[1] lists *lapsus calami*, typist's and printer's errors as inadvertent and these may be corrected. It does not accept errors made by the author, such as incorrect transliteration, improper latinization, or the use of inappropriate connecting vowels as inadvertent errors.

2. In bact. an apparently inadvertent error by de Bary occupied much of the time of the Judicial Commission for several years, and by a majority decision the Commission ruled[2] that the spelling of the epithet megaterium was intentional and to be preferred to megatherium. See **errors in names**.

[1, ZOO.32(a)(ii). 2, BAC.Ap5,Op1.]

-inae. 1. Bact. and bot. suffix used to form the name of a subtribe; BAC.[1] requires the suffix to be added to the stem of the name of the type genus, BOT.[2] to the stem of a legitimate name of an included genus.

2. Zoo.[3] suffix used to form a subfamily name, attached to the stem based on the name (valid at the time of publication) of a contained genus.[4]

[1, BAC.9T. 2, BOT.19. 3, ZOO.29. 4, ZOO.11(e).]

inapplicable character. A character that cannot be determined because it is dependent on another character that is negative (or not present) in the taxon. The shape of a spore, or the bulging of the cell wall are inapplicable characters in an organism that does not produce spores. Another example is the absurd assumption that has to be made to justify the inclusion of *Shigella* in the family Enterobacteriaceae, that the flagella of *Shigella* species would be peritrichate if they had any flagella. The site of insertion of flagella in *Shigella* spp. is an inapplicable character. See also **linked character**.

inappropriateness of a name is not sufficient reason to change or reject it in any of the biological disciplines.[1]

[1, BAC.55(1); BOT.62; ZOO.18(a).]

incertae sedis. Refers to organisms of uncertain taxonomic position;[1] when attached to a name it usually refers to the generic element of the binomial. Jeffrey (1977) uses a different spelling, *incertae cedis*.

Bergey's Manual (1974) lists *genera* and *species incertae sedis*, and also genera of uncertain affiliation. The suggestion that there might be a subtle difference is unintentional, and probably only reflects the preference of at least one member of the Editorial Board for English.

[1, ZOO.Gl.]

incidental mention. When an author does not state definitely that he is proposing a new name, but merely mentions it without the intention of introducing it,[1] or suggests it as a suitable or appropriate name, he has not fulfilled the requirements of any of the codes of nomenclature, and such a name will not be regarded as validly published.

[1, BAC.28b(3); BOT.34N2.]

incorrect Latin endings. A name that does not have the Latin ending (of a latinized word) in the form required by the nomenclatural code. BOT.[1] will accept such a name as validly published provided that in other respects its formation and publication comply with the provisions of the code; the error should be corrected without change in the author citation.

Differences in the endings due to gender are to be disregarded when homonymy is being determined.[2]

[1, BOT.32N1. 2, ZOO.57(b).]

incorrect original spelling. As names can be formed arbitrarily it may be difficult to decide what is an incorrect spelling of a new name, but ZOO. defines it as a spelling that is not in accordance with Art. 26 to 30, or is an **inadvertent error**, or is a name that had multiple spellings when first proposed and is not the one adopted by the first **reviser**.[1] Such an incorrect original spelling is to be corrected *wherever it is found.* The incorrect spelling has no status in nomenclature,[2] cannot be used as a **replacement name**, and does not enter into **homonymy**.

[1, ZOO,32(c). 2, ZOO.19.]

incorrect subsequent spelling. Any change in the spelling of a zoo. name other than an **emendation** (intentional change). It is without standing in nomenclature.[1]

[1, ZOO.33(b).]

independence of nomenclature. Bot.[1] and zoo.[2] nomenclatures are independent of each other, and a 1975 amendment of BOT.[3] excludes bacteria from the application of its rules; ZOO.[4] recommends the avoidance of duplication of generic names in all the biological disciplines.

BAC.[5] is independent of bot. (except algal and fungal) and zoo. (except protozoan) nomenclature, but is interdependent[6] with the nomenclature of algae, fungi, protozoa, and viruses.

[1, BOT.PrI. 2, ZOO.2. 3, BOT.(1975)PrI,Fo. 4, ZOO.R2A. 5, BAC.Pr2. 6, BAC.51b(4).]

index, *pl.* -es. 1. Alphabetically arranged detailed list of the contents of a book, based on the subject matter and giving reference to page (sometimes section or chapter) numbers.

2. A list of names; in taxonomy usually a list of scientific names arranged alphabetically. In zoo.[1] an Official Index deals with rejected (i.e. unacceptable, replaced) names; accepted names are recorded in Official Lists. Less official lists and indexes, such as *Index Bergeyana*, give both legitimate and illegitimate names of bacteria.

[1, ZOO.78(f).]

Index Bergeyana. A compilation of thousands of names that have been used in bact. This monumental work was published in 1966 and was the result of over fifty years' work in bact. by the first-named author, **R. E. Buchanan**. Unfortunately there are numerous errors, both typographical and of the literature citations, that seem inevitable in a book of this magnitude, and it bends the rules when it seems expedient to do so, as in *Clostridium tertium* (Henry) Bergey *et al.* 1923, which is declared to be legitimate (see **number**). Several addenda have been published.

indexing terms. Words or short phrases useful in indexing the title of a paper. The words will often be from the title itself but other words may lead readers to the paper and these should be added to the list. Retrieval systems depend on these words and the author should list words that will assure that his paper will not be overlooked.

index of dissimilarity. Term to describe differences in characters. A value of one indicates identity, and values greater than one indicate increasing dissimilarity. The index was used by Stanier in Hawkes (1968) in an attempt to express quantitatively serological reactions between anti-enzymes and cell-free extracts of bacteria. The index expressed the factor by which the antiserum concentration had to be raised to produce a precipitate equal to that produced by the homologous (immunizing) antigen.

Antonym, **similarity index** (cf. **peculiarity index**).

indicate, *v.* To point out, show, draw attention to something. Used in BAC.[1] and BOT.[2] to mean that an author should make clear his intentions; for example he should state that he is using a name (epithet) either as a new name or new combination by appending a qualifying phrase such as sp. nov. or comb. nov. in place of the author citation.

[1, BAC.33a; 34a. 2, BOT.33.]

indication. 1. In bot. the indication of a name is the **author citation** and any **qualifying phrase** appended to show the nature of the taxon.[1]

2. A term used in ZOO. for published information which, in the absence of a **description** or **definition**, makes a name published before 1931 **available**.[2] The details of what constituted an indication were given in ZOO.[3] but recent changes[4] have reduced their number; those that remain are (1) a reference to a previous description, (2) a new name proposed as a substitute for an established name. The following are not indications within the meaning of ZOO.:[5] (3) mention of a common name, type-locality, host, label or specimen in a collection, and (4) citation of a name in synonymy, and those excluded under the recent legislation.[4]

3. ZOO.[6] has another use for the word indication in *type by indication* in which the type-species is determined by the application of Art. 68(b) to (d) which cover monotypy, absolute tautonomy (see **tautonym**), and species named *typus* and *typicus* (q.v.).

[1, BOT.46. 2, ZOO.12; Gl. 3, ZOO.16(a). 4, ZOO.13(c). 5, ZOO.16(b). 6, ZOO.67(b).]

indicative epithets. Infraspecific epithets that supposedly indicate the taxon containing the type of the next higher taxon are not allowed[1] in bot. unless they repeat the specific epithet (to comply with Art. 26). The epithets that are of this nature are: *typicus* (q.v.), *originalis*, *originarius*, *genuinus*, *verus*, and *veridicus*. [1, BOT.24.]

indicator. 1. A substance which, by a change in colour, indicates a change in reaction such as pH value (hydrogen ion concentration) of a solution or a culture medium.

2. The word indicator is sometimes used for the abbreviated words referred to in this dictionary as **qualifying phrases** that are part of the authority for a name. These indicators must not be followed slavishly for as Ainsworth (1973, *Rev. Pl. Path.* **52**, 59–68) points out abbreviations such as *gen. nov.*, *sp. nov.* are used *only at the point of publication* by the original author; they are not repeated by subsequent authors who should replace them by the name of the author of the new genus or species name.

indicator medium. Nutrient medium containing a chemical indicator to show changes in reaction (pH value) of the substrate produced by the growth of an organism. Often combined with the substance that inhibits growth of unwanted organisms. See **selective medium**.

indirect reference. In nomenclature this term has a special meaning, viz. *a clear indication . . . that a . . . description or diagnosis* applicable to the newly named taxon has been effectively published earlier.[1] The clear indication is normally made in the author citation that follows the name but may be made as a reference in the text of the paper.

[1, BOT.32N2.]

individual. 1. The earlier editions of BAC. recognized individual as a category, but the newest edition[1] states that the term has little meaning in bact. and recommends that it should be avoided.

2. An individual organism (plant, animal) is considered to belong to a series of taxa of consecutively subordinate rank; of these, the species is basic.[2]

[1, BAC.Ap10A. 2, BOT.2.]

individual variation occurs among all forms of life and among non-genetic factors may be due to age, season, or habitat. It produced from Mayr the Orwellian sentence 'It is only rarely possible to say that this or that individual of a population is "more typical" than some others.' This variation between individuals of a population contrasts with **group variation**, which is variation between populations.

inducible enzyme is one that is produced in substantial amounts only when the inducer is present in the medium. The substrate (substance acted upon) may be the inducer but not all substrates are inducers.

industrial organisms. A loose but useful term used in a primary division of micro-organisms based on the interest they arouse in workers in different disciplines or occupations. Industrial micro-organisms are those of interest to manufacturers of chemical products made by microbiological processes, such as antibiotics, chemicals such as acetic acid, acetone, and of course, to brewers.

BAC.[1] clearly states that names of bacteria published in patent applications or in a patent itself will not be accepted as being effectively published.
[1, BAC.25b(5).]

-ineae. Suffix to be attached to form the name of a suborder; in bact.[1] it is attached to the stem of the name of the type genus; in bot.[2] it is attached to the stem of the name of a family, but when a taxon between genus and family is raised to the rank of suborder, the stem of that name must be retained. In bot.[3] the principles of priority and typification are not mandatory for names of taxa above family.
The suffix is not used in zoo.
[1, BAC.9T. 2, BOT.17. 3, BOT.(1975)16.]

infectious, *adj*. Readily transmissable from one organism to another; e.g. an infectious disease.

infective. *adj*. Able to infect another biological unit (man, plant, bacterium), e.g. an infective particle.

informal category. In bact. the categories *section*, *subsection*, *series*, and *subseries*, introduced into BAC in the 1966 revision, were thrown out at the next revision.[1] They are now said to be informal categories, and since their names do not compete with names of genera and subgenera they are without standing in nomenclature.
[1, BAC.11.]

infrageneric, *adj*. Refers to any or all ranks below genus.[1]
[1, BOT.60F0.]

infraspecific, *adj*. Below the rank of species; used to describe the names (epithets) of any taxa below species (but see also **infrasubspecific**). BAC. and ZOO. legislate for names of taxa down to the rank of subspecies but not below. BOT.[1] lists the infraspecific ranks as subspecies, varieties, subvarieties, forma, and subforma, and if these are not sufficient botanists may add supplementary ranks *provided that confusion or error is not thereby introduced*. But creators of new ranks should note that BOT. does not concern itself with the names of ***formae speciales***.
[1, BOT.4.]

infrasubspecific, *adj*. Below subspecies. **1. ~ form.** An organism of a taxon below the rank of subspecies, e.g. serotype, phage type.
2. ~ rank. Any rank below that of subspecies. Includes variety (equated with subspecies in BAC.[1]), subvariety, form, and subform. Since 1976 the use of variety in new bact. names has been without standing in nomenclature.[1]
3. ~ name. Name given to a taxon below subspecies. Not subject to the rules of BAC.[2] or ZOO.;[3] BOT.[4] regulates the names of taxa down to the rank of form, but not ***forma specialis***.
[1, BAC.5c. 2, BAC.5d. 3, ZOO.1. 4, BOT.4N.]

infundibuliform. Describes to funnel-shaped appearance of **gelatin liquefaction** in stab cultures.

inhibitory medium. A medium containing a substance(s) that prevents the multiplication of certain (unwanted) organisms. This kind of medium is particularly useful when intestinal contents (which always contain a mixed flora) are being examined for the presence of bacteria (such as salmonellas and shigellas) known to cause intestinal infections. However, caution must be exercised, and colonies growing on these media must not be assumed to be the growth from a single organism; often the colonies are mixed and contain a small number of bacteria whose growth is only inhibited by the substance in the medium, and in subculture to a non-inhibitory medium all the organisms making up the colony will multiply and continue as a mixed culture. It is essential that before carrying out tests to characterize or identify the organism(s), colonies from inhibitory or **selective media** are replated on non-inhibitory media, and only then can well-isolated colonies be assumed to be pure; even at this stage a further plating or platings may be necessary.

-ini. Zoo.[1] suffix added to the stem of the name of the type-genus to form the name of a tribe. Not used in bact. or bot.
[1, ZOO.R29A.]

initial letters of names and epithets. ZOO.[1] gives clear directions about the use of capital and lower case initial letters in names and species-group names; BAC.[2] and BOT.[3] are shy and hesistant, and what direction they give is scattered. However all the codes make similar recommendations, and the practice of workers in the discipline is essentially the same, which can be summarized as follows:
A *capital letter* is used at the beginning of the name of a genus and taxa of all higher ranks; a capital is also used for the epithet of a subgeneric name. Capital letters are *not* to be used for epithets (species-group names) formed from personal or geographical names, but BOT.[4] allows authors who wish to do so to use a capital for epithets that are formed from the name of a person, are vernacular (non-Latin) names, or are former generic names. Those tempted to use capitals should see **personal name**, and the caution contained in the paragraph on adjectival epithets (see **adjective**).
A *lower case letter* is used for species-group names (zoo.) and epithets of species and below; the permissible exceptions allowed only in bot. are listed above.
[1, ZOO.28. 2, BAC.7; 10a; 59; 61. 3, BOT.21. 4, BOT.R73F.]

initials. A characteristic of the late twentieth century is the greatly increased use of initials from names of organizations, national and international. Some of these make pronounceable words such as Anzac and Unesco, others have been made into words by additional letters, as Cospar from Committee on Space Research. There are so many of these sets of initials used as acronyms that a list is essential; some of those likely to be met in the literature of taxonomy and microbiology will be found on pp. 35–6.

intension. Meaning; comprehension of a notion. *A proper name has no intension...* (Heise & Starr, 1968, *Syst. Zoo.* **17**, 458 (460)).

interdependence of nomenclature. The 1966 revision of BAC.[1] stated, and the principle still applies, that the nomenclatures of bacteria, algae, fungi, protozoa, and viruses were interdependent in the sense that the name of a bacterial taxon must be rejected if it was a later homonym of the name of a taxon of algae, fungi, protozoa, or viruses. Bact. and bot. nomenclatures were also interdependent and the name of a bacterial taxon had to be rejected if it were a later homonym of the name of a plant. At that time BOT. did not contain a corresponding principle and it was theoretically permissible for a botanist to give to a plant a name previously given to a bacterium. BOT.[2] now indicates that its rules do not apply to bacteria, which implies the independence of bacterial and botanical nomenclatures.

The naming of algae and fungi should be in accordance with BOT. and protozoan names should follow ZOO. Virus nomenclature is still in an unsettled state, and the proposed VIR. has appeared only in draft; it has not been approved, and a starting date for virus nomenclature has not been fixed (see Viruses, Proposed Code, 2).

[1, BAC.(1966)Pr3. 2, BOT.(1975)PrI,Fo.]

interform. Strains showing characteristics intermediate between two (or more) typical members of a group.

intermediate. 1. The allocation of micro-organisms to different categories is not always as simple as it may seem, and because there is a great overlapping, but not identity of characters, the breakdown within a category may be difficult. The intergradations have been likened to a spectrum; Ainsworth (1962, *Symp. Soc. gen. Microbiol.* **12**, 251) describes it in these terms: 'For others, who are more analytical perhaps, all is reduced to shades of grey, one taxon fades into another, and there is a tendency for such systematists to give the intermediates they recognize the prestige of specific rank.'

2. The word intermediate is used in taxonomy for taxa that do not seem to fit into a recognized category, or that, in the opinion of the taxonomist who uses the term, fall between two groups. An example is seen in the coliform bacteria isolated from water supplies, which are not sufficiently different from recognized species to justify the creation of new species or even subspecies; they are often referred to as 'coliform intermediates' and designated by roman numerals, as *E. coli* I, *E. coli* II, *C. freundii* II, but species names fail and other 'intermediates' receive designations such as Irregular type IV.

3. In numerical identification an intermediate is an unknown organism that has a high absolute **affinity** with a given group, but does not have a high relative affinity with that group. LRH.

International Association of Microbiological Societies, *abbr.* **IAMS.** The organization responsible for the Microbiological Congresses and also for BAC. Nomenclature is organized in the International Committee on Systematic Bacteriology (ICSB), with its Judicial Commission, and many taxonomic subcommittees, each of which deals with a group of bacteria; the official organ is the *International Journal of Systematic Bacteriology* (*IJSB*).

International Association of Plant Taxonomy, *abbr.* **IAPT.** The governing body responsible for the various committees that deal with the nomenclature of plants. Of these, the General Committee receives, and the Editorial Committee disposes of proposals for amendment of BOT., conservation of names, and the rejection of names.[1] Between the two committees are others that can be called upon for advice on matters concerning particular groups of plants. The official journal of the Association is *Taxon*, published by the International Bureau for Plant Taxonomy and Nomenclature (q.v.).

[1, BOT. Div. III.]

International Bacteriological Code of Nomenclature. Official name given to the first Bacteriological Code, which was approved at the IV International Congress for Microbiology at Copenhagen in 1947; it was first published in the

Journal of Bacteriology, (1948, **55**, 287–306), then in the *Proceedings* of the Congress (not published until 1949) and reprinted in *J. gen. Microbiol.* (1949, **3**, 444). Translations in French, German, Spanish, and Japanese were published subsequently.

A revised version of this code was approved at the VI Congress at Rome in 1953, and was published in 1958 as the *International Code of Nomenclature of Bacteria and Viruses:*. The name was changed again in 1966 when 'and Viruses' was deleted from the title.

International Bulletin of Bacteriological Nomenclature and Taxonomy. Official publication of the International Committee on Bacteriological Nomenclature and of its Judicial Commission. Volume 1 appeared in 1951, and in 1966 the *Bulletin* name was changed to *International Journal of Systematic Bacteriology*, starting with volume 16. The *Bulletin* was published at Ames, Iowa, US, by Iowa State College (from 1959 University) Press.

International Bureau for Plant Taxonomy and Nomenclature, Tweede Transitorium, Uithof, Utrecht, Netherlands. Publishers of the *Botanical Code* (BOT. in this dictionary), *Taxon*, and *Regnum Vegetabile*.

International Code of Botanical Nomenclature (BOT. in this dictionary). The Code, adopted by the XI International Botanical Congress 1969, was published in 1972 by the International Bureau for Plant Taxonomy and Nomenclature. It superseded all earlier editions (1867, 1906, 1912, 1935, 1947, 1950, 1952, 1956, 1961, 1966).[1]

Proposals for changes at the XII Congress in 1975 appeared in *Taxon*, 1975, **24**, 201–54, and the decisions were reported in 1976. These decisions will be considered by the Editorial Committee, and after the necessary changes have been made a new edition of BOT. can be expected in 1978.

In this dictionary references to BOT. are to the 1972 publication, or where qualified (as BOT(1975)) to the modifications made at the XII Congress as reported in *Taxon* (1976, **25**, 169–74).

[1, BOT. Bibliographia, 394–7.]

International Code of Nomenclature of Bacteria. Official title of the Bacteriological Code adopted at the Moscow Congress in 1966. Retained as title of the revision intended to be known as *Bacteriological Code* (*1975 Revision*), which supersedes all earlier editions; in this dictionary abbreviated to BAC. The revision was the work of a drafting committee of which S. P. Lapage was chairman, and it was approved at the 1st International Congress of Bacteriology, Jerusalem, September 1973. After editing it was published on behalf of IAMS by the American Society for Microbiology, 1913 I Street, N.W., Washington, D.C. 20006 towards the end of 1975.

Rule 1a states that this code shall be cited as *Bacteriological Code* (*1975 Revision*) and that its rules apply from 1 January 1976.

International Code of Nomenclature of Bacteria and Viruses. Official title of the Bacteriological Code approved in 1953 at the VI International Congress of Microbiology. This code was published in 1958 with annotations by R. E. Buchanan and was a most useful document, unfortunately out of print. The annotations include comparisons with the botanical and zoological codes of the time, and illustrations of how the rules should be applied.

When, in 1966, provision was made by the IAMS Congress for an independent committee for virology, the name of the code was changed by deleting the words 'and viruses'.

International Code of Zoological Nomenclature. The official code of the International Commission on Zoological Nomenclature. The code adopted by the XV International Congress of Zoology was published in 1961; the second edition of this code (published in 1964) included changes approved at the XVI Congress, 1963, and it is to this edition that mention of ZOO. refers. Amendments to the code were made at the XVII Congress, 1972, and these were published in *Bull. zool. Nomencl.* 1972, **29**, 168–89, and have been noted whenever necessary.

International Commission on Zoological Nomenclature. A permanent body which derives its powers from International Congresses of Zoology.[1] The Commission considers requests to change ZOO. and makes recommendations to congresses for clarification or amendment of ZOO. Between congresses it issues **Declarations** (provisional amendments), **Opinions**, and **Directions** on points of nomenclature that do not need changes in the code. It compiles **Official Lists** of accepted names and also **Official Indexes** of rejected and invalid names. [1, ZOO.76; 77.]

International Committee on Bacteriological Nomenclature. Title by which the **Nomenclature Committee** was known in 1939 when it authorized the establishment of a Judicial Commission and the development of a Bacteriological Code.

International Committee on Nomenclature of Bacteria, *abbr.* **ICNB.** Title adopted in 1966 by the International Committee on Bacteriological Nomenclature when virologists decided to set up their own committee to draft an independent code of nomenclature for viruses.

International Committee on Nomenclature of Viruses, *abbr.* **ICNV.** Title of the committee formed in 1966 when it was decided that BAC. should not be applied to the names of viruses. A draft code of nomenclature was drawn up and appeared in the minutes of the 1966 meeting, but this provisional code was not generally accepted and has not been taken into use by virologists. Presumably as a temporary measure, eighteen Rules were approved by ICNV to serve as a guide and a basis for further discussion. (See also Fenner, 1976.)

In his presidential report Wildy (see Viruses, Proposed Code, 1) indicated that the Committee was more concerned with the recognition and classification of virus groups than with their nomenclature; it was not surprising, therefore, when in 1974 the Committee changed its name to the **International Committee on Taxonomy of Viruses**.

International Committee on Systematic Bacteriology, *abbr.* **ICSB.** Name adopted in 1970 by the International Committee on Nomenclature of Bacteria. Earlier designations used by the Committee had been the Nomenclature Committee for the International Society for Microbiology (1930), International Committee on Bacteriological Nomenclature (1939), and International Committee on Nomenclature of Bacteria (1966).

International Committee on Taxonomy of Viruses, *abbr.* **ICTV.** A committee of the Section of Virology of IAMS and from 1974 successor to the **International Committee on Nomenclature of Viruses** which was initially charged with the development of a code of nomenclature of viruses.

A Virological Code of Nomenclature has not so far been approved as the 1966 draft was not accepted and a further draft has not been published for discussion. Virologists work to a set of rules (abbreviated to VR. in this dictionary) of nomenclature to serve as a guide; these have been modified at Section of Virology meetings and those referred to are from the 1975 revision (Fenner, 1976).

ICTV has five main subcommittees which deal with the viruses of (1) bacteria (i.e. bacteriophages), (2) fungi, (3) invertebrates, (4) plants, and (5) vertebrates. Newly proposed names are first submitted to the appropriate subcommittee, then to the Executive Committee, and finally to the main ICTV, and names do not become 'official' until approved by ICTV. Reports of the Subcommittees and the Committee are published in *Intervirology*, the official organ of the Section of Virology.

International Journal of Systematic Bacteriology, *abbr.* ***IJSB.*** Name adopted in 1966 for the official journal of the International Committee on Bacteriological Nomenclature. As a successor to the *International Bulletin of Bacteriological Nomenclature and Taxonomy* (which was published from 1951 to 1965) the first volume of the renamed journal was numbered 16. Now published on behalf of IAMS by the American Society for Microbiology, 1913 I Street, N.W., Washington DC, 20006.

International Organization of Mycoplasmologists, *abbr.* **I.O.M.**

International Trust for Zoological Nomenclature. The publishing arm of the International Commission on Zoological Nomenclature; is responsible for the publication of ZOO. Its address is: c/o British Museum (Natural History), Cromwell Road, London, SW7 5BD.
[1, ZOO. Introduction.]

internode. Segment of a branch in a dendrogram between two nodes. LRH.

interpolated categories. ZOO. does not legislate for categories (ranks) above Superfamily, but Blackwelder (1967) listed 13 categories above Superfamily, and said that there was no theoretical limit to the number of ranks in these higher categories. As examples of these interpolated categories he cited cohort, series (between subkingdom and phylum), branch, division, legion, and phalanx.

Tribe has been incorrectly used above the level of family which, to the common man, would seem to be its natural place; however, in that position it is contrary to the accepted **relative order** or sequence[1,2] of ranks in a biological hierarchical system (shown in Tables R. 1, R. 2, p. 220).
[1, BAC.5a. 2, BOT.3; 4; 5.]

interpretation of rules (articles). When the application of the rules is doubtful the bodies responsible for the nomenclature of the discipline will give help in the interpretation of rules (articles) and recommendations. Requests for interpretation should be addressed to the secretary of the Judicial Commission (bact.[1]), the General Committee (bot.), or the Zoological Commission.
[1, BAC.S8c.]

interrupted publication of a name. A name may be proposed in a paper divided into parts and published in different issues of the journal. Such a name dates from the date of the part in which all the conditions required for **valid publication** are met; in zoo. this is when the conditions of Art. 11 are satisfied.[1] ZOO.[2]

advises editors not to allow interrupted publication of names of taxa below the rank of family-group.

[1, ZOO.10. 2, ZOO.R10A.]

interstudy. Numerical taxonomists term for comparison of data or groupings obtained in different investigations.

intraspecific, *adj*. Applied to categories within the species, but which themselves cannot be arranged in any order of precedence.

invalid. A difficult word in taxonomy that is not defined in BAC. or BOT. It is the antonym of **valid**, which has different meanings and uses in bact./bot. and in zoo.

1. In bact. and bot. it refers to publication, and invalid means publication that does not meet the requirements of BAC.[1] or BOT.[2] which deal with **effective publication**, the formation of names, descriptions, diagnoses, the designation of types, and the author's acceptance of his own proposals. When these requirements are not met a name is not validly published and therefore must be invalid. Valid is not a synonym of legitimate, so invalid cannot be a synonym of illegitimate.

2. In the Glossary of ZOO. invalid is defined as *Any name for a given taxon other than the valid name.* A valid name is one of many available names and is the correct name for a taxon; consequently an invalid name must be an incorrect name.

[1, BAC.23aN5; 25; 27–32. 2, BOT.29; 32–44.]

in vitro. A biological reaction that takes place outside the body, e.g. the reaction of a micro-organism (or one of its products) with a biological product such as antibody, as in the Ramon flocculation reaction between diphtheria toxin (or toxoid) and antitoxin, a reaction that is carried out in small tubes. The corresponding test *in vivo* is to give a susceptible animal (guinea-pig) passive protection by the inoculation of diphtheria antitoxin, and some hours later test resistance by intradermal inoculation of varying doses of diphtheria toxin.

In vitro should be used adverbially after the noun;: 'an experiment *in vitro*'; if it must be used adjectivally then it should be printed in roman and hyphenated: 'an in-vitro experiment'.

in vivo. Action in the body, or the reaction between the microbe and the defence mechanisms of an animal (and man is only one of the animals) or plant. (See second paragraph of ***in vitro***.)

-iorum. Suffix added to the name of men (presumably brothers, father and son, or other near relatives with the same name) to form a substantival specific epithet in bot.,[1] as ***millsiorum*** for the Mills brothers. BAC. does not legislate for the unlikely event of an author wanting to name a bacterium after more than one person.

[1, BOT.(1975)R73C.]

-ious. Characterized by; full of.

-ism. A suffix, usually added to the name of an individual, to form an abstract noun indicative of some thought, principle, or discovery by the person thus honoured. Common in medicine as the name of a sign; in microbiology used to indicate succinctly the views for which a man was (or became) famous, e.g. Adansonism. The adjectival form is **-ian** is seen in Adansonian.

isogenotype, -ic, -y. Two generic names that have the same type species. A term used by those zoologists who use genotype for the type of a genus. In microbiology such usage is undesirable, and so probably would be the use of isogenotype.

isolate. 1. *n.* (US variant, isolant). The equivalent of the word **strain**, which is commoner in bact. A micro-organism grown on lifeless media and maintained by serial transfer (subculture) on culture media.

2. *v.* The action of growing a micro-organism from a **specimen 2**. Often implies that the culture obtained is pure and free from contaminating organisms, but this is not necessarily so.

Methods used to isolate a micro-organism may be (1) simple plating on nutrient medium; (2) enrichment in media specially suitable for the organism (expected) to be isolated, followed by (1), (3) or (4); (3) growth on differential medium on which different organisms produce different kinds of colonies or colour changes; often combined with (4) inhibitory (seletive) media which prevent multiplication of unwanted organisms. The use of such selective media assumes that one expects a certain kind of organism, and the success of the method depends on the experience of the worker and a good deal of luck. Bacteria isolated on (4) are frequently not in pure culture as the inhibitory material does not kill the unwanted organisms, which multiply when a subculture is made to a non-inhibitory medium.

isonym. One of two or more names (objective synonyms) based on the same type. An isonym may be a name (epithet) or a combination (generic name+specific epithet).

isosyntype. Used in bot. for a duplicate of a **syntype**.

isotype. A bot. term for a duplicate of a **holotype**, the specimen being one of those collected at the same gathering;[1] an illustration or description cannot be an isotype, and a bot. isotype is always a **specimen 1**.

Although not used in microbiology, the term might be useful for subcultures (or strains) of different colonies (one of which was designated the holotype) from a single **specimen 2**, but see **isosyntype**.

[1, BOT.7.]

italic type. What appears in italic depends largely on journal style, but in nomenclature certain conventions are followed by most journals. The codes do not rule on the subject but recommend that the names of genera and species should be printed in a distinctive type, or the letters spaced. There are critics of this style, and Savory regards the use of italic as the uncritical following of a convention that has nothing to recommend it.

There are occasions when italic should *not* be used: in the names of taxa above the rank of genus; for abbreviations of qualifying phrases such as gen. nov., sp. nov. – these should be printed in a contrasting type and if the name is in italic the qualifying phrase should be in roman or in boldface.[1]

[1, BAC.33aN2.]

-ive. Having a tendency to; having the nature, character, or quality of; given to (some action).

J

j. A letter not found in classical Latin, but used in scientific names especially those made from the names of people, e.g. *Clematis jackmanii*, *Actinomyces janthinus*.

When there are alternative spellings in *i*/*j* BOT.[1] accepts *i* before a consonant and *j* before a vowel (*Saurauja*, not *Saurauia*).

[1, BOT.(1975)73.]

Jaccard similarity coefficient. The similarity coefficient first used by Sneath for use in numerical taxonomy; it counts only positively scored matches as similarities. (See **S**). LRH.

Jordanon. Term suggested by Brierley for use in mycology for microspecies. Ainsworth thinks that Brierley's suggestion is more often quoted than adopted.

journal. 1. A serial published at regular intervals. Papers on taxonomic subjects are not accepted by all biological journals and authors should choose an appropriate journal for each paper. *Taxon*, the *International Journal of Systematic Bacteriology*, the *Bulletin of Zoological Nomenclature*, and *Systematic Zoology* are some of the journals that publish papers on taxonomy and matters concerning the nomenclatural codes; in addition some other, less specialized journals publish papers of interest to taxonomists. Authors of papers on numerical taxonomy, particularly ones that include large triangular tables of figures (data matrixes) should not be surprised if they are asked to deposit the mass of information at some storage centre, and confine their publication to information essential to the understanding of their paper by the general reader.

2. Diary or day-book in which dates of gatherings of plants, isolations of micro-organisms, and dates of freeze-drying may be recorded.

journal style. Each scientific journal has its own **conventions** in regard to the use of words, and the manner of printing scientific names and abbreviations. Journal style changes from time to time (often with a change of editor) and authors are advised to consult a recent issue of the journal of their choice and to study the general layout, division into sections, and particularly the manner in which earlier literature is cited. Information on these points is often given in the end papers under a title such as Advice to Authors or, more realistically, Instructions to Authors. These should be studied by a prospective author before he writes his paper, and if he is not prepared to accept them (e.g. American spelling in an American journal) he should look for a journal that has a style of which he approves.

Judicial Commission. 1. Title of the subcommittee of the International Committee on Systematic Bacteriology[1] that is responsible for considering between congresses any requests for opinions on the interpretation of the Bacteriological Code, on the validity or otherwise of names, and proposals for amendment of the rules of nomenclature. The Judicial Commission may request the setting up of Subcommittees on Taxonomy of different groups of bacteria with the object of establishing **minimal standards** for descriptions of taxa. It may also receive from subcommittees recommendations for accepting lists of names as valid and applicable; it may make exceptions to Rule 23a (dealing with priority), and may

correct the **Approved Lists** that will come into effect in 1980. It may also recommend names to be made available for **reuse**.
2. Zoologists also have a judicial interpretative body to deal with nomenclature and the problems that arise between congresses, but the Zoological Commission is not authorized to investigate or pass judgement on breaches of the principles of the *Code of Ethics*.[2]
[1, BAC.S8c; S9. 2, ZOO.ApA.]

K

k. Although not found in classical Latin, the letter *k* has its place in latinized scientific names, generally in names derived from those of people or places, as in *Klebsiella*, *Kluyvera*, *Salmonella kauffmannii*, and *Serratia kiliensis*.
In zoo.[1] names differing in spelling only by *c* or *k* are regarded as homonyms.
[1, ZOO.58(3).]

karyology. Study of the nucleus, its structure, and influence on the organism.

Kauffmann, Fritz. German bacteriologist who specialized in the antigenic analysis of the coliform bactera. He was one of the first outside England to appreciate the importance of Bruce White's work on salmonellas, which he extended into an elaborate scheme of classification (see **Kauffmann–White scheme**). He applied the same principles to the serological classification of *Escherichia*, *Klebsiella*, and to other groups outside the field of the enteric bacteria.
Kauffmann is a firm believer in putting (what he thinks are) first things first and treating succeeding things as inferior; in taxonomy this is shown by the great weight he attaches to certain characters such as antigens, and the lightness with which he treats (or discards) other characters. As the leader of a school opposed to Adansonism, he has given his name to Kauffmannism.

Kauffmannism. A school of thought in which unequal weighting of characters forms the first principle of classification. Named after Fritz Kauffmann, the distinguished German serologist who attaches much wieght to antigenic structure and little, or none, to other characters. This is exemplified in the phrase *higher groups are only biochemically defined and therefore badly defined*.

Kauffmann–White scheme. A system of classification based on the antigenic structure of the salmonellas. First worked out by P. Bruce White (1891–1949), it was developed by F. Kauffmann, who labelled it the Kauffmann–White scheme. Originally Bruce White had labelled the O (somatic) antigens by Roman numerals, but with the recognition of many new O antigens the complex Roman numbers became laborious and sometimes confusing; after 1955 Arabic figures were substituted. H (flagellar) antigens in two (later more) phases were labelled by lower case letters and Arabic numerals. In the antigenic formulae the O and H antigen labels are separated by a colon, individual antigens by commas (see **antigenic formula** for examples). When the first (specific) phase H antigen labels exhausted the alphabet, newly described antigens were added to the schemes by labelling them with the letter *z* and a subscript (inferior) number, e.g. z_{23}.

key. **1.** An old English use of the word key meant something that disclosed or explained the unknown or mysterious, and this seems to be the sense in which the word has been used in biology. As a verb it means the construction of a series of statements of the 'black–not black' kind that will enable a biologist to work out the identity of an organism from the characters that have been found by observation and experiment. The noun relates to the step-like series of questions which, as the answers should be of the 'yes–no' kind, are called **dichotomous keys**. The constrasting propositions which form the leads are not exclusively of the yes–no kind, but must show definite contrast as black, white, red, blue, etc. or sphere, spiral, rod-shaped. The successful applications of keys to microbiology are few, mainly because the constructor soon runs out of known contrasts and only too frequently we have entries such as this (from *Bergey's Manual*, 7th edn):

aa.	No acid from sucrose.		
b.	Lipolytic.	92	*Pseudomonas polycolor*
bb.	Not lipolytic.	93	*Pseudomonas viridiflava*
bbb.	Lipolytic action not reported.	94	*Pseudomonas ananas*
		95	*Pseudomonas bowlesiae*
		.	
		.	
aaa.	Acid from sucrose not reported.	102	*Pseudomonas barkeri*
		.	
		.	
		108	*Pseudomonas xanthochlora*

The only successful key for the identification of bacteria (to the level of genus) is that developed by V. B. D Skerman. Some keys attempt to summarize a classification, and these are labelled *natural keys* because of an assumed indication of evolutionary relations. Most successful keys are less pretentious and are frankly artificial.

2. ~ **character.** Key may be used to qualify constant characters that are useful to the taxonomist either because they are always present (or absent) in a taxon, or because they are used as differentiae to subdivide a taxon and thus, finding their places in dichotomous keys, become key characters. Positive characters of this kind may also be termed diagonostic or marker characters.

3. ~ **words.** Words useful for indexing and indicating the nature of the material in a publication. Also known as indexing terms and descriptors.

kingdom. The highest rank, category, or level of a hierarchical classification. For many centuries only two divisions of living organisms were recognized, plants and animals, but there is now a tendency to increase the number of kingdoms from the traditional two to four or five. A useful review of the subject, with a presentation of a five-kingdom scheme was published by R. H. Whittaker (1969, *Science*, **163**, 150), and this should be consulted by microbiologists who want more details. Bacteria were included in the kingdom Monera, divided into 5 phyla; algae were in Protista (10 phyla), and the kingdom Fungi had 8 phyla. Whittaker found it convenient *not to treat the viruses as organisms*.

In the 8th edition of *Bergey's Manual*, bacteria are placed in the kingdom Procaryotae.

Kluyver, A. J. 1888–1956. Like many continental bacteriologists, Kluyver was trained as an engineer; he was appointed to succeed M. W. Beijerinck at Delft in 1920. Kluyver remained at Delft as Professor of Bacteriology for the rest of his life, and although he built up a school of biochemically orientated microbiologists investigating the unity and diversity of microbial metabolism, he developed a great interest in taxonony and with C. B. van Niel produced in 1936 a scheme for the classification of bacteria based primarily on their morphology and their catabolic processes. For several years he was Chairman of the International Committee on Bacteriological Nomenclature.

Koch's postulates. A bacteriological legend, but a worthy one. Two of the problems of medical bacteriology are to determine whether an organism isolated from a clinical specimen (i) is present in the body or is a contaminant picked up in taking the specimen, and (ii) is the cause of an infective process; this is really two questions: did it cause the disease, and is it a pathogen?

Facts pointing to a causal role between an organism and a disease are labelled Koch's postulates, which he made in 1884 in connexion with his work on tuberculosis: (1) the organism should be isolated from all patients suffering from the particular disease; (2) after several serial subcultures the organism should be capable of reproducing the disease in a susceptible animal; and (3) the organism should be recoverable from the tissues of the infected animal.

It is needless to point out that postulate (2) can seldom be successfully proved, for few organisms reproduce in a laboratory animal a disease that resembles the clinical syndrome seen in man.

Králschen Sammlung von Mikroorganismen. The world-famous Král Collection, founded towards the end of the last century by Franz Král of Prague. When Král died in 1911 his collection came under the care of Professors Rudolf Kraus and Ernst Pribram in Vienna. Apparently it stopped distributing cultures during the Great War, and the Král Collection was one of the victims of the break up of the Austro-Hungarian Empire, when the scientific life of Vienna came to a temporary halt. However, all was not lost, and when Pribram moved to Chicago he took with him half of the Král Collection; on his death some of the cultures went to the ATCC.

KWIC. In computer retrieval systems, Key Word In Context. A method of listing titles of papers according to key words, without altering the sequences of words in the titles. LRH.

L

label. *v.* To place a name or number on a specimen. In bact.[1] labelling of specimens in a collection does not constitute valid publication. LRH.
[1, BAC.25b(2).]

laboratory data. The form of words, figures, or signs that are the basis of descriptions or characterizations of organisms. Described by Quadling & Martin as the *raw material for classification and identification.*

Lamarck, J. B. P. A. de Monet, Chevalier de, 1744–1829. French biologist. Intended for the Church, he became a soldier; pensioned from the army, he decided to study medicine but became a botanist. He published a Flora of France that ran to three volumes, this brought him fame and election to the Academy and the appointment of Royal Botanist. However, he is best known for the development of a much-disputed theory that acquired characters could be transmitted to succeeding generations. He is less well remembered for his useful invention of the **dichotomous key**.

Lamarckism. The theory put forward by Lamarck that an acquired character can be transmitted, and was stated in two laws. I, that organs develop and are useful in proportion to the use made of them; and II, characters that are acquired during lifetime may be passed to the progeny. The theory of the inheritance of acquired characteristics is also called Neo-Lamarckism.

language of nomenclature. The adaptation of words to the special needs of nomenclaturists. Scientific names of taxa are treated as Latin, regardless of their derivation[1] which may be Latin, Greek, or a modern language. Names not in Latin form are treated as indeclinable. The use of Latin and the latinization of names, even when formed arbitrarily, has provoked criticism and accusations that much of nomenclature is ritualistic. The problem was discussed by Yochelson in relation to the difficulties of handling scientific names of animals in automatic data processing.

Nomenclature, according to Yochelson (1966, *Syst. Zool.* **15**, 88), should be treated purely as a written language (even in one country the pronunciation of Latin varies); as a start he suggests a statement that a name is an arbitrary combination of letter characters; that generic names must differ from each other by at least one letter character; specific names may be identical as long as they are not combined with the same generic name. Names must always be spelt as originally written, any deviation to be treated as an error. Grammatical niceties such as gender, agreement of cases and endings would be ignored.

Why not forget all 'latinization' says Yochelson, *and its accompanying sterile scholarship and get on with the study of biology*?

[1, BOT.PrV.]

lapsus calami. A slip of the pen. In nomenclature accepted as an inadvertent error which, together with a copyist's or printer's error, is to be corrected by an author or editor who detects it; the correction does not involve a change in the author citation.[1]

[1, ZOO.32(a)(ii).]

Latin. In the nomenclatural codes Latin means ancient, mediaeval, or modern (Neo-)Latin;[1] latinized words are modern words that are given a Latin ending and treated as if they were Latin.

Letters not found in classical Latin (*j*, *k*, *v*, *w*, *y*) may be used or retained in modern words that are latinized.

[1, ZOO.29(a)(ii).]

Latin diagnosis. For publication to be valid BOT.[1] requires that after 1 January 1935 a **diagnosis** in Latin should accompany the publication of a new name of a taxon of plants, and after 1 January 1958 a Latin diagnosis is required to ensure the valid publication of a new name of an alga.[2]

Before 1 January 1935 for most plants and fungi, and after 1 January 1958 for algae, a diagnosis in any language is acceptable for **valid publication**.

The Latin diagnosis may be either a translation of the whole description or a summary which includes only those characters that distinguish the taxon from similar taxa.

BAC. and ZOO. do not require Latin diagnoses.

[1, BOT.36. 2, BOT.39.]

Latin letters. The letters *j*, *k*, *v*, *w*, and *y* were not used in classical Latin. BOT.[1] allows the consonants *k*, *w*, and *y* to be used in plant names; but in pre-1800 names the letter *i* and *u* should be used before a consonant for *j* and *v*; but *j* and *v* should be used before a vowel (e.g. *Taraxacum*, not *Taraxacvm*, and *Jungia*, not *Iungia*).[2]

In forming new generic names[3] and specific epithets[4] from personal names BOT. requires changes to be made when the names contain *letters foreign to Latin*.[2] BOT. also treats the letter *y* as a vowel when it occurs at the end of a personal name, and a specific epithet formed from Keay is shown as *keayi*.

The other codes allow use of letters *j*, *k*, *v*, *w*, and *y*.

[1, BOT.73N2. 2, BOT.(1975)73. 3, BOT.R73B. 4, BOT.(1975)R73C.]

latinization of names. Biological nomenclature presents words in Latin form; words derived from Greek or from modern languages are either translated into Latin (as Lindum for Lincoln) or given an ending of the kind used in ancient Latin. Latinization of the names of people does not follow a uniform practice and the different disciplines do it in different ways. Thus ZOO.[1] treats the patronymic Morgan as if it were Morganus in Latin, but BAC. and BOT. assume the latinized form to be Morganius; the genitive of the first is *morgani* and of the second *morganii*, and that is how they would appear (without capitals) as specific names of animals or specific epithets of bacteria or plants. Names that contain letters not used in classical Latin, such as Jackway, retain these letters, but the letters of ligatures are separated, umlauts and other diacritic signs are removed, and diaeresis marks are generally suppressed. But see **Latin letters** for usage in bot. of letters not found in classical Latin.

[1, ZOO.ApD16, 17.]

law. Rule or regulation under a disciplinary code. Zoologists use the terms Law of Priority[1] and Law of Homonymy.[2] Under recent revision[3] the Zoological Commission has been given additional guiding principles to help it deal with disputes that arise about the strict application of these laws.

[1, ZOO.23. 2, ZOO.53. 3, ZOO.79(b).]

LB system. Designation used by Gibbs *et al.* (1966, *Nature*, **209**, 450) for a nomenclature system using latinized binominals (LB), in contrast to their own **VAC** scheme for virus nomenclature.

lead. One of the leading questions or contrasting statements of a character couplet in a dichotomous key.

lectotype. A specimen (strain, culture) chosen as nomenclatural type from the original material studied by the author of a new name who failed to designate a **holotype.**[1] In status it follows holotype, and it is more important than, and senior to, a neotype. A bot. lectotype should be chosen from an **isotype** or, if none exists, from one of the **syntypes**.

When a bacterial species cannot be grown in culture, a description or plate may have to suffice for the lectotype; if at a later date the bacterium can be cultivated the type strain should be designated by the person who isolated the strain or, if he did not do so, by a later author. This strain will replace any non-living type.[2]

[1, BAC.18d; BOT.7; ZOO.74. 2, BAC.18h.]

Leeuwenhoek, Antonie van, 1632–1723. Pioneer microscopist and microbiologist; described by Dobell as the first bacteriologist and the founder of bacteriology. His best lenses gave him magnifications of about 300 diameters, and with them he discovered protozoa in 1674, and bacteria in 1675. A man of little education, he did not add to scientific confusion by premature naming, except in the vernacular, of his 'little animals' or 'beasties'.

Dobell, who described him in the words *This old Hollander... belongs to a genus of which he is the type and only species*, found nineteen different spellings of his name in English versions of his writings, and attributed most to misprints or misreadings of his signature. Dobell uses the spelling Antony; the Dutch journal named after him uses Antonie.

legitimate name or **epithet** is one that is in accordance with the rules of nomenclature.[1] Not used in zoo. in which the equivalent term is **available**.[2, 3] Only legitimate names or epithets published in combination with a generic name are considered in establishing the priority of a name.[4] This legitimacy requirement seems to have been interpreted too rigidly by excluding epithets that were legitimate when published. *Bacterium freundii* Braak 1928 was regarded by some workers as illegitimate since the generic name *Bacterium* was rejected by Opinion no. 4 (revised) in 1954. A 1966 amendment to BAC. (Rule 25, Note) recognized the injustice of this interpretation and allowed the legitimacy of an epithet of a species name in which the generic name was illegitimate. BAC.[5] says that Rule 23 (dealing with priority) must be applied separately to the generic name and the specific epithet of a combination. An epithet originally published as part of an illegitimate name can be legitimate in another combination.[6]

[1, BAC.23aN5; BOT.6. 2, BOT.45F0. 3, ZOO.Gl. 4, BAC.23b; BOT.11. 5, BAC.23aN1. 6, BAC.51a; BOT.66N2.]

leptospire. Common or vernacular name for an organism of the genus *Leptospira;* these organisms have tightly wound spirals and show movement by flexion of the organism.

letter. ZOO.[1] allows the use of a letter of Latin alphabet as the first component of a species-group name; it should be joined to the remainder of the name by a hyphen as in *c-album*. This is the only use of a hyphen permitted by ZOO.[2] VR.[3] also allow letters to be used in names of viruses, but BAC.[4] will not accept a letter as a substitute for a specific epithet.

[1, ZOO.26(c). 2, ZOO.27. 3, VR.12. 4, BAC.52(3).]

level. In taxonomy refers to **rank** in a hierarchical system.

L form. The L stands for Lister and was the designation given to aberrant forms of bacteria by Emmy Klieneberger who was, at the time, working at the Lister Institute, London. For a time there was confusion between the L forms and mycoplasmas (then called PPLO or pleuropneumonia-like organisms and now organized in the class Mollicutes).

L forms were found in natural material, and can be produced in the laboratory by growing the organism under certain conditions unfavourable for normal growth. The morphology of the organism changes, large, swollen forms appear and the rigid cell wall is lost. Thus the L form corresponds to the sph(a)eroplast or protoplast. L forms are now called L-phase variants.

ligature. In typography the term for the union of two letters as *æ*, *œ*, *ffi*. In all the codes of nomenclature such ligatures are incorrectly described as diphthongs; in scientific names the letters should be separated,[1] but BOT.[2] allows them when they are part of **journal style**.
[1, ZOO.ApE3. 2, BOT.73Fo.]

limitation of priority. See **Statute of Limitation**.

linkage. In numerical taxonomy, the joining together of two or more branches of a dendrogram. The similarity level at which the linkage is made may be that corresponding to the highest, average, or lowest similarity between all pairs of **OTU**s (one OTU from each branch), according to the technique used: single linkage, average linkage, or complete linkage respectively. LRH.

linked characters, *syn.* **dependent characters.** Some characters can only be observed when others are present, and they are known as linked characters. A good example is a positive VP test for acetoin; this can only be obtained when the organism is able to break down glucose, for an organism that cannot attack glucose cannot produce acetoin. The characters glucose breakdown and acetoin production are linked, and the tests for them are linked tests. The phenomenon is known as character nesting.

Linnaean system. A system of classification named after C. von Linné who systematized and popularized binominal names, usually called binomials. Linné was not the first biologist to use binominals but when, in the 18th century, he introduced his *Systema Naturae* and *Species Plantarum* the binominals were found to be much more convenient than the long polynominals, which he retained as diagnostic descriptions. Binominals were consistently applied in *Species Plantarum* and because of this the work became the start of botanical nomenclature.

Linné, Carl von, 1707–1778. Swedish biologist whose name is latinized to Linneus or Linnaeus, from which are derived the adjectival forms Linnean (the spelling used by the Linnean Society, London) and Linnaean, which has a wider usage.

lists that concern taxonomists are usually alphabetical lists of scientific names of algae, animals, bacteria, fungi, plants, protozoa, and viruses, and these are labelled *nomenclators*. Some such lists are named indexes and in zoo. there is a subtle difference between a List and an Index; Official Lists give accepted (approved) names, whereas Official Indexes give rejected (not approved, not accepted) names.

Lists of conserved and rejected names of bacteria and plants are published in BAC.[1] and BOT.[2]
[1, BAC.Ap4. 2, BOT.ApII; ApIII.)

literacy. In scientific writing clarity of expression comes first and is usually attained by good English. Careless writing confuses both the writer and the reader. Apart from the choice of words (which should always be the fewest and shortest to make the sense clear), the order of words and phrases needs more careful study than any other part of writing. Compare 'The term for a species in bacteriology is...' with 'In bacteriology the term for a species is...' and the greater clarity of the second is at once apparent; again in 'The epithet should not be used alone in writing about a species' and 'In writing about a species the epithet should not be used alone', although both make sense the second is the more logical order. In a proof a good example of incorrect word order was found (and corrected): 'cultures were suspended in distilled ice-cooled water'; it is technically difficult to distill ice-cooled water and what the author meant was 'suspended in ice-cooled distilled water'.

Too great a use of scientific names (binomials) when **common names** would be better is a fault found more often in US than in English journals; the binomial is needed only for increasing the specificity of the sense, and is seldom needed when writing about organisms at the generic level. Plurals of *generic names* cause trouble and the plural of the Latin form should never be printed in italic. The formation of the plural is discussed under **plural forms**, but the point to be discussed here is Buchanan's dictum that a generic name (which is singular) should not be used with a verb in the plural; no one, for example, would write 'staphylococcus are...' but many use 'salmonella are...'. But Buchanan (1958) does not discuss the elliptical use of the generic name in italic, when standing alone it means all the organisms (species) of the genus; then it is not only permissible but correct to use '*Salmonella* are...' because this is a shortened version of 'organisms of the genus *Salmonella* are...'.

An error too often made by bacteriologists is to treat flagella as the singular and to make a plural ending in *-ae*. Those who have lapsed should see **flagellum**.

Number. This word is a source of difficulty to authors and editors alike. It is a singular noun and if you write 'A number of strains...' the verb should be singular, which although correct is not euphonious and is not used in speech; when writing there is a temptation to put the verb into the plural. Insistence on the singular is regarded as pedantic, and *Fowler* advises us to treat it as plural when the word number follows an indefinite article. The difficulty can be avoided by using 'many' or 'several' instead of the words 'a number of'.

Total. Popular with US author and editors as 'A total of...', an unnecessary phrase which lands them in the same trouble as the word number. The solution is to delete the needless words 'a total of'.

Do not be obsessed by use of *elegant variation* (*Fowler*); make your meaning clear even if it means repeating the same word twice or three times in a sentence or twenty times in a paragraph.

Typescript. Instructions to the printer, shown on manuscript by underlining or marginal notes, are hardly matters for discussion here, but some literate authors spoil typescripts they submit to editors by too liberal of use of underlining; they should remember that it is easier for an editor to add underlining than to delete it. Do not underline unless you are *sure* the word should be set in italic type. The use of italic and other type is dealt with under **conventions** and **typography 2**.

literature citation. References to published work of earlier authors that are embodied in the text or tables of a scientific paper or book. The manner in which the citation is made in the text and in the list of references (**literature cited**) at the end of the paper will be determined by the editorial **conventions** of the particular journal.

1. Most journals adopt a **name–date system** (Harvard system); for details of punctuation (in which journals differ) see a recent issue of the journal of your choice. In this system the text reference gives the name of the author followed by the year of publication of the paper or book, but the page number is not usually included in the text, e.g. Smith (1984) or (Smith, 1984). When dealing with the names and descriptions of new taxa nomenclators sometimes include the actual page number in the text reference in this manner, Smith (1984, 123), meaning that Smith, on page 123 of his paper published in 1984 (journal details in the list of references) first used the new name and/or gave its description. This usage is now required by BAC.[1] for the full citation of the authority for a name.
2. In long review articles or in journals that do not use the Harvard system, the text references may be numbers in brackets or printed just above the text (superior or superscript figures); these numbers may be sequential from the beginning of the paper when they will be used in conjunction with a numbered list arranged in the same order.

More often the list of references at the end of the paper is arranged alphabetically (as in the Harvard system) and when complete the alphabetical sequence is numbered 1 to *n*. In preparing a manuscript by this method it is essential until the last minute to use the names of authors and dates in the text; deletions and additions cannot be made once the numbered list has been prepared and the numbers transferred to the text. Because of these limitations and the greater likelihood of error, this system is cursed by authors and editors alike; it does not have anything to commend it, and the space saved is negligible.
[1, BAC.33bN2.]

literature cited. Heading at the end of a paper for the list of references to authors and their published works cited in the paper. For details of the preparation of the list see **alphabetization** and **name–date system** (**2**).

The list should include all published work cited in the text, but it should not include any papers not cited. References to unpublished work or personal communications should *not* be included in this list; nothing annoys a reader more than a reference such as Jones (1976) which, in the Literature Cited list is given as 'Jones, A. (1976) Personal communication'; that information should have been given, in brackets, in the text.

In reviews or books that give lists of other books for more details the inclusion of titles not referred to in the text is permissible under a heading such as 'Further reading'.

lithography. Literally printing from smooth and porous stone, but used to describe printing from blocks as in photolithography, in which a photographic process is used for the preparation of the block.

Lithography, when it reproduces what was originally printed, is a form of publication accepted by the codes of nomenclature, but when the original was in manuscript or autograph, the lithograph is acceptable only in cases in which the original would be acceptable for effective publication, namely only to names of plants published before 1 January 1953.[1]
[1, BOT.29.]

lithotroph. An organism whose carbon needs are satisfied by the fixation of carbon dioxide. See **chemolithotroph**.

loco citato, *abbr.* ***loc. cit.*** In the place cited. Formerly used to avoid repeating a reference, but it sometimes left the reader wondering which reference. It is not allowed by editors of scientific journals and should not be used in books or publications that escape the eagle eye of a critical editor.

locomotion. Movement from one place to another. In microbiology taxonomists must distinguish between **motility** due to swimming in a liquid medium, and **mobility** by creeping or gliding movement over a solid surface.

The rapid movements of motile bacteria are due to the whip-like action of the flagella, each flagellum probably using the rigid cell wall as a fulcrum. Spirochaetes, on the other hand, do not have flagella, but make rotary and vibratory movements of the spiral-shaped body and these seem to be partly responsible for propulsion; some have fibrils wound spirally around the cell, and these, too, may act as propelling agents. In treponemata the whole cell may act as a motor organ when it makes its flexing movements.

logical classification. A classification based on the correlation of characters without any pretence that it represents phylogenetic relationships, though possibly it may do so.

logomachy. Hair-splitting about words; a controversy turning on the choice and use of words. This is a favourite pastime of nomenclaturists, particularly in dealing with priority, as in the searching out, and attempted re-introduction of a name that has seldom been used except by its original author. In bact. this problem should be avoided when the **Approved Lists** become effective in 1980.

lophotrichate, -ous. Having a tuft of flagella at one or both poles of a rod-shaped bacterium. Lautrop & Jessen think that the term should be limited to those polarly flagellated rods that have a **flagellar index** $\geqq 25$, in contrast to monotrichous organisms with an index $\leqq 10$.

L phase variant. A term for **L forms** of bacteria, which must not be confused with mycoplasms.

luminescent. Descriptive of a body or material that emits light but which is not itself incandescent or burning. It is the correct adjective to describe photobacteria which, in aerated cultures, emit light.

lumper (sometimes spelt **clumper**). A taxonomist whose inclination is to combine several taxa of the same rank (usually species or subspecies) and so reduce the number of taxa in that category. A lumper may apply this principle to taxa of more than one category; e.g. Shaw *et al.* combined the genera *Staphylococcus* and *Micrococcus*, and then combined dozens of species into five. In contrast to (c)lumper is the **splitter**.

lyophil, lyophilize, lyophilization. US terms for drying by sublimation of water vapour, = **freeze-drying**.

lyse, *v.* To undergo lysis (q.v.).

lysis. Dissolution of the cell. *Bacteriolysis*, dissolving of the bacterial cell; *haemolysis*, rupture of the red-cell envelope and liberation of the contents.

lysogen, -ic. A lysogenic bacterium is one that is carrying a bacteriophage (often as prophage) to which it is not itself susceptible. The products of the prophage genes (phage-specific enzymes and structural proteins) may appear only after vegetative growth of the bacterium has started, as in the production of toxin by *Corynebacterium diphtheriae*, or in the synthesis of surface antigens, so that the phage itself seems to be taking part in the formation of the bacterial phenotype.

A lysogen cannot be infected (or superinfected) by the temperate phage it is carrying, a phenomenon known as lysogenic immunity.

lysotype. Term recommended by 1966 edition of BAC.[1] for infrasubspecific forms based on reactions to bacteriophages. Replaced in BAC.[2] by **phagovar**.
[1, BAC.(1966)R8a(3). 2, BAC.Ap10B.]

lysozyme (muramidase). A lytic substance found in tears by Alexander Fleming which could dissolve the cell walls of many Gram-positive bacteria. It is found also in egg-white, saliva, and other substances. It is used to form **protoplasts** and **sph(a)eroplasts**. It was the first enzyme to be synthesized.

M

M. A matching coefficient which allows comparison of both positive and negative characters in the strains being compared.

$$\% \mathrm{M} = \frac{n_c}{n_c + n_d} \times 100,$$

where n_c = number of similar characters (whether + or −) and n_d = number of dissimilar characters.

macrobiology, -ist. Useful terms in the jargon of microbiology, for the study of the so-called higher forms of life (plants and animals). From macrobiology is derived macrobiologists, an inclusive term for both botanists and zoologists.

macrotaxonomy. Taxonomy that is mostly concerned with the theory and practice of classifying higher taxa; a term coined by Mayr. Not to be confused with the taxonomy of macrobiology.

maintain, *v.* To maintain a name in nomenclature is to support its use; it is the equivalent of the verb to *retain*, used in connexion with the names of taxa that are remodelled, divided, or translated.

Manhattan distance, or 'city block distance', is the taxonomic distance between two **OTU**s reckoned like the moves of a rook on a chess-board (Sneath) rather than the shortest, direct straight line (Euclidean distance). Derived, of course, from the famous borough of New York City, Manhattan distance has uses in statistics where each 'street' and 'avenue' can be regarded as a character and the progression of a lineage has to follow the street and avenue route. The distance therefore between start and stop points may well be longer than the Euclidean distance. LRH.

manuscript, *abbr.* **MS.,** *pl.* **MSS.** The material, usually typewritten, submitted by an author to the editor of a journal. Before having a MS. typed an author (and his typist) should study the **style** of the journal, pay attention to editorial instructions on lay-out, underscoring, tables (which must *always* be typed on separate sheets), and use only the abbreviations accepted by the particular journal.

When checking the typed MS. before sending it to an editor, errors or omissions may be found. These should not be corrected by using proof-reader's marks, but the corrections should be typed or written clearly between the lines and the place of insertion indicated. Do not pin additional typed lines to the sheet; it is better to cut the page and insert the new material, attaching it to the main by one of the dull-surfaced invisible adhesive tapes.

Finally make sure that all tables are numbered, that they are referred to by table number (and not as 'this Table'), and that the places of insertion are indicated; that all references in the text are also in the references cited list, and that all those in the list are mentioned in the text. A check list such as that found in *CBE Manual* and in O'Connor & Woodford's book will be found useful.

'man versus machine' controversy. Some workers have expressed a belief or suspicion that by giving greater roles to the computer in all parts of taxonomy (but especially in numerical classification and identification) the taxonomist may become little more than a press-button mechanic. While it is true that some developments in computer technology are towards a goal of creating 'artificial intelligence', in taxonomy the role of computers is simply that of a calculating machine (see **computers**), and the value of computers to taxonomy is governed by the programmes and the quality of the input data, expressed in the adage 'garbage in, garbage out'. It can be claimed with some justification that the taxonomic 'horizon' has itself been considerably widened by computers. LRH.

marker. Indicator, identifier, pointer. Applied to (1) strains, particularly in numerical taxonomic surveys, put in to pin-point or identify known taxa, and (2) characters that are constant and perhaps unique to a taxon.

matching coefficient. In a comparison of the characters of two strains the similarity or dissimilarity can be expressed as the number of similar characters divided by the total number of characters being compared (i.e. positive and negative matches). See **coefficient of association** and **M**.

matching strategy. Term used by Sneath for the comparison of characters of the unknown strain with those of known (named) taxa in diagnostic tables.

matrix. 1. In biology used for the substance between cells.

2. Substratum on which a fungus is growing.

3. Used in mathematics for a rectangular arrangement of figures or symbols. Sokal & Sneath describe it as a synonym for table.

meaning of a name. The name of a taxon is a label intended primarily for recognition and as a means of communication. Although the names used by the older botanists and zoologists might have been descriptive, a meaning is not a requisite for the legitimacy or availablity of a name. Consequently, a name may not be rejected or replaced because it has lost any meaning intended by the original author.[1]

[1, BAC.55; BOT.62.]

medical bacteria. 1. A useful term found in a primary division of bacteria made on their interest to different groups of workers. Used loosely to describe bacteria of interest in clinical medicine; these are the pathogens (to man and animals) and organisms with which they may be confused, and also those bacteria (e.g. *Clostridium botulinum*) which, although not pathogenic, can produce substances that are toxic to man and other animals. Thus medical bacteria may be isolated from clinical specimens (faeces, urine, pus, cerebrospinal fluid, etc.) and from food and other substance that may have caused either infection or intoxication (or both) in mammals, birds, or fishes.

The term has been criticized in the phrase *identification of medical bacteria* but that, surely, is being pedantic and overlooks the usefulness of the term. The alternative, bacteria isolated in a clinical or public health laboratory, or from food and water, is too cumbersome to contemplate.
2. In parts of US a usage (not found in the UK) of the word medical for medicinal; thus medical bacteria could be used for bacteria that produce substances used in therapy, e.g. antibiotics. In the UK antibiotic-producing organisms would be listed under industrial as they are of primary interest to manufacturers of antibiotics.

medium, *pl.* **-ia** (US *pl.* **-s**), *n.* In microbiology, nutritional substances in liquid or solid (solidified by agar) form suitable for the growth of micro-organisms. Formerly made from infusions of meat or plant tissues, these are gradually being replaced by better **defined** (so-called synthetic) media made up from chemical substances of known composition. By the use of defined media the nutritional requirements of an organism can be worked out, and classifications based on these requirements have great taxonomic value.

Many different kinds of media are used and these can be classified into (1) *media for isolation*; straightforward non-inhibitory nutrient media, differential, enrichment, and selective media; and (2) *media for characterization*; which are non-inhibitory, and may be defined media or have a nutrient broth made with animal- or plant-tissue infusion as a basis.

meristic variation is an expression of a variation by a number, e.g. the number of flagella on a bacterium. It has not been used widely in microbiology, but could have an obvious appeal to numerical taxonomists.

mesocaryote, -ic. Describes a cell form with a membrane-covered, histoneless skein of DNA; this kind of cell seems to span the difference between eucaryotic and procaryotic cells, and is found in the Dinoflagellates, a widely distributed group of protozoa.

mesophil(e), -ic. Applied to organisms that grow over a wide range of temperature but not below 8° or above 45 °C. Not every organism will grow between the limits 8–45 °C; individual strains may have a narrower temperature range, with the optimum near the maximum. Many **psychrophilic** bacteria grow at temperatures within the range of mesophilic bacteria, but the mesophils are not tolerant of cold. Some mesophils, such as *Neisseria* spp. have a narrow temperature range; others, such as *Pseudomonas* spp. have a wide one.

metabolite. A small molecule product of metabolism essential to the life of an organism. The idea that these metabolites were essential to all organisms was put forward independently by Lwoff and by Knight. They are intermediate and end products of metabolism produced during growth, and they may be used to form larger molecules or converted to coenzymes.

Secondary metabolites are produced after the growth phases of a culture, and they include microbial toxins, antibiotics, and growth factors.

method. 1. Manner or way of doing something, hence **technique**. In microbiology can be used as an abstract noun as in method of classification, or in a concrete form when technical details of a method are described so that others may repeat the work, or at least know how it was done.

2. Type method is a bot. term relating to the application of names to types of names (specimens, strains in bact.).

methodology. 1. The science of scientific language concerned with the use and meaning of words, and their relations to ideas and things. Woodger used the term for that branch of science that concerns itself with the critical analysis of hypothesis and theory. This could be defined briefly as the philosophy of scientific statement.

2. In common use as a synonym for several methods, for which the simple plural, methods, is obviously a better choice.

metonym. A later name given to a specimen (other than the type specimen) of a taxon that has a valid name. Generally the result of an inadequate search of the literature.

microaerophil(e), -ic. Microaerophils may grow feebly in air or under conditions described as anaerobic, but they grow much better (i.e. give a greater cell yield) under conditions of reduced oxygen tension. The terms should not be used for organisms whose growth is benefited by carbon dioxide; for these **capneic** or **carboxyphilic** may be used.

microbe. Common name for any living unit too small to be seen by the unaided eye; includes algae, bacteria, bacteriophage, fungi, protozoa, viruses.

microbiology, -ist. An unfortunate name for the study of (and those who study) microbes. Microbiology means little biology, and little biologist is an appellation that few bacteriologists would appreciate, however much algologists, mycologists, and molecular biologists may like the term. A better word for the discipline would have been microbology, but it is almost certainly too late to hope for the adoption of this sensible suggestion made in a light moment by a former editor of a microbiological journal.

microbiota. Microbial flora; microbial population.

microcard, microfilm, microfiche. Reproduction on card, on roll or sheet film; photographic methods used for storing information (journal papers, books) in a small space. Such reproductions do not constitute **publication** or **effective publication** in the senses of the nomenclatural codes, but if the material photographed was printed matter (as distinct from typescript), the original could have been effectively published.

Technically the differences between the processes are (1) *microcard*, a trade name for an opaque photographic card; a reader is needed; (2) *microfilm*, three standard widths (16, 35, and 70 mm) of roll film normally in 30 m lengths; (3) *microfiche*, sheet film in different formats (e.g. 9×12 cm, or 5×3 in) on which several pages of a book or document are reproduced. Many microfiche readers will also handle microfilm.

micrococcus. 1. In roman; common name for cocci in clusters. Medical bacteriologists often reserve the name staphylococcus for the potentially pathogenic Gram-positive, catalase-positive cocci, and use micrococcus for the saprophytes and commensals. This may be convenient but it is a practice to be avoided; the distinction between staphylococci and micrococci should be made (if at all) on scientific grounds, such as differences in the characters of the organisms, rather than on hypothetical pathogenicity or otherwise.

2. In italic, with an initial capital; a generic name.

microcolony. Intracytoplasmic microcolonies is the term used by Page for inclusions (of *Chlamydia psittaci*) in tissue cells.

micromethod. Method of carrying out a test in which the reagents are used in small volumes, often measured in drops. Biochemical tests on bacteria can be made by these methods, and two quite different types of test have been devised: (1) a heavy inoculum is made into a small volume of nutrient medium, which is then incubated and the test carried out after a relatively short growth period; (2) non-multiplying suspensions are added to the substrate and buffer; in some tests (e.g. sugar reactions) chemical indicator shows when acid has been produced; in other tests (e.g. indole production) the test reagents are added and a colour change may be seen.

The chief advantage of these tests is that the results can be obtained in a much shorter time than by using the standard techniques; an additional advantage of the suspension method is that, in the absence of growth medium, the enzymes of the bacterial cells can be tested against one substrate at a time.

micrometre, symbol **µm.** SI unit of measurement = 10^{-6} metre. Replaces the obsolete unit micron, which should not now be used.

micromorphology. Term used for the description of the microscopic morphology, staining reactions, capsulation, sporulation, and motility of a species.

micron. An obsolete unit of measurement, one thousandth of a millimetre, represented by the Greek letter μ; used particularly by microscopists. The term micron and the symbol μ will be found in microbiological literature published up to about 1970. It is now replaced by the SI unit, the **micrometre** (10^{-6} metre), the abbreviation for which is μm.

The subunit, millimicron (mμ) = 10^{-9} metre, used in measurements of viruses, is also obsolete and has been replaced by nanometre (nm).

microspecies. In bot., a unit of a **species aggregate**; according to Davis & Heywood the differences between microspecies, though small, are usually more constant than those between subspecies.

microsubspecies. A small but distinct population limited to a small geographical area; it is not customary to give it a scientific name. In higher plants may be recognized as a **variety**, sometimes as var. (geogr.), but the abbreviation geogr. is not sanctioned officially.

mihi. Latin, of me. In the past used in place of his own name by the author of a new name or combination. This usage is not recommended by BOT.[1] which advises authors to use their own names in all author citations.

[1, BOT.R46F.]

minidefinition. Term for brief descriptions of genera that contain only (but generally all) the essential information for distinguishing one genus from another. When introduced by Cowan & Steel (1965) it served the purpose of drawing attention to the clumsiness of the fuller description necessary to characterize an organism essential for sound taxonomic work, but unnecessarily detailed and complex for the **diagnostician**. Before 1965 bacteriologists did not have a word for these brief descriptions limited to ***differentiae***, but botanists and zoologists used **diagnosis**, inappropriate in clinical bacteriology with its close relation to medicine, where diagnosis has a different, but well-established meaning.

Some numerical taxonomists use *character-set minimization* for a minidefinition, but this is neither euphoneous nor lucid.

minimal standard descriptions. In an attempt to obtain adequate descriptions bacteriologists have decided that certain minimal standards should be laid down for each group of bacteria by workers who have made special studies of the group.[1] For example, the description of *Vibrio cholerae* would include the following reactions: Gram negative, non-sporeforming rod; oxidase positive; ferments glucose without producing detectable gas; acid produced from mannitol but not from inositol; H_2S not produced; lysine and ornithine decarboxylated; arginine dihydrolase not produced; grows in 1% tryptone broth without NaCl; G+C ratio 40–50 mole %.
[1, BAC.R30b.]

minimum spanning tree. In numerical taxonomy, a two-dimensional graphical representation showing all the **OTUs** linked together through their nearest neighbours. Neither the axis or abscissa is scaled, but the lengths of the links are made proportional to the number of differences betrween the linked OTUs. LRH.

minireference, *abbr.* **miniref.** Brief reference used in this dictionary to the codes of nomenclature. Minirefs are in form so simple that a key is hardly needed (but one will be found on p. *xii*) by anyone who knows that the codes consist of provisions, rules (articles), recommendations, notes, examples, and appendixes.

minutes of meetings and author citation. ZOO. has a rule[1] which states that when the name of a taxon is established by publication in the minutes of a meeting, the individual who proposed the name, and not the secretary or *rapporteur* of the meeting, is to be cited as the author. But this rule is followed immediately by a recommendation[2] which states that *reporters of meetings should not include in their published reports new names of taxa or any information affecting nomenclature*. In other words, what has been done, let it be done – but don't do it again!

BAC.[3] does not accept new names reported in minutes as being effectively published.
[1, ZOO.50(a). 2, ZOO.R50A. 3, BAC.25b.]

misapplication. The application of a name to an organism that is not included within the range of variation of the named taxon. Misapplications, unlike misidentifications, may be a consequence of a change in classification; e.g. the name *Escherichia freundii* could be correct in a classification that did not include the genus *Citrobacter*, but it would be a misapplication in a classification that included both genera, *Escherichia* and *Citrobacter*.

misidentification. 1. Unintentional errors in the identification of a micro-organism are most often due to contamination of the culture examined. Diagnosticians whose examination of a culture suggests a new species should pause and made certain that they are dealing with a pure culture before publishing the characters of their unusual (exotic) microbe.

2. Use of a wrong name for a species; better terms for this sense would be misnaming, mislabelling, or **misapplication** of a name. Misidentifications of this kind should not be included in the synonymy of a name; they should be

indicated by adding the abbreviation auct. non, followed by the name of the original author and the literature reference to the author who made the misidentification is to use the qualifying phrase *sensu. . . non*, with the name of the misidentifying author between the two words, and after the word non the name of the original author of the name.[2]
[1, BOT.R50D. 2, BOT.R50C.]

misidentified type species. A misidentified type species seems to be an unlikely event, but should it occur it would put a taxonomist in a quandary. When he detects that a type species (or type specimen or strain) belongs to an established species should he apply the first principle of the type concept, that a name is attached to a type and that name must stick to the type specimen (or strain)? This has been bact. practice, and if the type is shown to be a misidentification and the strain belongs to an established species, the name attached to the misidentified strain sinks into oblivion as a synonym of the name of the species to which the organism belongs.

Zoologists require a misidentified type-species to be referred to the Zoological Commission.[1]
[1, ZOO.70(a).]

mis-matching. Dependent on the technique used, levels of DNA molecular hybridization *in vitro* may be overestimated due to non-matching of sections between, or at the ends of, matching sections. The more stringent the experimental conditions, the less mis-matching will occur; the level of mis-matching can be estimated from the decrease, if any, in the thermal stability of the duplex *in vitro*. LRH.

misplaced term. The rank of a taxon is misplaced when it is not in the approved sequence (see Tables R. 1 and R. 2, p. 220); an example would be a species that contained genera. The name of such a taxon is not validly published,[1] and is a misplaced term.
[1, BOT.33.].

misplacement. Allocation of a category to an incorrect place in the sequence of ranks in a **hierarchy**, e.g. a tribe placed between order and family.

mixed culture. Term used in bact. for impure or contaminated culture; the original culture (supposedly pure) may have been a mixture of two or more kinds of organism or, if pure, has been contaminated – and possibly overgrown – by another organism.

Names given to mixed cultures, described as new species or subspecies, are not validly published.[1]
[1, BAC.31a.]

mobile, mobility. The ability to move on a solid surface, as distinct from the ability to swim in liquid (**motility**). This type of movement of bacterial cells is described as gliding motility, and is characteristic of myxobacteria.

A different phenomenon is the mobility of colonies of certain spore-forming rod-shaped bacteria; it is not confined to one species although a new species based on this characteristic (*Bacillus rotans*) was created by one observer. The movement is generally circular, and counter-clockwise rotation predominates (2:1); all the species and individual strains that have been shown to have mobile colonies had peritrichous flagella (Murray & Elder, 1949, *J. Bact.* **58**, 351).

molecular biology. The branch of biology that attempts to explain inheritance, development, and behaviour of organisms in terms of chemical structure, and as the interactions between enzymes and nucleic acids (DNA and RNA). It has been described as *the child of biochemistry and genetics*. . . [which] *broke away from its parents under the influence of its godmother physics.*

molecular biology approach to taxonomy. A cumbersome phrase to mean the use of gross genetic characters, separately from phenetic characters, in classification. For example, the use of DNA base compositions, genome sizes, and levels of in-vitro molecular DNA/DNA or DNA/RNA hybridization. LRH.

molecular taxonomy. Molecules in relation to taxonomy; patterns of compounds and their biosynthetic origins. Since identical compounds may be formed by different biosynthetic pathways, these patterns are more informative, taxonomically, than lists of chemical end-products. Erdtman, who seems to be the originator of the term molecular taxonomy, points out that it is not the structural complexity of a compound that makes it taxonomically valuable, but the complexity and number of biosynthetic pathways involved in its formation. This kind of taxonomy is believed by Stanier to supplement the more elementary studies on bacterial nutrition, and to be essential if we are to make full use of biochemistry in taxonomy.

Monera. Used by many systematists who, adopting a hierarchical system of categories, recognize and name a kingdom of bacteria. Replaced by Procaryotae when it is thought that a name other than Bacteria is needed for a kingdom (see R. G. E. Murray in *Bergey's Manual*, 1974).

mongrel combination. A literary hybrid made up of elements from two languages, e.g. *television.* In making a name for a new taxon, authors are instructed by the codes of nomenclature to make names in Latin form, but the stem may be of Latin, Greek, or any origin; indeed, the name may be composed quite arbitrarily.[1] Words made from the stems of two different languages are frowned upon; as Savory (1962) remarks such words 'produce in the minds of some a faint distaste, in others a real distress, and in others again no reaction at all'.

[1, BAC.10a; 12c; BOT.20; 23; ZOO.11(b).]

Monilia. Common name for *Candida.*

monobasic, *adj.* Applied to a genus based on a single species. The term is an attempt to get away from monotypic, which is inappropriate in this sense.

monogenic, *adj.* Determined by one gene.

monomorphic. 1. Having one characteristic shape and size; may be of taxonomic value.

2. Lacking an ability to change shape (cf. **pleomorphic 2**).

monomorphism. Name for the theory of the constancy or fixity of the species.

monophyletic. Applied to a taxon whose members are all part of a single line of descent. It is a relative term, dependent on how far back into the evolutionary history it is applied. The extreme case would be if life had originated only once in the 'primordial soup', then all life today is one (large) monophyletic taxon. LRH.

monothetic. With **polythetic**, a term based on Backner's monotypic and polytypic. A monothetic group is formed by a series of rigid and successive logical divisions in such a way that the possession of a unique set of features is both sufficient and necessary for membership of the group so defined. The end results of dichotomous keys are monothetic units.

monotrichate, -ous. Having one or a very few polar flagella. Lautrop & Jessen introduced a more precise definition of monotrichous by limiting it to polarly flagellated bacteria with a flagellar index of $\leqq 10$.

monotypy, -ic. Based on one type. This is the Linnaean concept of species, and these species are not broken down into races or subspecies. Also used for taxa at other levels, such as genus and family.

When an author describes a species on the basis of one strain (an unwise thing to do) and does not designate it as the type strain, it automatically becomes the type strain. This process is *designation by monotypy.*[1]

A description of a new species assigned to a new monotypic genus is acceptable as a combined generic and species description;[2] in such a case the generic name and specific epithet must be published together.

[1, BAC.Ap7. 2, BOT.42.]

monstrosity. In earlier editions of both bact. and bot. codes a name was to be rejected when it was based on a monstrosity. The problem, unresolved in microbiology and bot., was what constituted a monstrosity? Monsters have now been deleted from the codes of both disciplines.

morphospecies. Term used by zoologists and botanists for species defined on morphological features alone. Obviously applicable to mycology, but in bact. morphospecies are likely to be found only among the spirochaetes and in the Actinomycetales, and even in that order they would be regarded as 'poor' or 'bad' species.

morphovar. BAC.[1] recommends that morphovar be used in place of morphotype for infrasubspecific forms distinguished by differences in morphological characteristics.

[1, BAC.Ap10B.]

motile, motility. The ability to swim and move independently from place to place. In bacteria rapid movement may be due to flagella, or to thrashing movements (flexion) of the bacterial body (in spirochaetes); a slower creeping movement over moist surfaces is shown by cytophagas (see **mobile, mobility**). Bacteria that are motile by flagella may produce non-motile, aflagellate variants; other bacteria may have functionless flagella which, although non-motile, have H antigens and can be agglutinated by H antisera.

multidisciplinary. Drawing on the resources of more than one scientific discipline. Bact. taxonomy is multidisciplinary since it utilizes data and methods derived from chemistry, physics, molecular biology, computer technology, mathematics, statistics and so on. The organization of science into 'different' disciplines is itself an arbitrary action, which may be useful for teaching purposes, but in a research context the description of something as multidisciplinary is really uninformative. LRH.

multifactorial inheritance. Characters controlled by several genes (polygenic), which in practice gives rise to continuous variation.

multi-entry system, *syn.* **polyclave.** A key in which the user is not restricted in his choice of characters or starting point. Punched card identification systems fall into this category, especially those which use a separate card for each character.

multiple authorship of a paper in which a new name or a name change is proposed raises problems in dealing with **author citation**. The obvious but clumsy method is to write out all the authors' names, but BAC.[1] and BOT.[2] recommend that the name of the first author should be followed by *et al.* when more than two authors are to be cited.

Committee reports come into the multiple-author category; Buchanan *et al.* in *Index Bergeyana* (p. 620) stated that it was 'customary to ascribe names given by a committee to the chairman of the committee and "others"', and they used this form of citation for one that would, in the UK, probably be cited as 'Report 19XX' or as 'XYZ Committee, 19XX'.

[1, BAC.AdB(1). 2, BOT.R46B.]

multiple-entry key. An identification system in which the user chooses the starting point; it will be a non-sequential method such as a punched-card system or a peek-a-boo system.

multiple original spellings. When a name is spelt in more than one way by the original author, the spelling adopted by the first revising author should be regarded as the correct original spelling.[1] The codes do not define what actions the revising author should take in connexion with multiple spellings, but ZOO.[2] makes recommendations on what the *first reviser* should do when dealing with multiple names for a single taxon are published simultaneously; in that case if none of the names has any particular advantage, the one that appears first in the manuscript should be given preference. If this reasoning is applied to multiple spellings of names then the first spelling should be adopted.

[1, ZOO.32(b). 2, ZOO.R24A.]

multi-point inoculator. Device for making several inoculations simultaneously on solid medium in a Petri dish. The apparatus can be used for spot inoculation of several cultures on one medium, or for spotting typing phages on a medium seeded with the organism under test.

multistate. A simple character such as the ability to hydrolyse urea depends on the presence of a single enzyme; in a bacterium the enzyme may be present (+) or absent (−). and the character is described by numerical taxonomists as a two-state character.

Other characters, such as the consistency of a colony, may show a greater number of different states (e.g. friable, granular, butyrous), and these are multistate characters. Characters that can be measured may be treated as multistate rather than as continuously quantitative characters.

multitest media are those media on which several tests can be made simultaneously. One of the earliest and best known was Russell's double sugar medium; media have now become more complex and more biochemical characters can be determined on a single medium. Normally these multitest media are suitable only for use with pure cultures, but in some the base is selective and can be used for cultures from the primary (isolating) plate.

multitrichate, -ous. Having a tuft of several flagella at one pole of a rod-shaped bacterium.

multivariate. Term used to describe a comparative analysis of units which may or may not share several characters, each of which can vary independently of the others. Used for the kind of taxonomy that employs this approach, which today implies numerical taxonomy.

muramidase. A biochemist's name for lysozyme.

Murray, E. G. D., 1890–1964. One of the great 'characters' of bacteriology, Murray was born in South Africa, and from this derived his nickname, Jo'burg, by which he was known to all his English students and colleagues. He was a teacher, and while at Cambridge trained several men who became outstanding bacteriologists. While there he isolated and described with Webb and Swann the bacterium that became known as *Listeria monocytogenes*, to which he devoted great affection for the rest of his life. In 1930 he went to Canada as Professor of Bacteriology at McGill University. He helped Breed with several editions of *Bergey's Manual*, and was one of the original members of the Bergey's Manual Trust when it was formed in 1936. He was Chairman of the International Committee on Bacteriological Nomenclature from 1953 to 1962.

mutation. An abrupt change in some hereditary determinant (such as the nucleotide sequence) which is not induced by the presence of a substrate, and is shown by a change in character or function. The change may be the loss of a character or the acquisition of resistance to a bacteriophage or the ability to grow in the presence of a deleterious substance. Because the change is permanent, the newly-acquired character is passed on to the progeny; in adverse surroundings and by a process of selection, the mutant may become the dominant and even the only form (cf. **adaptation**).

mutatis characteribus, *abbr.* **mut. char.** The abbreviated form is appended to the name of a taxon when the circumscription has been altered (but without exclusion of the type), and this is followed by the name of the author making the change.[1]

[1, BOT.R47A.]

mutualism. In biology used for the state in which two organisms are associated and contribute to each other's well-being; the benefit is mutual and is symbiotic rather than synergic (see **symbiosis**).

mutually exclusive grouping is a principle of orthodox (classical) taxonomy, namely that an organism can belong to only one species, and a species to only one genus and, as a general statement, that a taxon can be a member of only one taxon of the next higher rank. This concept contrasts with the **overlapping of taxa**.

mycelium. Ainsworth (1961) defines this as *a mass of hyphae*. Bacteriologists use the term differently, for long filaments of segmented cells which may be on the surface or buried in the solid medium.

-mycetes. Suffix used in bot. to indicate the name of a class of fungi.[1] The rules of priority need not be applied to names of classes,[2] but the names of all taxa above the rank of family that are based on the name of a contained genus are **automatically typified** by that genus.

[1, BOT.R16A. 2, BOT.(1975)16.]

-mycetidae. Suffix used to indicate the name of a subclass of fungi.[1] If the name is based on a genetic name it is **automatically typified**; if it is the nomenclaturally typical subclass it must be based on the same generic name as the class.[2]

[1, BOT.R16A. 2, BOT.(1975)16.]

mycoplasma. **1.** Printed in roman, the word means a member of the group now organized in the class Mollicutes. There are two families, each of which has one genus: Acholeplasmataceae with its genus *Acholeplasma*, and Mycoplasmataceae, with *Mycoplasma* as its genus. Mycoplasma-like bodies have been found in the tissues of plant hosts previously regarded as infected by viruses (1971, *CMI Phytological Papers*, no. 9, Suppl. 1), and other genera described recently are: *Spiroplasma*, *Ureaplasma*, *Anaeroplasma* and *Thermoplasma*.
2. Printed in italic and with an initial capital, the name of a genus of organism which, unlike bacteria, lack a rigid cell wall and have a triple-layered 'unit membrane'. Morphologically heterogeneous, these organisms grow on lifeless media but, since they cannot synthesize lipid, use what is available in the medium. They are among the smallest free-living organisms (others are the bacterial parasitic bacteria, *Bdellovibrio*). They have both RNA and DNA and in this they resemble bacteria; they should not be confused with **L forms** derived from bacteria.
3. Common name for mycoplasma and mycoplasma-like organisms; in forming the plural the terminal *a* seems to be optional, mycoplasmas, mycoplasms (Andrewes *et al.* 1965, *Nature*, **208**, 332). The former laboratory shorthand PPLO has been replaced by mycoplasm(a).

mycoplasmology, -ist. Words coined for the study of and for those who study mycoplasms.

mycorrhiza, pl. **-s.** An association of fungus and the roots of plants. Formerly regarded as an example of symbiosis, it is now believed that the association is mildly parasitic. There are two main types: *ectotrophic*, in which the fungus is on the root surface, and *endotrophic*, when the fungus lies within the root, as in orchids and heaths.

-mycota. Suffix indicating the name of a division of fungi. The stem of the name should indicate the nature of the division. Priority need not be applied to divisional names, but when the name is based on that of a contained genus it is **automatically typified.**[1]
[1, BOT.(1975)16.]

-mycotina. Suffix indicative of the rank of a subdivision of fungi.[1] Not subject to the rules of priority but when the name is based on the same genus as the divisional name, the subdivisional name is subject to **automatic tautonomy**, and has the same stem as the divisional name but with the different (subdivisional) suffix.[2]
[1, BOT.R16A. 2, BOT.(1975)16.]

N

name. **1.** A name is a simple word of more than one letter or a **compound word.**[1] Names are words attached to objects or organisms by which they can be referred to, but they need not be descriptive as the name is simply a label and a number (not in the form of a meaningful code, see **code 2**) would serve equally well.

Scientific names of taxa are treated as if they were Latin and usually have Latin endings. The name of a genus is a noun in the singular and names of higher ranks are adjectives in the plural used as nouns.

2. The word name in the codes[2] means a name that has been validly published (bact., bot.) or is available (zoo.), irrespective of its legitimacy.

3. The name of a taxon below genus is a **combination** consisting of the generic name and one or more epithets or names. The name of a species is a binary combination of the generic name and specific epithet (bact.[3] and bot.[4]) or a binomen (zoo.[5]) which is the generic name plus the specific name.

4. *v.* The act of coining a word to form the name and applying it to the object. The formation of scientific names for micro-organisms is subject to the rules and recommendations of the various codes of nomenclature; they can be formed arbitrarily[6] and need not have any meaning. See **formation of names.**

[1, ZOO.11(g). 2, BAC.Pr1N; BOT.6N1; ZOO.11–15. 3, BAC.12a. 4, BOT.23. 5, ZOO.5. 6, BAC.10a; BOT.20; ZOO.11(b).]

name-bearer. Name of a specimen to which a species or subspecies name is permanently attached: =**nomenclatural type**=**basionym**=**onomatophore.**

name change. Although the codes of nomenclature were drawn up to discourage and eliminate name change, new work may bring to light facts that point to a need for change in the taxonomic position of an organism, and in the new position a change in name will be required.

At the species level the transfer of a taxon from one genus to another is shown by altering the generic name and putting the name of the original author of the specific name (or epithet) in round brackets (parentheses) immediately after the binominal name, and this in turn is followed by the name of the author who made the transfer to the other genus.[1] When the consequent combination (binomial) already exists based on another type, a new name must be sought. As an example, chemical analysis of cell walls and various serological tests indicate that the organism *Corynebacterium pyogenes* (which causes abscesses in cattle) may be more closely related to the streptococci than to the corynebacteria. If the species were moved to the genus *Streptococcus* the resulting combination *S. pyogenes* would be a homonym of the name of the type species of the genus *Streptococcus* and, probably because of the nomenclatural problems, no author has yet had the courage to make a definite proposal for the cattle species to be moved from *Corynebacterium* to *Streptococcus*. If such a proposal is ever made a literature search will have to be made to find the second (i.e. the next) binomial proposed for the species; according to *Index Bergeyana* this was *Bacterium hyopyogenes* L & N 1907; if this confirmed the resulting new combination would be *Streptococcus hypopyogenes* (Lehmann & Neumann) followed by the name of the author who proposed the transfer.

Changes in rank follow much the same procedure, and the name of the original author retains its association with that name and is shown in brackets after the scientific name. The annotations to BAC.(1958) give a good example of the absurd situation that can arise, as in *Coxiella* (Philip) Philip, when Philip raised a subgenus of his own creation to genus.

Name changes may seem desirable when the circumscription of a taxon is altered but a change should not be made unless the taxon, as redefined, excludes the nomenclatural type to which the name is permanently attached.

When the type is excluded a new name will be needed for the redefined taxon, but the original name must be retained for the remnant that contains the type of the name.

[1, BAC.34b; BOT.49; ZOO.R51B.]

name–date system. A method of making text references to previously published work by the author's name and the date of publication, combined with the arrangement and layout of the details of these works in a list at the end of the paper. The advantages of the system, which is followed with minor modifications by most scientific journals, is that it avoids references in footnotes, and the equally objectionable use of numerals, superior or in brackets, to refer to a numbered list of the authors' names and literature cited. British and US usage differs slightly, but the general practice is shown below.

(1) *References in the text* are made as follows:

(*a*) at first mention the surnames of all the authors are given in full, but *et al.* may be used for the first mention if the number of authors is more than four or five, followed by the year of publication of the paper or book, as Smith, Jones & Robinson (1984).

(*b*) At second and subsequent mention of paper (or book) by three or more authors the name of the first author is followed by *et al.*, as Smith *et al.* (1984).

(*c*) Where there are multiple authors the use of the word *and* or the ampersand (&) depends on journal style. Punctuation, particularly in the use of a comma between an author's name and the date, also varies; that used here is the style I use most often.

(*d*) When the author's name forms part of the sentence the date of publication is enclosed within brackets, as 'Smith (1984) showed...', but when it is not part of the sentence both the author's name and the date are enclosed, as 'pigment was produced (Smith, 1984; Jones & Brown, 1985)'.

(*e*) When more than one paper is cited at the same point in the text citations are arranged in chronological order starting with the earliest, as 'pigment was produced (Smith, 1984; Jones & Brown, 1985; Jones, 1986, 1989; Andrews, 1987)'.

(*d*) When reference is made in one paper to several papers by an author (or group of authors with the same first author) published in a particular year, the different papers referred to are distinguished by adding italicized letters, *a*, *b*, *c*, etc. to the year of publication as in Smith, 1999*a*, *b*; Brown *et al.* 1988*a*. But when a group of workers published in one year a series of papers in which the author-sequence was changed, each sequence must be treated separately, as A, B, C & D, 1990*a*, *b*; B, C, D & A, 1990*a*; C, D, A & B, 1990 (only one paper in that author-sequence referred to in the paper) and so on.

(2) *Arrangement in a list* of references of **literature cited**:

(*a*) The list is arranged alphabetically by authors. See also **personal name 3**.

(*b*) Works by a single author are listed before those he wrote in collaboration with others, i.e. Smith (1900) precedes Smith & Brown (1899).

(*c*) Papers by more than one author may be arranged in any of the following ways. (i) Alphabetically by second author (irrespective of the number of authors), i.e. Smith, Brown & Jones (1900) precedes Smith & Jones (1899). (ii) Author with one other, in alphabetical order of second author; author with two others and so on. The order would be: Smith (1899); Smith & Brown (1900); Smith & Jones (1900); Smith, Jones & Brown (1899). (iii) Where there are a large number of multi-author entries it may be helpful to arrange two-author entries as (ii) above and those by three or more chronologically, i.e. Smith & Brown (1900); Smith & Jones (1899); Smith, Jones & Brown (1899); Smith, Brown & Jones (1900); Smith, Brown, Jones & Andrews (1901); Smith, Brown & Jones (1902).

(*d*) The initials of given names follow the surname, but in US this practice applies only to the initials of a single author or the first (senior) author when there are two or more authors; e.g.
Smith, A. B. (1984). (British and US usage.)
Smith, A. B., Jones, C. D. & Brown, E. (1981). (British usage.)
Smith, A. B., C. D. Jones & E. Brown. 1981. (US usage).

(*e*) The year of publication is enclosed in brackets as (1984) or (1899*a*, *b*), but in US brackets are less often used.

(3) The *title of the paper of book* should follow the year of publication in the list of references cited.

(*a*) Titles of papers in journals are printed in roman except for scientific names, which should be printed as in the original journal (note that italic for family names is common in US journals).

(*b*) Titles of books are printed in italic (roman in US).

(*c*) Titles should be reproduced exactly; errors in titles should be followed by [*sic*].

(*d*) Titles of theses may be printed in roman.

(4) The name of the journal is normally abbreviated unless it consists of a single word (e.g. *Taxon*) or is the name of a person (*Antonie van Leeuwenhoek*). The abbreviation of titles has been a journal characteristic but is slowly becoming more standardized (see *CBE Manual*, 1972, and O'Connor & Woodford, 1975). In English journals the abbreviations are printed in italic, in US journals in roman. See **abbreviation 4**.

(5) The *volume* (and in US occasionally the part) *number* in arabic figures follows the abbreviated journal title. Bold face figures are usual for these numbers, but ordinary roman type is found in some US journals.

(6) The *number of the first page* follows the volume number; this is followed by a dash and the number of the last page of the paper; these are arabic figures in roman type.

(7) *Books* should be listed under the name of the author; chapters and sections written by a named author in a book published under an editor's name should be listed under the name of the actual author.

The sequence of the place of publication and the name of the publisher differs in journal style; in US the publisher's name generally comes before the place of publication.

Sometimes the number of pages of a book is recorded, but this is unusual; reference to a particular page in a book is better cited in the text of the paper.

name group. Name groups (hyphenated as species-group) are recognized by ZOO. and are named after the basic category[1] (not necessarily the lowest category). The names within such a group are **nomenclaturally co-ordinate**, which means that the same rules apply to all members of a name-group, and that the names within the group are competitive for priority.

In bact. and bot. specific and subspecific epithets within the same species group (without a hyphen) could be competitive; epithets of species and subspecies within the same genus must be different unless they are based on the same type.[2] BOT.[3] allows infraspecific taxa within different species to bear the same epithets as other species, and then advises against doing so in practice.

[1, ZOO.Gl. 2, BAC.12b; 13b; BOT.64. 3, BOT.24; R24B.]

nanometre, *abbr.* **nm.** SI unit of measurement = 10^{-9} metre. Replaces the obsolete millimicron.

napiform. Term used to describe gelatin liquefaction in a stab culture that has a turnip-like appearance.

National Collection of Dairy Organisms, *abbr.* **NCDO.** A collection of bacteria used in cheese making, kept at the National Institute for Research in Dairying, Shinfield, Reading, Berkshire RG2 9AT.

National Collection of Industrial Bacteria, *abbr.* **NCIB.** A collection that specializes in actinomycetes and maintains other bacteria that are likely to be of use to chemists and biochemists. Located at the Torry Research Station, P.O. Box 31, 135 Abbey Road, Aberdeen AB9 8DG, Scotland.

National Collection of Marine Bacteria, *abbr.* **NCMB.** A collection maintained at the Torry Research Station, Aberdeen AB9 8DG, Scotland.

National Collection of Plant Pathogenic Bacteria, *abbr.* **NCPPB.** A collection started by the well-known plant pathologist, W. J. Dowson, at the Botany School, Cambridge, this collection is now maintained at the Plant Pathology Laboratory, Hatching Green, Harpenden, Hertfordshire AL5 2BD.

National Collection of Type Cultures, *abbr.* **NCTC.** A collection of bacteria of medical and veterinary significance maintained at the Central Public Health Laboratory, Colindale Avenue, London NW9 5HT.

Started in 1920 at the Lister Institute, London, to fill the gap left by the Král Collection which had ceased to function during the Great War. Its first curator was Ralph St John-Brooks, and until 1946 the Collection was financed and run jointly by the Lister Institute and the Medical Research Council. By 1947 the Collection had enlarged its interests beyond the medical field, and these extensions were used to form the nuclei for other specialized collections such as NCIB, NCDO, and NCYC.

Now part of the Public Health Laboratory Service of England and Wales.

National Collection of Yeast Cultures, *abbr.* **NCYC.** A collection kept at the Brewing Industry Research Foundation, Nutfield, Redhill, Surrey RH1 4HY.

natural. A word that causes difficulty because, like species, it means different things to different people. It was originally used as the opposite of supernatural, and indicated something open to reason or rational. It is used as an adjective to describe a kind of classification, and in English bot. usage it means (*a*) a classification based on many characters, i.e. overall resemblance, as opposed to an artificial classification; (*b*) phenetic as opposed to phyletic; or (*c*) evolutionary or phylogenetic system of classification.

In bact. taxonomy it does not have an accepted meaning as most bact. classifications are regarded as artificial, and bacteriologists seldom use the word natural with the (*c*) meaning. Orla-Jensen's classification, based on metabolic products, and its extension by Kluyver & van Niel, were described as natural. Although these classifications are based on the end-products of bacterial metabolism (e.g. the lactic-acid bacteria), they do not entirely ignore morphological differences.

In US natural often has the meanings of (*a*) and (*c*) combined.

natural key. A summary of a classification arranged in key form; such a key is supposed to indicate the evolutionary relations of the taxa. Some authors provided keys to species of genera (*Bacteroides, Fusobacterium*) in their contributions to *Bergey's Manual* (1974); these could be described as natural keys.

naturalness. For non-Linnaean taxonomy defined by DuPraw as bringing together specimens that are similar or identical in properties that are *not* used in making a classification. This implies that the common characters (as distinct from the *differentiae*) indicate the naturalness of the specimens.

NC. Abbreviation used by numerical taxonomists for ***no comparison***; shown when it is not possible to compare a character in two different strains. The NC symbol is used when information is not available and normally it is not counted when totals of similarities and differences are made. The NC symbol may indicate that data are missing; when characterizing a bacterium that produces a violet pigment (e.g. *Chromobacterium violaceum*) one cannot detect oxidase production by a method dependent on the appearance of a purple colour; the NC symbol is then appropriate. But it is not appropriate when a character appears to be absent or missing, and the negative state should be recorded.

Jišin & Vašíček (1969, *J. gen. Microbiol.* **58**, 135) think that it is illogical to omit NC from the **coefficient of association**, and that for each character marked NC there is the same (equal) probability that the two strains being compared will agree or disagree in this character.

NCDO, *abbr.* National Collection of Dairy Organisms (q.v.).

NCIB, *abbr.* National Collection of Industrial Bacteria (q.v.).

NCMB, *abbr.* National Collection of Marine Bacteria (q.v.).

NCPPB. *abbr.* National Collection of Plant Pathogenic Bacteria (q.v.).

NCTC. *abbr.* National Collection of Type Cultures (q.v.).

NCYC. *abbr.* National Collection of Yeast Cultures (q.v.).

nearest neighbours. 1. In molecular biology: dinucleotide sequences in DNA. 'Doublet analysis' is the term applied by Subak-Sharpe *et al.* (1974, *Symp. Soc. gen. Microbiol.* **24**, 131) for the estimation of the frequency distribution of the sixteen possible (GpG, GpC, GpT, GpA; CpG., CpC, etc.) nearest neighbour base sequences. Not surprisingly, it has been found that actual frequency distributions of natural DNA samples differ from frequencies that would be expected if the bases were distributed along the DNA randomly. Of considerable taxonomic potential, however, is the study of the way in which actual frequencies differ from random.

2. In numerical taxonomy: see **minimum spanning tree**. LRH.

negative. The opposite of positive; absence of a feature; a failure to obtain a positive result in a test.

negative acts, possessives, and qualities. The addition of 'not' to a verb indicates that the action does (or did) not take place, as 'he did not speak', but often in scientific papers 'no' is substituted for 'not' and a negative becomes the subject of the verb as in 'no indole was produced' for 'indole was not produced'. Sometimes 'no' is substituted for 'did not [verb] any', as in 'he filled no tube', an impossible negative act for although a bowler may deliver a 'no ball' in cricket, a laboratory worker cannot fill a 'no tube' (see **no for not...any**). An unsuccessful attempt to avoid this construction is seen in the phrase 'ability of strains to invade host plant tissue directly is lacking'; this could have been rephrased to read 'strains are unable to invade host plant tissue directly'. Other good (or bad) examples of this construction: 'All the potential inhibitors had no effect on the rate of lysis...', and 'No growth,...or NH_4^+ utilization...occurred...and use of...resulted in no increase.'

Negative possessives occur in phrases like 'it had no proteolytic enzymes' for 'it did not have any proteolytic enzymes', and 'it had no flagella' for 'it was aflagellate' or 'it did not have flagella'. These constructions are acceptable in speech but in written work indicate a careless use of words by the author; this may not be important in itself but it may indicate that his bench work is equally slipshod.

A taxonomist often wants to indicate a *negative state* or negative potential, and he does this in listing the characters of taxa in terms such as Gram negative, catalase negative and so on. These qualities (or characteristics) are hyphenated when used as adjectives (a Gram-negative bacterium) but as a predicate (the bacterium was Gram negative) the hyphen is not used.

A *multiplicity of negatives* makes for obscurity of meaning, and a phrase such as 'not infrequently' is not as clear as 'often' or 'sometimes'. Too many negatives make a part of Art. 34 of BOT. ambiguous, and it can be interpreted differently according to the mood of the reader. It states that a name is not validly published (*1*) *when it is not accepted*..., and continues that *Provision no. 1 does not apply to names or epithets published with a question mark...yet published and accepted by the author*. This would be clearer if rephrased (assuming my present interpretation is what is intended): A name is validly published (1) when it is accepted by the author, even if he adds a question mark...

negative matches. Characters scored as negative for two organisms being compared are said to match, but the reasons for the negative state may be different. For example, absence of motility (a negative state) may be due to either (1) a lack of development of flagella because of adverse cultural conditions, or (2) there may not be a genetic factor to produce flagella; in the first the negative state of the character is environmental, in the second it is innate.

neighbour. 1. In numerical identification, a term applied by Gyllenberg & Niemelä (1975) to an unknown which has high absolute affinity, but not exceeding a previously defined linit, with a given group.

2. See also **nearest neighbour**. LRH.

neo-Adansonian, -ism. Term applied, usually with derisory or derogatory intent, to those who think that equal weighting of characters is a sound principle of taxonomy; it is applied by those who believe in selecting important characters and giving greater emphasis to them. See **Adansonian**.

Kiriakoff (1965, *Syst. Zoo.* **14**, 61) says that neo-Adansonism is unacceptable because it is *the substitution of the dogma of overall similarity for the plain fact of evolution by descent.*

neontology. The study of living things.

neotype. A specimen (strain, culture) chosen to be the nomenclatural type of a taxon for which a **type** was not designated by the original author, and of which none of the specimens originally studied has survived. Both BOT.[1] and ZOO.[2] state that the neotype first chosen must be followed unless material for a **lectotype** (which always takes precedence over a neotype) should be rediscovered. ZOO.[2] also states that a neotype should be designated only in **exceptional circumstances** and not *for its own sake*, or as matter of curatorial routine; when a neotype is designated, deposition in a museum or institution is mandatory.

BAC.[3] describes a new procedure for proposing a bact. neotype and authors should consult the original to make sure that they fulfil all the requirements. Briefly, the proposal *must* be published in the *International Journal of Systematic Bacteriology* (*IJSB*); it must be accompanied by a description or reference to a previously published description with author citation of the name, and the name of the culture collection (and catalogue number) in which the strain has been deposited; when all this information is given in the publication the strain will be a *proposed neotype.*

Provided that objections are not received by the Judicial Commission the neotype becomes an *established neotype* two years after the publication in *IJSB*. A strain suggested as a neotype but not formally proposed in accordance with Rule 18e will not have any standing in nomenclature until it has been formally proposed and has become established.[4]

[1, BOT.7; 8. 2, ZOO.75. 3, BAC.18e. 4, BAC.18f.]

neotypology. 1. A term used to indicate a philosophy that sees in numerical taxonomy a new tool for typologists.

2. Used by Davis & Heywood for the accumulated experience of taxonomists who build up a typological picture of a species (the **hypodigm** of Simpson).

nested table. Another term for a table of distinguishing characters used particularly in the identification of bacteria. See **diagnostic table** and Table D. 1, p. 96.

neutral term. Word used for a category made up of units whose place in a taxonomic hierarchy is uncertain. Examples are form, group, complex.

new name. An author can publish a name as new only once; he should not repeat publication of the same paper in another journal or in another language. After the first publication he should refer to the original paper as the source of the name and in the second and succeeding papers he should not use a qualifying phrase indicative of a new name or new taxon.[1] A new name should not be published for the first time in (1) the summary of the first paper; (2) an abstract; (3) a table of contents; (4) a key; (5) an index; or (6) the introduction or preface to a book.[2]

Publication of a new name has been possible in any scientific journal and was acceptable as long as the requirements of the codes (on matters of publication)

were met. *Bacteriologists* should note that after 1 January 1976 publication of a new bacterial name *must*, to be valid, be made in *IJSB*.[3]
[1, ZOO.ApE22. 2, ZOO.ApE23. 3, BAC.27(1).]

New Systematics. A book published in 1940 by Oxford University Press and edited by Julian Huxley. It contained articles by various authors which showed that a new concept of systematics had made its appearance, a concept into which material from allied disciplines could be fitted. A term widely applied to the concept outlined by the book.

nexus hypothesis. In numerical taxonomy theory, the nexus hypothesis states that each phenetic character is determined by more than one gene and, at the same time, most genes affect more than one character. There is therefore a complicated nexus (*SOED*: bond, link) of cause and effect. LRH.

nineteenth-century bacteriology. Descriptions of bacteria based on morphology, colony form, and cultural characteristics such as the different kinds of liquefaction seen in gelatin stab cultures; the simple characterizing tests that were the only distinguishing features available to the first bacteriologists. The expression nineteenth-century bacteriology is not used in a derogatory sense, but implies that such simple descriptions are inadequate for present day taxonomy.

nitrification. Bacterial oxidation of ammonia to nitrate.

nitrogen fixation is a characteristic of certain soil bacteria and blue-green algae that are able to use atmospheric nitrogen in the synthesis of organic compounds. Characteristic of *Rhizobium*, the genus of root nodule bacteria, and of *Azotobacter* spp.

nocardioform. Bacteria morphologically and culturally resembling *Nocardia* spp. As the circumscription of the genus is debatable, there is much to be said in favour of this useful term for borderline organisms that might be species of *Nocardia*, *Mycobacterium*, or *Actinomyces*.

no for not any. Phrases such as 'it was no mean feat...', 'There is no doubt that...' are rhetorical figures of speech and colloquial English, but what may be called the 'negative possessive' is becoming too common in scientific writing. 'Control with no antibiotic...' when 'without antibiotic' is meant rival the familiar reporting phrase 'no tubercle bacilli seen', which makes the reader wonder how one sees a no thing, or nothing. The height of this absurdity occurred in a manuscript with the phrase 'tubes showing no agglutination were centrifuged'.

nomen**, pl.* ***nomina. Name, names.

nomen ambiguum**, abbr.* **nom. ambig.** A name having different meanings, or applied to more than one taxon. The tag applies only to the name; it should not be added to the name of a taxon when the taxonomic position of the organism is in doubt, or where the placement is made with reservations; for these the appropriate qualifying phrase is ***incertae sedis.

***nomen approbatum**, abbr.* **nom. approb.** Approved name. After 1979 may be applied to a bact. name that is included in one of the Approved Lists. Probably it will be more usual to specify the list number, when the citation will read[1] 'Approved List no...., 1980'.
[1, BAC. Prov.A1(2), (3).]

nomenclator. In biology has two meanings: **1.** one who creates scientific names to be attached to organisms.

2. A list or index of scientific names.

nomenclatorial. Adjectival form of *nomenclator*; sometimes incorrectly used for the adjectival form of nomenclature, but according to Blackwelder (1964) this usage is not acceptable American. An example of such incorrect usage is seen in the heading *Taxonomic and nomenclatorial problems.*

nomenclatural. Adjectival form of *nomenclature*; concerned with, referring to, or relating to nomenclature.

nomenclaturally co-ordinate taxa. Taxa whose names are governed by the same rules and recommendations; e.g. in bact. generic and subgeneric names are to be formed in accordance with Rule 10 and its subsections, and names of species and subspecies are regulated by Rules 12 and 13. On subspecific epithets the rules of BAC. are clearer and less ambivalent than those of BOT.;[1] BAC.[2] lays down that a subspecific epithet of one species should not duplicate a subspecific epithet of another species within the same genus, a rule that should prevent confusion and homonymy when two species of a genus are united.
[1, BOT.24; R24B. 2, BAC.13c.]

nomenclatural synonym. Synonyms are different words for the same thing, and nomenclatural synonyms are different *names* for the same organism. In terms of biological nomenclature, nomenclatural synonyms are names based on the same type; if they are generic names they are based on the same species, and if they are names of species they are based on the same specimen to type strain. Thus in the binary names *Malleomyces mallei* and *Loefflerella mallei* the two generic names are nomenclatural synonyms. They are also called *objective synonyms*, and in bot. are known as *homotypic synonyms*. In lists of synonyms they are represented by ≡. For the synonymy of the adjectives qualifying the noun see **synonym**.

nomenclatural type (typus). Nomenclatural types are types of names of taxa.[1] Type is a term applied to a strain of bacterium that has been chosen and designated as the representative of a species bearing the name to which it is attached. The important relation is the attachment between the *name* and the *strain* for the application of a name is determined by its nomenclatural type. After 1 January 1976 the type of a named taxon of bacteria[2] must be designated by the author of the name when he proposes it.

The nomenclatural type of a species of bacteria is the same as the *type strain*, and the code lays down procedures for determining the different kinds of type (see **holotype**, **neotype**). The nomenclatural type of a rank between Subtribe and Order is the genus that provides the **basionym**, i.e. the name-bearing stem.

The nomenclatural type is used by typologists as the centre or reference point of a species, and is that element to which the name is permanently attached: it is not essential that the name should be the correct name (it may be a synonym), or that the species has a valid binomial to enable it to serve as a type. Except in species and subspecies (where the type is a specimen or strain), the types of taxa above species are, by extension, taxa which, by their subordinate taxa, are themselves typified (Stafleu & Voss, 1975, *Taxon*, **24**, 207).
[1, BAC.15; BOT.7. 2, BAC.16.]

nomenclature. The scheme (believed to be a system) by which names are attached to objects, including micro-organisms. In the biological sciences there are Codes of Nomenclature which consist of Principles, Rules (Articles), and Recommendations but, since they cannot be enforced, they are no more than codes of good behaviour, or the ethics of nomenclature. The names attached to micro-organisms are no more than labels, and even when originally intended to be descriptive (e.g. *Staphylococcus aureus*) should not now be regarded as informative.

Some workers regard nomenclature as an end in itself, to others it is just a bore; in fact it plays only a small part – but an important one – in microbial systematics. The nomenclature of a group of organisms does not depend on the correct latinization of words, but on the thoroughness of the preceding work to define and classify them in their appropriate taxonomic position; when this has been done it is a relatively simple matter to apply the rules of nomenclature. Unfortunately, the rules seem to be unnecessarily involved, trying as they do, to anticipate every possible contingency.

The tyro often seeks advice for an appropriate name; he should be firmly told that he is the best judge of that, and that the main pitfalls are two: the first is classificatory, namely he must be sure that he has created a new taxon and is not merely giving a new name to an old-established taxon (i.e. coining a later synonym). The second is nomenclatural, and is the avoidance of creating a homonym, i.e. using a name that has been attached to another taxon of the same rank. To many this is the more difficult part involving extensive library research to discover if the name favoured has been used before for an organism of the same rank (within the same kingdom) or, if a specific epithet, within the same genus.

Nomenclature Committee for the International Society for Microbiology. The original title of the Committee, formed in 1930, to deal with the nomenclature of bacteria; the name was not abbreviated to a set of initials as that practice was not then fashionable. Subsequently it was referred to as the Nomenclature Committee and later renamed the International Committee on Bacteriological Nomenclature (1939); by the late 1940s the word Permanent preceded that title on the letterheads used by officers of the Committee, presumably because this committee was the only continuing body of the International Association of Microbiologists, as the committee responsible for organizing the congresses changed at each congress.

nomenclature for bacterial genetics. A proposal for a standardized nomenclature for the use of geneticists was made by Demerec *et al.*; this so-called nomenclature is a series of abbreviations (e.g. *lac* for genes determining and regulating lactose utilization; (Rtf) means harbouring resistance-transfer-factor), and recommendations for their use.

Recommendation 9 is of interest to more than geneticists as it states that strains should be designated by serial numbers, and that different laboratories should use different letter prefixes. This is a system familiar to taxonomists, especially those working in cultures collections, where the letter-number combinations are known as accession or catalogue numbers.

nomenclature of viruses does not follow any of the other codes; a set of rules serves as a guide to the naming of viruses but, unlike the other codes, a name formed in accordance with the rules still requires official approval. For a new name this involves at least three steps: (1) recommendation by one or more of the subcommittees of the International Committee on Taxonomy of Viruses (ICTV), (2) consideration by the Executive Committee of ICTV and, if approved (3) submission to ICTV itself. Only after approval by ICTV does the name of a virus become official. It is intended that names (nomenclature is the word used in the rules) shall be applied to all viruses but the usual principles of priority in the selection of names, and the author citation of the names are not to be observed.

ICTV also organizes the arrangement of viruses into groups and some of the groups of viruses of vertebrates have been designated as families and subfamilies. The 1975 set of Rules of Nomenclature (VR.) allows existing latinized names to be retained, but in forming new names binomials are neither required nor desired. While an effort is to be made towards a latinized nomenclature,[1] numbers, letters, and combinations of these may be used in forming the *names of species*,[2] and these may be preceded by an *agreed abbreviation of the latinized name of a selected host genus*.[3] [The word host refers to the natural host (plant, animal, or bacterium) in which the virus multiplies; the rule applies to virus groups in which the host range may be a distinguishing feature (H. G. Pereira, personal communication).]

Sigla are allowed as virus names *provided that they are meaningful... and recommended by international virus study groups*.[4] Other points about the naming of viruses will be found under **cryptogram**, **rank**, **suffix**.

[1, VR.4. 2, VR.12. 3, VR.13. 4, VR.7.]

nomen confusum, *abbr.* **nom. confus.** A confusing name; a name applied to a taxon of dubious integrity, e.g. an impure culture.[1]

[1, BAC.56a(3).]

nomen conservandum, *abbr.* **nom. cons.** A name conserved (i.e. may be used with official sanction) in place of another, usually earlier name; a conserved name must be used in preference to all synonyms and homonyms.[1] Conservation is made by the Commission[2] or Committee set up for each discipline to judge the need for action to set aside the rules of nomenclature. Lists of *nomina conservanda* are published in BAC.[3] and BOT.[4] and accepted names (zoo.) are published in Official Lists.[5]

[1, BAC.56b. 2, BAC.23aN4(i). 3, BAC.Ap4. 4, BOT.ApIII. 5, ZOO.77.]

nomen conservandum propositum. A name proposed for conservation. Until action has been taken by the Commission or Committee existing usage should be followed.[1]

[1, BOT.15; R15A.]

nomen dubium, *abbr.* **nom. dub.** A name the application of which is uncertain; it is to be rejected,[1] but only the Commission or Committee can do this. It can be argued that *Serratia marcescens* Bizio 1823 is a *nomen dubium* because the organism named might have been a yeast or a bacterium, as either can produce 'bleeding polenta'. Bizio probably saw and named the yeast; the bacterium is much smaller, so small that it was at one time used to test the efficiency of bacterial filters.

[1, BAC.56a(2).]

nomen hybridum. A hybrid name; a name formed by combining words from different languages.

nomenifer. Term proposed by Schopf for a **holotype** that may not be characteristic of the species but is the name-bearer. It was suggested that it should follow the word holotype and be enclosed in brackets, e.g. holotype (nomenifer).

nomen illegitimum, *abbr.* **nom. illegit.** An illegitimate name; although validly published, a name that is not in accordance with all the rules. It need not be considered when priority is being determined, but must be noticed when questions of homonymy arise. When an illegitimate name is included in the synonymy of the name of a taxon it is useful to make a reference to the rule against which the name offends, as in the citation *nom. illegit.* (Art. 64).

nomen invalidum, *abbr.* **nom. invalid.** A name that has not been validly published. In mycology could apply to a name that, when published, was not accompanied by a Latin diagnosis or description (or by a reference to such a Latin diagnosis or description) as required by BOT.[1]

[1, BOT.36.]

nomen legitimum, *abbr.* **nom. legit.** A legitimate name; a name formed and validly published according to the rules (bact., bot.). Not used in zoo.

nomen non rite publicatum, *abbr.* **nom. non rite public.** A name not published in accordance with all the rules concerned with **valid publication**.

nomen novum, *abbr.* **nom. nov.** New name; a **substitute name**, which must be typified by the type of the older name.[1] It is not often used as it is better to specify the rank of the new name, which in subsequent publications is replaced by the name of the author of the name. In zoo.[2] it should be used only to indicate that the new name replaces a **preoccupied name**.

[1, BOT.7. 2, ZOO.ApE21.]

nomen nudum, *abbr.* **nom. nud.** Naked name; a name that, when proposed, was not accompanied by an adequate description (or reference to an adequate description) of the taxon.[1] In zoo. a name that did not meet the requirements of ZOO.[2] for availability at the time of publication.

[1, BAC.AdB. 2, ZOO.12; 13(a); 16; Gl.]

nomen oblitum, *abbr.* **nom. obl.** A forgotten name; a name that has not been used as a senior synonym within the last fifty years. The stability of nomenclature can sometimes be upset by the application of the law of priority, as when a long-established name is rejected under the law and replaced by an unused senior synonym. The procedure for dealing with this situation has changed recently, and an appeal should now be made to the Zoological Commission[1] to reject the unused senior synonym that, in current usage, is not applied to a particular taxon. *Nomen dubium* is applied to names rejected by the Zoological Commission under the **plenary powers**[2] conferred on it.

[1, ZOO.23(a–b). 2, ZOO.79(b).]

nomen perplexum. Perplexing name. A name which causes uncertainty in bact. though its application is known. BAC.[1] gives as an example *Bacillus limnophilus* Bredemann & Stürck *in* Stürck 1935 and *Bacillus limophilus* Migula 1900, which differ only in having the letter *n* in the epithet of the first name and not in the second.

[1, BAC.56a(4).]

nomen provisorium. A name proposed provisionally. To propose a name provisionally is a unforgivable action as such a name has no standing in nomenclature and is merely an unnecessary encumbrance.

nomen rejiciendum, *abbr.* **nom. rejic.** Name rejected by the appropriate Commission or Committee of the discipline.[1] Rejection is often made at the time of conservation of a junior synonym; the reasons for the rejection may or may not be stated in the Opinion rejecting the name. Lists of *nomina rejicienda* are published in Appendixes to BAC.[2] and BOT.[3] but in zoo.[4] they are published in Official Indexes. A rejected name should not be used to designate any taxon.
[1, BAC.23aN4. 2, BAC.Ap4. 3, BOT.ApIII. 4, ZOO.78(f).]

nomen revictum, *abbr.* **nom. rev.** Revived name. After 1 January 1980 bacteriologists will be able to reuse a name that is not on one of the **Approved lists**. When a name or epithet is reused or revived and applied to what the author regards as the same taxon to which it was orginally attached, he can show this by adding nom. rev. to the abbreviation which indicates the rank.[1] BAC. gives as an example to Provisional Rule B2 *Bacillus palustris* sp. nov. nom. rev.
[1, BAC.28a.]

nomenspecies. The naming species; a term coined by Ravin for bacteria that share a group of common characters. In general these are species delineated by the **nomenclatural type** which acts as the **name-bearer**.

nomen superfluum, *abbr.* **nom. superfl.** A superfluous name which, when first published, labelled a taxon that included the type of another name which should have been adopted.[1] The phrase 'included the type' means included the type strain, or cited the name of the type species, or an illustration that acts as type specimen.
[1, BOT.63.]

nomen triviale, *abbr.* **nom. triv.** A trivial name. At one time applied in zoo.[1] to the specific name, and also used by some authors for a common name.
[1, ZOO.Gl.]

nominal, *adj.* Used to qualify the name of a category or rank to indicate that the rank has been fixed or defined by the designation of its type as a particular taxon of a lower rank.[1] Thus a nominal genus (e.g. *Escherichia*) is that to which its type species (*Escherichia coli*) belongs, and the nominal species (*E. coli*) is that to which the type strain (NCTC 9001) belongs.
[1, ZOO.Gl.]

nominalism. The second of Mayr's five basic theories of taxonomy, and although he would probably not agree, the one most applicable to bacteria and bacteriology. According to this theory only individuals exist (cf. N. W. Pirie's *only individuals are real*) and all groupings of them are artificial and created in the mind of man. Numerical taxonomy is a modern example of nominalism; numerical taxonomists deliberately make groups (taxa) on the basis of calculated overall similarity. Because the groups so formed often coincide with those created by intuitive methods, Mayr believes that it is demonstrably false to claim that the groups are man-made rather than the products of evolution. He believes that this numerical approach is most successful when applied to groups in which classification is immature, by which he means micro-organisms.

nominalistic. A concept of a category defined by its formal relations and position in a hierarchical system of nomenclatural taxonomy.

nominate, *adj.* **1.** Applied to a taxon that has the same name (except for the suffix) as a taxon of higher rank, and contains the *type* of the subdivided taxon. The nominate taxon is a subordinate taxon (cf. **nominal**).

2. Applied to the first named of a series of taxa within one superior taxon.

nominate subordinate taxon. Term used in zoo. but of wider application for the subdivision of a taxon that contains the type (genus, species, specimen) of the subdivided taxon.[1] The subordinate taxon has the same name (modified when necessary by a different suffix) as the higher taxon.

Nominate subordinate taxa are created automatically under certain circumstances: (1) when a species is divided into subspecies, the one that contains the type of the species becomes the nominate subordinate taxon and must, for its subspecific epithet, repeat the specific epithet; (2) when several species are made subspecies of another species, the subspecies that contains the type of the species becomes the nominate subordinate subspecies and repeats the specific epithet as its subspecific epithet; e.g. if *Bacillus anthracis* and *B. mycoides* are treated as subspecies of *B. cereus*, this action automatically creates another subspecies, *B. cereus* subsp. *cereus*, which is the nominate subordinate subspecies, although this term is not used in bact.

[1, ZOO.37; 44; 47.]

nominifer. Nomenclatural type to which a name is permanently attached. The word appeared with this spelling (cf. **nomenifer**) in bact. literature with the publication of the 1966 edition of BAC.

nominotypical taxon. The taxon that contains the nomenclatural type of the next higher rank.

non. Not. Used particularly in a **synonymy 1** to indicate that the name which follows is a homonym of an earlier author whose name is given. Sometimes only the names of the authors separated by non followed the name of the taxon and in such cases the date of publication of each author is helpful.[1] BAC.[2] indicates the use of non after 1980 in connexion with **reused names**.

[1, BAC.AdB(3)(c); BOT.R50C. 2, BAC. Prov.B4.]

non-Linnaean taxonomy. Any classification that departs from the Linnaean system can be described as non-Linnaean, but the name was used deliberately by DuPraw for a system which eliminated hierarchies of mutually exclusive categories, and was intended to simplify recognition of individual organisms (as opposed to categories of organisms) representative of populations of unknown specimens. In practice, quantitative measurements were made of measurable characters (qualitative characters were avoided) and these were submitted to multivariate analysis and recorded as two-dimensional distribution points. Classification was regarded as a statistical procedure, and the naturalness of the classification was tested through its 'recognition function'; revision was thought to be simpler than with any Linnaean system.

non . . . nec. Neither . . . nor. Used in citing the homonym of a name.[1]

[1, BOT.R50C.]

nonoperational. 1. Impracticable; unworkable.

2. Jargon for not provable by experiment.

non-pathogen. A micro-organism that is not known to be capable of causing infection in an animal or plant. It is *not* synonymous with **saprophyte**.

nonspecificity hypothesis. A numerical taxonomy hypothesis that one class of characters is not exclusively under the influence of a distinct class of genes. LRH.

non vide, *abbr.* **n.v.** Not seen. When the synonymy of an organism is given the names and references should (but seldom are) seen and checked in the original publications. When they are not all so checked the abbreviation n.v. should be appended to the author citation of any synonym not seen.

notation. 1. The representation of numbers, quantities, qualities, and measurements by signs, and the symbols used. See **symbolic notation**.
2. ZOO.[1] uses notation in the sense of **qualifying phrase** (q.v.).
[1, ZOO.ApE13.]

note. Explanatory notes are appended to rules to clarify the wording and to illustrate good usage; they have the same force as the rules to which they apply.

note indicators. Typographical terms for signs normally used for footnote references; in biological works used frequently in tables to direct attention to additional information in footnotes. The signs should be used in the following sequence: * (asterisk), † (dagger), ‡ (double dagger), § (section mark), || (parallels), and ¶ (paragraph mark); when more than six such marks are required in one table (or page of text) they may be doubled (**, ††, ‡‡, etc.). Note that in tables footnotes are set immediately below the table, so that if there are two tables on one page, the marks used in the first table may be used again in the second. The sequence of marks should be read across the table, and then down, as in reading the lines of text.

nucleoid. Structure containing nuclear material (DNA) which, during multiplication of the bacterial cell, divides into two and passes to the daughter cells.

nucleotide. A molecule made up of three parts: a purine (adenine, guanine) or pyrimidine (thymine, cytosine, uracil) base linked to a 5-carbon sugar and a phosphate group.

DNA contains many nucleotides the bases of which are adenine, guanine, thimine, and cytosine; the sugar is deoxyribose. In RNA, uracil takes the place of thymine, and the sugar is ribose.

nucleotide sequence homology. The relative nucleotide sequence homology can be measured by so-called DNA–DNA hybridization experiments.

number. A number or figure cannot be used as a substitute for a specific epithet and under BAC.[1] and BOT. would be illegitimate. Adjectives formed from numbers (ordinal adjectives) such as primus, secundus, etc. are not acceptable as epithets and they should be rejected.[2] However, *Clostridium tertium* (Henry) Bergey *et al.* has survived and was declared legitimate in *Index Bergeyana*, though that does not accord it official recognition. It is not included in any list of conserved or rejected names[3] and it will be interesting to see whether it reaches one of the Approved Lists.

VR.[4] accepts numbers, letters, or combinations thereof in the formation of names of species of viruses.

ZOO.[5] allows for numerals and numerical adjectives forming part of a compound name, but they must be written in full and joined to the rest of the name; the example is *decemlineata*, not *10-lineata*.

[1, BAC.52(3). 2, BAC.52(2); BOT.23. 3, BAC.Ap4. 4, VR.12. 5, ZOO.26(b).]

number (grammatical) in nomenclature. The name of a genus is treated as a singular noun, and the binomial name of a species is also treated as singular.

Names of higher categories are nouns or adjectives treated as nouns in the plural number, so that in writing we say 'Enterobacteriaceae are interesting bacteria'. See **literacy**.

numbers, printing of. Long numbers are often split into groups of three figures. Past British and American practice separated the groups by commas, but some continental European printers used stops (periods) as spacers, and commas were reserved for decimal points.

It is now advised that *neither commas nor stops (periods) be used as spacers in either typescript or print.*

numerical classification. A term applied to methods of classification in which the similarity of characters is presented quantitatively and expressed as **S** (similarity) values or **M** (a matching coefficient).

The term has a much older use in bact., and occurred as the title of a paper by Bergey & Bates in 1906 for a coding of the characters by Chester's (1905) method. The term numerical classification, like numerical taxonomy, has become well established in the sense that it is used by Sneath, and an attempt to restrict its use now to the sense of Bergey & Bates would be an example of **logomachy** and has nothing to recommend it.

numerical identification. Term used by Tsukamura and other numerical taxonomists for methods of identification made by comparisons of the S values of the unknown and that of **hypothetical mean organisms** of various species. (*N.B.* Tsukamura's S value takes into account both positive and negative matches, and corresponds to the matching coefficient, M, of other workers.)

numerical phenetics. Used by Mayr for a school of systematists which minimized the importance of information obtained by means other than the calculated resemblance (phenetic distance) of taxa. Mayr preferred this term to numerical taxonomy which he regarded as unfortunate because good taxonomists had been using numerical data for about 200 years.

numerical taxonomy. An approach to classification in which it is hoped that human weaknesses and bias are eliminated, and the large subjective element removed from taxonomy. Numerical methods have been applied for many years to classification but a new and more systematic approach to the subject was made by Sneath to bact. classification, and by Sokal to zoo. systems. Briefly, they attempted to make comparisons of large numbers of characters and to pick out groups (phena) with the greatest similarity of characters. One of the main points made was the absence of *a priori* weighting of characters because of the lack of a known logical basis on which to do so. This principle was attributed to Adanson, hence numerical taxonomy is sometimes referred to as Adansonian.

Numerical taxonomy as we now know it has only been made practicable by the development and wider use of computer techniques, and it has been described by Sneath (1968 *Class. Soc. Bull.* 1, 35) as the *automatic processing of information so as to give classifications.*

Heiser thinks that one of the chief virtues of numerical taxonomy is that it makes taxonomists look for and use more characters; at the same time he is thankful that not all taxonomists work with paper chromatography, computers, *or even compound microscopes*!

numericist. One who practices numerical taxonomy; a term used more by critics than by supporters of numerical taxonomy.

numericlature. The use of numbers instead of names. In microbiology these numbers can be (1) arbitrary, such as accession numbers given to cultures in collections (reference numbers); (2) indicative of rank or category, as in the systems proposed by Michener (1963, *Syst. Zool.* **12**, 151) and by Little (1964, *Syst. Zool.* **13**, 191) (classification numbers), also referred to as biological numerical nomenclature; or (3) coded descriptions of the characters of the organisms, as in the examples given by Cowan (1965*a*, *b*) and characters indicated by Chester type codes.

O

o. 1. A personal name ending in *-o* can be used to form the name of a genus in bact.[1] and bot.[2] by adding *-a*, e.g. *Lysenkoa* from Lysenko. In zoo.[3] a personal name ending in *-o* can be made into a generic name by adding the appropriate suffix *-us*, *-a*, *-um*.

A specific epithet or species-group name can be made by adding *-i* to the name of a man or *-ae* to the name of a woman ending in *-o*.[1,4,5]

2. The prefix O in surnames such as O'Brien should be united with the main part of the name when it is to be used to form the name of a genus or a species, as *obrieni*.[5]

[1, BAC.Ap9B. 2, BOT.R73B. 3, ZOO.ApD37. 4, ZOO.R31A. 5, BOT.(1975)R73C; ZOO.Ap.D21.]

O antigens. Antigenic complex of the body (or soma) of a bacterium, which may be masked by more labile antigens overlying the soma. O antigens are heat stable and resist 100 °C for several hours. Called 'O antigens' (from *Ohne Hauch*) by Weil & Felix. The chemical nature of O antigen(s) determines their specificity; the different O antigens of salmonellas are different polysaccharides.

objective synonyms are different names based on the same organism (type), e.g. in *Pseudomonas campestris*, *Phytomonas campestris*, and *Xanthomonas campestris*, the generic names are all objective synonyms. Although objective synonym is the term most used in bact.,[1] other disciplines use different adjectives, and the **synonymy** *of synonyms* (*q.v.* and Table S. 3, p. 253) is quite complex.

-odes, -oides. Bot. and zoo. confer different genders on names with these suffixes. BOT.[1] says that generic names with these endings should be regarded as feminine, whatever gender was assigned by the first user of the name. ZOO.[2] makes generic names ending in *-odes* or *-oides* masculine. Annotations to BAC.[3] say custom has fixed the generic name *Bacteroides* as masculine. (See also **-oides.**)

[1, BOT.R75A(4). 2, ZOO.30(a)(ii)Ex. 3, BAC.(1958), p. 118.]

Official Index. The International Trust for Zoological Nomenclature issues indexes of rejected and invalid names,[1] of family-groups, genus-groups, and

species-groups. Names in these Official Indexes have been rejected and made invalid by the Zoological Commission.
[1, ZOO.78(f).]

Official List. A list of names accepted (although usually contrary to a rule or rules) by the Zoological Commission;[1] separate lists deal with family-group, generic, and specific names. Official lists (though not so named) of conserved and rejected names appear in BAC.[2]
[1, ZOO.77(5); 78(f). 2, BAC.Ap.4.]

Official Opinion, see **opinion.**

offprint. Reprint of part of a larger publication, e.g. a single paper from one issue of a serial publication of journal. A **separate**.

O F test. Test of Hugh & Leifson (1953, *J. Bact.* **66**, 24) to distinguish between oxidative and fermentative breakdown of glucose. These characters are of great importance to bact. taxonomists, and form the basis of a division of bacteria into two major groups.

-oidea. Suffix added to the stem of the name of the type-genus to form the name of a superfamily in zoo.[1] Not used in bact. or bot.
[1, ZOO.R29A.]

-oideae. Suffix added to the stem of a legitimate name of an included genus in bot.[1] or the type genus in bact.[2] to form the name of a subfamily.
[1, BOT.19. 2, BAC.9T.]

-oides. 1. A generic name ending in *-oides* should be treated as feminine irrespective of the gender given to the name by the original author.[1]
2. A bot. subgenus epithet formed by adding *-oides* to the name of a genus the subgenus resembles, but it must not be attached to the name of the genus to which the subgenus belongs.[2]
[1, BOT.R75A(4). 2, BOT.21; R21B.]

oligo-. Prefix to indicate a small(er) number; e.g. oligosporogenous means an organism producing a small number (or an abnormally small number) of spores.

oligogenic, *adj.* Applied to character determined by only a few genes; a character controlled by one gene is **monogenic**.

oligonitrophil(e), (-ic). Term used by Beijerinck for microbes able to grow on media which, although not submitted to processes to remove the last traces of nitrogen, had not received any nitrogen-compound supplement. Carbon was supplied as carbohydrate and the oligonitrophilic organisms were able to fix free nitrogen and use it for their nutrition. In a shallow layer to create aerobic conditions, Beijerinck isolated *Azotobacter* spp. from garden soil.

omega classification. The ideal classification that will not, and cannot, be achieved while evolution continues. It is the ultimate stage in the development of classifications, the first of which is the **alpha classification**.

omnispective system of classification. A term used by Blackwelder (1964) to replace the vague 'classical taxonomy'. The method uses all available data *so far as necessary*. The taxonomist uses his experience to discard differences due to sex, age, disease, and from a workable number of characters, selects (Blackwelder uses the word considers) those necessary to show groupings and distinctions.

onomatophore. Strain or specimen that acts as the name-bearer.

ontogenesis, ontogenetic. Pertaining to the origin and development of the individual organism.

operational. 1. Practicable, workable.

2. Jargon for capable of proof by experiment.

operational taxonomic unit, *abbr.* **OTU.** A convenient unit for the purpose in hand; it may be an individual, a population, a species or even a genus. In microbiology it is used by numerical taxonomists as a useful substitute for **strain.**

A useful OTU suggested by Hill (*IJSB*, 1974, **24**, 494–9) is U-OTU, the initially unidentified OTU in identification work.

opero citato, *abbr.* ***op. cit.*** In the work cited earlier. an imprecise term to be avoided by all authors. Editors of scientific journals seldom allow its use.

opinion, *n.* Bac. and zoo. make provision for **commissions** that are empowered to consider between congresses questions that involve the application or interpretation (but not the alteration) of the rules of their codes. The commissions issue Opinions[1,2] for the guidance of bacteriologists and zoologists, and the effective date of each Opinion is that of its publication. Names may be **conserved** (*bact.*), **accepted** (zoo.) or **rejected.**

The method of preparation and presentation of a **request** for an Opinion on a bact. name is described in BAC.[3] The Chairman and the Editorial Secretary of the Judicial Commission prepare the Request for publication in *IJSB*; after six months the Request and any appeals against it will be submitted to members of the Judicial Commission and put to the vote.[4] An Opinion will be issued when 10 or more Commissioners vote in favour, and it will be published in *IJSB*. An Opinion is submitted to ICSB and must be approved by a majority of members voting. In bact. Opinions become final after approval by ICSB; all Opinions (no. 4 as revised) are printed in Appendix 5 to BAC.

An Opinion in zoo. must be reported to the next congress; a name that is accepted or rejected is entered in the appropriate Official Index or List, and the Opinion is then regarded as repealed.[2]

[1, BAC.23aN4(ii). 2, ZOO.78. 3, BAC.Ap8. 4, BAC.S8a(5); S8c(2).]

-ops. In zoo.[1] a genus-group name ending in *-ops* is to be treated as masculine, irrespective of its derivation or its treatment by the original author.

[1, ZOO.30(a)(i)(2).]

-opsida. Suffix to denote the name of a class in the Cormophyta.

[BOT.R16A.]

-opsis. In bot. indicates the name of a section. It must not be attached to the name of the genus to which the section belongs,[1] but may be added to the stem of the name of a genus that the section resembles.[2] The lesson to be learnt here is the suffixes *-opsis* and *-oides* are to be avoided. (See **-oides.**)

[1, BOT.21. 2, BOT.R21B.]

order (and **suborder**). Ranks between **class** and **family** recognized by BAC. and BOT. The names of orders in bact. should be based on the stem of the name of the type genus[1] and should end in *-ales*; names of suborders should end in *-ineae*.

In bot.,[2] where typification is not mandatory above the rank of family, the rules are less restrictive. But when the name of an order (or suborder) is based on the name of one of its families (which it need not be) then the suffix *-ales* (*-ineae* for suborder) is added to the stem of the family name; it is then automatically typified.[3]

[1, BAC.9. 2, BOT.17. 3, BOT.(1975)16.]

ordinal adjective. Adjective denoting number or used for enumeration. Such adjectives are not acceptable as specific epithets by either BAC.[1] or BOT.[2] but *Clostridium tertium* is well established by common use.

Since numbers and letters can be used in the formation of a name for a species of virus, ordinal adjectives will be acceptable.[3]

[1, BAC.52(2). 2, BOT.23. 3, VR.12.]

organelle. The microbial counterpart of an organ in a higher plant or animal. In higher plants and animals refers to subcellular units such as Golgi-apparatus; in lower forms, membrane-bound units such as the mitochondria of eucaryotic cells, but in procaryotic cells the plasma membrane forms the most important organelle.

organ-genus. A genus assignable to a family (cf. **form-genus**). Used in bot.[1] for fossil plants when the relation between specimens can rarely be proved.

[1, BOT.3N1.]

organize a genus, *v.* The act of classifying the intrageneric members; sorting organisms that make up a genus into orderly groups and sub-groups, and arranging them in categories of different ranks (levels), such as species, serotype, phagotype. Used more in US than in English.

original author. In taxonomy generally, and in nomenclature in particular, the term original author refers to the first author who used and published a name or epithet; it does not refer to the **discoverer** of the organism, an individual without importance to the nomenclator, even when the discovery was made long before thoughts had turned to a code of nomenclature for bacteria.

Some workers who discover new organisms, and who should be original (or publishing) authors, are too modest to name them, incorrectly thinking that they are not entitled or knowlegeable enough to propose a new name, or that the naming process should be left to a committee or to so-called experts. Sometimes a worker uses only a common name, or proposes a scientific name so tenatively that the **proposal** is unacceptable. For example the phrase 'the name *Bacillus stinkus* would be appropriate' is not strong enough to be a proposal; it should be either 'I name this organism *Bacillus stinkus*' or 'I propose the name *Bacillus stinkus* for the organism'.

originalis, originarius. Epithets professing to indicate the taxon containing the nomenclatural type of the next higher taxon. These epithets, together with *typicus*, *genuinus*, *verus*, and *veridious* cannot be accepted as validly published infraspecific names unless they repeat the specific epithet because BOT. requires their use (see **typus**).

[BOT.24.]

orthographia mutata: *abbr.* **orth. mut.** A changed spelling. Can be used as a qualifying phrase after a name and would be followed by the name of the author who made the change.

orthographic error. In bact. and bot. an error in spelling or transliteration which, in a scientific name, may be corrected by a subsequent author; such correction does not affect the attribution of the name to the original author, or the date of its valid publication. Among the errors that may be corrected are the wrong use of endings such as *-i*, *-ii*, *-ae*, *-iae*, *-anus*, *ianus*,[1] and, in bact.[2] and bot., the use of wrong connecting vowels or omission of a connecting vowel (compounding form[3]) in compound words.[1]

Zoo.[4] does not regard an inappropriate connecting vowel as an orthographic error, and consequently does not authorize its correction, neither does it allow correction of faulty transliteration or improper latinization. A name incorrectly spelt, or with diacritic marks or a hyphen should be corrected;[5] in the original form it is not an **available name** in zoo., and does not enter into homonymy.

[1, BOT.73. 2, BAC.Ap9A. 3, BOT.(1975)73. 4, ZOO.32(a)(ii). 5, ZOO.26; 27; 30; 32(c).]

orthographic variants. 1. Two or more names of the same rank, or specific epithets in the same genus, that differ in spelling so slightly that they are likely to be confused; e.g. compound names that differ only in the connecting vowel[1,2] or incorrect compounding form.[3] If these are based on different **types** they are **homonyms**.

2. Alternative spellings of the same name, e.g. *B. megaterium*, *B. megatherium*.

3. A name that differs from another only in its transliteration into Latin or in its grammatic correctness.[4] The example given is *Haemophilus* and *Hemophilus*.

4. Names from a language other than Latin or Greek, one of which has been latinized and the other has not, e.g. london, londonensis.

Orthographic variants that are errors should be corrected[2,5] but special care should be taken if the correction affects the first syllable of the name.[6]

[1, BAC.Ap9A. 2, BOT.73. 3, BOT.(1975)73. 4, BAC.57b. 5, BAC.57cN1. 6, BOT.73N2.]

orthography. Conventional spelling. In nomenclature transliteration not in accordance with convention can be corrected without prejudice to the date of publication and validity of the original publication.

The original spelling should be retained except for correction of typographical and orthographic errors.[1]

Although the letters j, k, w, and y were not used in classical Latin, they can be used in the Neo-Latin of nomenclature.

Diacritic signs, and certain letters used in Scandinavian languages, are not used in nomenclature. **Ligatures** seem to be allowed in bot., but not in zoo. names;[2] BAC. does not mention them, but the custom has been to separate the letters.

[1, BOT.73. 2, ZOO.ApE3.]

-orum. Suffix recommended for the formation of a species-group name in noun form from the personal name(s) *of men or of man (men) and woman (women) together*. Which piece of fine, clear English is from ZOO.[1]

It must be a rare occasion for an organism to be named after more than one person. It would be possible to form a name (or epithet) from the names of brothers, sisters, and husbands and wives, but more often multiple authors who are the object of such nomenclatural honour will have different names and the mind boggles at the idea of an epithet such as *smithgordonclarkorum*.

BOT.[2] adds *-orum* to the name of a man and woman (example given is *hookerorum* for Mr and Mrs Hooker) when the name ends in a vowel or the letters *-er*.

BAC. wisely leaves this problem alone and does not legislate for the formation of names (epithets) from the personal name(s) of more than one person.
[1, ZOO.R31A. 2, BOT.(1975)R73C(b).]

OTU. Operational taxonomic unit (q.v.).

-ous. Suffix indicating full of, characterized by.

outlier. Term used by Gyllenerg & Niemelä, in numerical identification, for an **OTU** whose absolute and relative affinities (see **affinity**) are low for all groups; in effect, the outlier falls outside reference system (cf. **identified**, **intermediate**, **neighbour**). LRH.

overall similarity. Defined by Sneath as *the proportion of agreements between . . . two organisms over the characters being studied*, but often used in a less precise sense by other taxonomists.
Usually represented by **S** or **M**.

overlapping of taxa. A system in which a species can be a member of more than one genus is an example of the overlapping of taxa. This is a realistic approach that is not widely accepted but may in time supersede the present nested hierarchical system of non-overlapping taxa. Most taxa are members of only one next higher taxon.

oxidation, oxidative type. In taxonomy these terms refer to the method by which micro-organisms attack carbohydrates. When the Hugh & Leifson medium and method are used (see 1953, *J. Bact.* **66**, 24), organisms that are called oxidative produce acid only in the open tube.

oxybiont, -ic. Bacteria capable of using atmospheric oxygen during growth.

oxygen relations and needs may be important in the characterization of micro-organisms. In addition to the well-known terms **aerobe, microaerophil** and **anaerobe** which McBee, Lamanna & Weeks thought should be limited to the description of actual cultural conditions, they introduced new terms to bacteriology such as **aerotolerant**, **anoxybiontic**, **capneic** and **oxybiontic** to describe the gaseous needs.

P

paper. 1. In science refers generally to an article in a journal or book.
2. The presentation of scientific matter at a meeting of scientists. Distinguished from a lecture which may be presented to a mixed assembly, a group of students or even to a wholly non-scientific body.

para-, (par- before a vowel or h). Prefix meaning by the side of, beyond, or faulty, irregular, subsidiary relation, perversion, and simulation. Used often to form specific epithets and to indicate likeness to another species, as in *Clostridium parasporogenes*, *Bordetella parapertussis*, *Haemophilus paraphorphilus* (see **elision**). Examples where the para- form is retained before a vowel or h are: *Haemophilus parainfluenzae* and *H. parahaemolyticus*.

paracolobactrum. Name of genus set up by Borman, Stuart & Wheeler, (1944, *J. Bact.* **48**, 361) for what had been previously called the **paracolons**, an assortment of coliform organisms that differed from strains typical of species established at that time by the delayed (or even absence of) fermentation of lactose.

As a taxon, the genus *Paracolobactrum* is made up of species of different genera; it is, therefore, made up of **discordant elements** and under BOT.[1] the name would have to be rejected. BAC. does not have a rule to deal with such a situation as presented by *Paracolobactrum* unless it be considered a **consortium**, q.v. (*Paracolobactrum* is unlikely to be placed in a Approved List under BAC.). [1, BOT.70.]

paracolon. A word to be avoided because its meanings are almost as numerous as its users. Formerly used for late lactose-fermenting coliforms or non-fermenters that could not be classified by the user of the word. An attempt was made to introduce **paracolobactrum** to replace it, but the new name, like paracolon itself, is not used by the enterobacteriologists.

parallelism. 1. Simpson uses the word to imply similar evolutionary changes in the characters of related organisms; hardly applicable to microbes.

2. Sokal & Sneath's use of the term is entirely different, and they apply it to unrelated organisms which show constant differences in the evolution of characters. This sense, also, can have little meaning in microbiology.

parameter. A word borrowed from statistics and used in microbiology for points on or near the limits of a boundary; these variable points or boundaries represent known values rather than estimates.

Described as a favourite of Ph.D. candidates, it is a popular word for the circumscription of a taxon, but sometimes it is used to indicate measures or measurement. Because its meaning is vague and understood by so few, parameter is a word to be avoided. Alternatives suggested by O'Connor & Woodford (1975) include index, criterion, factor, characteristic, value.

paramorph. Any **variant** of a species, especially one that, for lack of information, cannot be more accurately defined.

paranym. A name so similar in spelling to another that confusion may arise. Examples in bact. are *Gaillonella* and *Gallionella* both used by Ehrenberg; in the 1950s the Judicial Commission wasted much time discussing the two spellings. Eventually, *Gallionella* was conserved.[1] Significant in nomenclature only when both names are validly published, *and* they label taxa based on different types. [1, BAC.Ap5.]

parasite, -ic. 1. An organism that lives on the surface or in tissues, from which it derives its nourishment, but does not benefit the host. The word implies that the cells on which an organism is parasitic are living; if the tissues are dead or decaying the correct word to use would be **saprophyte, -ic**. It should not be confused with **commensal** which is an organism not deriving its nutrition from its host.

2. In describing a parasitic plant or micro-organism scientific names should be used for the host species as the application of common names may vary in different parts of the world.[1]

3. Epithets for parasitic fungi may be formed from the generic name of the host; the accepted spelling of the generic name of the host should be used (with an initial lower case letter) and orthographic variants should be avoided.[2]
[1, BOT.R32E. 2, BOT.R73H.]

paratype. Every specimen (strain, isolate) cited in the original description (**protologue**) other than the **holotype** or **isotype**. However, if the author does not designate a holotype, the strains cited are **syntypes**, not paratypes.

parenthesis, *pl.* **-es. 1.** In literary work parentheses may be shown by brackets (round or square), dashes, or commas. Square brackets are generally reserved for material inserted into a quotation by an author using material of another author.

2. In nomenclature the plural form is used for the round brackets used to enclose (1) the name (epithet in bot.[1]) of a subgenus,[2] e.g. *Salmonella* (*Arizona*) *arizonae*; (2) the name of the original author of a specific name (epithet) when the species has been transferred to another genus.[3] The name of the author (reviser) when made the change is not enclosed but follows the closing bracket. See **author citation**.

[1, BOT.21. 2, BAC.10c; BOT.R21A; ZOO.6. 3, BAC.34b; BOT.49; ZOO.51(d), R51B.]

parsimony of evolution. A theory that evolution takes the shortest possible pathways. Though speculative, it does permit possible cladogenies to be drawn. LRH.

partition. Taxometrical term for classification; also used for division.

passage, *v.*, *n.* The act and result of inoculating an animal with a culture, the intention being to enhance the virulence of the culture, which, if the passage has been successful, will be recovered from the blood or tissues of the moribund or dead animal. This is a technique that is used not only to restore virulence, but also to maintain virulence in organisms that rapidly become avirulent in culture. The word passage is firmly established in the literature of bact. and immunology, especially in relation to the inoculation–reisolation–inoculation... sequence; the verb, to passage, has a more specific meaning than the verb to pass though. The pronunciation follows the French and is derived from the works of Devaine.

patent name. Commercial organizations, particularly manufacturers of antibiotics brought pressure to bear on nomenclature committees to recognize the names of micro-organisms used in patent applications. BAC.[1] clearly states that names in published patent application or in the issued patent are not acceptable and are not regarded as effectively published names.

[1, BAC.25b(5).]

pathogen. A micro-organism capable of causing disease. This disease is normally an overt infection but could be tumour formation.

Pathogenicity has always been difficult to determine and the so-called **Koch's postulates** were formulated in an attempt to define it. But, like insanity, it continues to defy description. Sometimes it has to be assumed, and some character may be taken as an index of potential pathogenicity (e.g. production of coagulase by staphylococci). Pathogen is not a word synonymous with virulent (but see also **virulence**).

pathognomonic, *adj.* A word taken from medicine via pathology (where it means specially characteristic of a disease) to bacteriology for characters of a bacterium that are peculiarly characteristic of or specific for a taxon. Examples are coagulase production by *Staphylococcus aureus*, and the bile solubility of cells of *Streptococcus pneumoniae*.

pathotype. Term used by plant pathologists for infrasubspecific divisions whose members have limited infectivity. Lelliot, discussing *Erwinia carotovora* var. *chrysanthemi* says *There is evidence of differentiation into pathotypes*; and quotes a 'carnation' strain that can infect chrysanthemum, but strains from other hosts cannot infect carnation.

BAC.[1] recommends **pathovar** in preference to pathotype.

[1, BAC.Ap10BT.]

pathovar, *syn.* **pathotype.** Term recommended by BAC.[1] for a strain pathogenic to one or more host.

[1, BAC.Ap10BT.]

patristic. Defined by Cain & Harrison (1960) as *similarity due to common ancestry, not to convergence.* Since we lack any knowledge of microbial ancestry, microbiologists do not have a use for this term (cf. **cladistic**). STC. (Also used by Silvestri & Hill (1964) to mean similarities in DNA base sequences, since these must be due to common origin. LRH.)

patronymic. A name derived from that of a parent or ancestor; in nomenclature refers to a name of a person it is intended to honour. Patronymics can be used in the formation of generic names and specific epithets (names). See **personal name.**

patronymic prefixes. When using a patronymic to form a scientific name or epithet the Scottish Mac, Mc, and M' should be spelt as *mac* and joined to the main part of the name.

The Irish prefix O' should either be joined to the main part of the name or omitted, as *okellyi* from O'Kelly.[1]

The Dutch *van* and the German *von* may be included when normally part of the name, but otherwise omitted.

Prefixes of an article (le, la, l', les, el, il, lo) or containing an article (du, dela, des, del, della) are joined to the name.

Nobiliary particles are omitted, as are prefixes indicative of sainthood; in this, practice differs from that of the formation of scientific names from **geographical names** (q.v.).

(1, BOT.R73C; ZOO.ApD21.]

pattern. 1. Pattern has an old but loosely defined use in taxonomy; it is used for a listing of characters (irrespective of particular sequence) which describe, say, the enzymic make-up, or the sensitivity (resistance) to various antibiotics (but for this a more common expression is antibiotic **spectrum** or **antibiogram**).
2. Sneath used the comparative terms **vigour** and pattern to describe the behaviour of strains influenced by growth rate, incubation temperature, and variation in results due to differences in time of reading or carrying out the tests. Pattern is the difference between two strains of equal vigour. The Total Difference (D_T) between two strains is the sum of two components, Vigour Difference (D_V) and Pattern Difference (D_P).

PCNV. Provisional Committee for Virus Nomenclature, which became ICNV, not ICTV (q.v.).

peculiarity index, or **PI.** The converse of a similarity index. For each character the common state (positive or negative) is found, the number of strains with the rarer state is subtracted from half the total number of strains to give the PI of the taxon. If 3 of 40 individuals had the rare state, the PI would be $(40 \div 2) - 3 = 17$, a high figure. Taxa with lowest PIs have most features in common, and strains in which the PIs are high should probably be in different taxa.

peek-a-boo system. A punched-card system in which holes are punched in spaces specified for particular (individual) characters. All the cards are held in register between the observer and a light source; when a character (quality) is shared by all units (taxa) represented by cards the light will be visible to the user. See **sorter** for a practical application.

percentage. Bact. taxonomic literature is frequently plagued with figures quoted as 'percentage positive' when the number of organisms tested is nowhere near 100, indicating the author has forgotten the meaning, number per one hundred. LRH.

perfect state. That part of the life cycle of Fungi Imperfecti in which sexual spores (ascospores, basidiospores) are produced. Separate names may be given to the organism in the different states, but the name of the perfect state has precedence over the name of the **imperfect state.**[1]
[1, BOT.59]

peritrichate, -ous. Used to describe the type of flagellation of rod-shaped bacteria in which the flagella are inserted at any point around the body of the organism. Sometimes the insertion is confined to a region near one pole, when it may be difficult to decide whether flagellation is peritrichate or **lophotrichate**. Another difficulty arises when all the peritrichate flagella on a cell appear to twist around each other and sweep to one end of the rod, the free ends giving the appearance of a tuft of polar flagella.

permissive philosophy. Expounded by the International Working Group on Mycobacterial Taxonomy (IWGMT), the basis of the permissive philosophy is that if organisms *A* and *B* are really different then this should be apparent in anyone's laboratory. What may differ from one laboratory to another is the precise similarity level calculated between *A* and *B*. Since any calculated similarity is only an estimate of the 'true' resemblance, it is therefore permissible to pool data for numerical taxonomy analysis from different laboraties. The IWGMT use this philosophy with common sense and, prior to pooling data, examine the data to delete obvious duplication of results. LRH.

perplexing name, (*nomen perplexum*). Bacteriology has many names whose application is known but whose spelling is uncertain, or causes difficulty:[1] examples are *Serratia keilenses* and *Serratia kilienses*. If these are treated as orthographic variants (as in *Index Bergeyana*, pp. 10–14) they may be corrected by any author, but if they are regarded as perplexing names one of them must be rejected and this can only be done by the Judicial Commission.[2]
[1, BAC.56a(4). 2, BAC.57c.]

personal communication. Information received in conversation or private correspondence. Such information must be regarded as confidential and, when the other worker is known to be alive, should not be quoted without permission. Reference to it should be confined to the text of the paper and its nature indicated by giving the other worker's name and initials followed by the words personal communication. It should not be cited in the text as Smith (1984) or included in the references (literature cited) as Smith, A. B. (1984) personal communication. This is a method designed to irritate the reader, and no author can afford to to that.

personal name. **1.** Except in virus nomenclature (see **2** below) the name of a person may be used to form either a generic name or a specific epithet (zoo. specific name). A name that is not in Latin form must be latinized, but the letters *j*, *k*, *v*, *w*, and *y*, not used in classical Latin are retained, and a terminal *-y* is treated as a vowel.

Accents, umlauts, and other **diacritic signs** are suppressed[1,2,3] and the modified letters are transcribed into acceptable form as follows: *é*, *è*, and *ê* become *e*; the letters *ä*, *ö*, and *ü* become *ae*, *oe*, and *ue*; Scandinavian *ø* (*ö*) becomes *oe*, and *aê* (*ä*) and *æ* become *ae*, but *å* becomes *aa* (BAC.[1]) and *ao* (BOT.[2]). The **diaeresis** sign is allowed in BOT.[4] but not in ZOO.;[3] it is not mentioned in BAC. Ligatures are suppressed in zoo. (ZOO.[5]), are allowed in bot. (BOT.[4]) in journals that normally use them; they are not mentioned in BAC. but in practice are invariably suppressed and the vowels separated.

Patronymic prefixes are dealt with similarly in BOT.[6] and ZOO.[7] and the same methods can be used in bact. The Scottish Mac, Mc, M' should be spelt *mac* and joined with the rest of the name; the Irish O' drops the apostrophe and is joined to the name, or it may be omitted. The articles (le, la, l', les, el il, lo) and prefixes containing an article (du, de le la, des, del, della) are united with the name. Nobiliary particles and prefixes denoting canonization (De, St) are generally omitted. A German or Dutch prefix that is normally treated as part of the name (as is the custom in US) is united with the rest of the name, but unless regarded as part of the name these prefixes are omitted.

Generic name (*or subgeneric epithet*). BAC.[8] and BOT.[9] make similar recommendations. (1) a personal name used as a generic name should be given a feminine form, whether it commemorates a man or a women.[10,11] (2) If the personal name is in Latin and ends in *-us*, drop the *-us* and add the suffix appropriate to the new ending. (3) If the name ends in *-a* add *-ea*, but if it ends in *-ea* it does not take a suffix.[9] When it ends in any other vowel or in *-y* add *-a*. (4) If the name ends in *-er* add *-a* but when it ends in a consonant other than *-er* add *-ia*. (5) In bact. a diminutive ending such as *-ella* may be added to a personal name that does not end in a sibilant, or *-iella* to a name ending in *-s* or *-x* (*Klebsiella*, *Coxiella*).[8,12] (6) In bot. a prefix or a suffix may be added to a personal name, or an anagram or abbreviated form of the name may be used; all these are regarded as different names from the simple patronymic.[9]

ZOO.[13] frowns on creation of compound genus-group names from personal names; nevertheless *names of modern persons* with appropriate suffixes can be used. When the name ends in a consonant the suffix will be *-ius*, *-ia*, *-ium*; when it ends in *-a* the suffix will be *-ia*; if it ends in any other vowel it should be *-us*, *-a*, or *-um*.[14]

Generic names and subgeneric epithets are printed with initial capital letters.

Table P. 1 summarizes the method of formation of generic names.

Specific (*and subspecific*) *epithet*. May be either a substantive in the genitive or an adjective (not used as a substantive); the two forms made from the same personal name are treated as different epithets and not as orthographic variants

Table P. 1. *Suffix to be added to a patronymic to form a generic name.*

Name ends in	BAC.	BOT.	ZOO.
-a	*-ea*	*-ea*	*-ia*
-ea		no addition	
-e, -i, -o, -u, -y	*-a*	*-a*	*-us, -a, -um*
-er	*-a*	*-a*	*-ius, -ia, -ium*
other consonant	*-ia*	*-ia*	*-ius, -ia, -ium*
to form a diminutive			
consonant (except *-s, -x*)	*-ella, -illus*		
-s, -x	*-iella*		
latinized ending in *-us*	delete *-us* and treat stem as above		

Table P. 2. *Suffix to be added to a personal name to form a substantival specific epithet or name*

Code	Name ends in	Sex of person(s) named				
		♂	♀	♂♂	♂♀	♀♀
BAC.[1]	*-a*	*-i* or *-e*	*-e*	—	—	—
	-e, -i, -o, -u	*-i*	*-ae*	—	—	—
	-y	*-i*	*-ae*	—	—	—
	-er	*-i*	*-ae*	—	—	—
	other consonant	*-ii*	*-ae*	—	—	—
BOT.[2]	*-a*	*-e*	*-e*	*-rum*	*-rum*	*-rum*
	-e, -i, -o, -u	*-i*	*-ae*	*-orum*	*-orum*	*-orum*
	-er	*-i*	*-ae*	*-orum*	*-orum*	*-orum*
	other consonant*	*-ii*	*-iae*	*-iorum*	—	*-iarum*
ZOO.[3]	any letter	*-i*	*-ae*	*-orum*	*-orum*	*-arum*
BAC. BOT. ZOO.	Latin or latinized name	Appropriate Latin genitive†				
		-i	*-ae*	—	—	—

* BOT. treats *y* as a consonant.[4]
† Modern names should not be treated as if they were third declension nouns (e.g. not *munronis*, not *richardsonis*).[2]
[1, BAC.Ap9B. 2, BOT. (1975) R73C. 3, ZOO.R31A. 4, BOT.73.]

unless they are so alike as to cause confusion,[15] but within one genus the future creation of a second epithet derived from the same personal name is not recommended.[16]

Substantival epithets are formed from personal names in the following manner:[8, 17] names that are Greek, Latin, or in latinized from (pastor from Pasteur) take the appropriate genitive ending (*Streptococcus pastorianus*) and names of women have feminine endings (*Rosa beatricis*, *Scabiosa olgae*). There are relatively few of these ancient Greek and Latin names, and the commoner method of forming substantival epithets from names in modern form is determined by the sex of the person after whom the organism is named (cf. *adjectval epithet* below).

Personal names ending in *-a* take *-i* or *-e* (♂) or *-e* (♀) in bact.[8] or *-e* (♂, ♀) in bot.;[17] names ending in any other vowel (including *-y*) or in *-er* add the genitive inflexion appropriate to the sex and number of the person(s) named,

Table P. 3. *Suffix to be added to a personal name to form an adjectival specific epithet or specific name.*
(The adjectival epithet must agree in gender with the gender of the generic name)

Code	Personal name	Gender of the generic name		
		m	f	n
BAC.[1]	whole name	*-ianus*	*-iana*	*-ianum*
BOT.[2]	name ends in *-a*	*-nus*	*-na*	*-num*
	ends in *-e, -i, -o, -u*	*-anus*	*-ana*	*-anum*
	ends in *-er*	*-anus*	*-ana**	*-anum**
	ends in other consonant	*-ianus*	*-iana*	*-ianum*
ZOO.[3]	whole name†	*-ianus*	*-iana*	*-ianum*

* The *-er* exception (*Taxon*, 1976, **25**, 172) mentions only *-anus* instead of *-ianus*. The full effects of this amendment from the floor of the 1975 Congress will probably not be known until publication of a new edition of BOT., timed for 1978.
† Permitted but not recommended by ZOO., which prefers use of the genitive singular.
[1, BAC.Ap9B. 2, BOT. (1975) R73C. 3, ZOO.ApD.16.]

as *-i* to the name of a man, *-ae* to the name of a woman[8, 17] (see Table P. 2, p. 198). To personal names ending in a consonant (but not in *-er*) add *-ii* (♂) or *-ae* (♀),[8] *-iorum* (♂♂), or *-iarum* (♀♀).[17]
Adjectival epithets must agree in gender with the gender of the generic name;[8, 17] the sex of the person named does not influence the ending of the adjectival epithet. The suffix added to the whole name is *-ianus* (m), *-iana* (f)., or *-ianum* (n).[6, 8] Thus the name Gordon takes the adjectival forms *gordonianus*, *gordoniana*, or *gordonianum* depending on the gender of the generic name and not on whether the epithet refers to John or Ruth Gordon. (See Table P. 3, above.)

Specific epithets formed from personal names should be written and printed with initial lower case letters,[18] but BOT.[19] permits initial capitals if an author prefers them; however, it is *never incorrect* [in any discipline] *to decapitalize the specific epithet* (Ainsworth, 1973, *Rev. Plant Path.* **52**, 59). Commenting on this point, Stafleu & Voss (1975, *Taxon*, **24**, 230) point out to those who use capitals that it must be *Lobelia Cardinalis*, but *Mimulus cardinalis*. Thus if capitals are used each epithet that might be formed from a patronymic becomes the subject of library research.
Specific names (*species-group names*) are formed from personal names in a different way; ZOO.[20] does not recommend latinization by the addition of *-ius* to the modern name (as in bact. and bot.) but prefers *-us*, to give genitives ending in *-i*. The *-us* ending of the latinized name of a man[21] is replaced by *-i*, and a feminine form (*-ae*) is used for the name of a woman; the final *-a* or *-e* of the patronymic may be elided for euphony, as *josephineae* or *josephinae* from Josephine.[22] ZOO.[23] makes recommendations on the endings to be used in the unlikely event that a species-group name is to be formed from the personal name of men, or of man (men) and woman (women) together, when the suffix will be *-orum*; it will be *-arum* when the personal name is that of women. Adjectival endings (see Table P. 1) are allowed in ZOO. but are not recommended.[20]

2. Virologists are adamant that personal names shall not be used in virus nomenclature, but the interpretation of Rule 8 of ICTV, *No person's name shall be used*, is far from clear. It refers both to the use of patronymics in scientific names, and to author citations, which virologists do not attach to virus names (see **depersonalization**). Dr Helio Pereira (personal communication, 1976) says that the intention is both to prevent the use of personal names in the formation of the name of a virus and to rule out author citations. The rule led to the rejection of *Heliovirus* (proposed for a virus with regular projections from its surface) because it might have been presumed to be formed from Dr Pereira's first name.

3. In lists of references or **literature cited** personal names are arranged alphabetically by the surname of the author or, where more than one, by the surname of the first author. This procedure is normally straightforward but some names may cause difficulty, especially in those languages in which the surname is not the last name. In Spanish the last name is that of the mother, and the father's name (surname) precedes this; thus A. Lopez Moreno on the byline should be shown in the list of references as Lopez-Moreno, A. In Portuguese the maternal name comes before the paternal, so that Silvo do Amaral becomes, in the alphabetical list, Amaral, S. do. The Spanish *hijo*, and the Portuguese *filho*, meaning son, are the equivalents of *junior*, and like it should follow the initials of the individual in lists of references, e.g. Niven, C. F. Jr; da Fonseca, O., Filho.

The position of prepositions and articles, e.g. da, in Olympio da Fonseca, may be treated differently in different journals, and O'Connor and Woodford recommend that journal style should be studied and, if determinable, should be followed. When journal style is not apparent, the names can be written (1) in the style customary in the country in which the authors quoted live, or (2) in a consistent system in which do, da, etc. are listed under D, and van, Van under V, and so on, irrespective of capitalization (see **alphabetization**).

Scottish and Irish names beginning with M', Mc, Mac, meaning 'son of', should be treated as Mac and listed accordingly. Names beginning with O', such as O'Hare, O'Meara, are listed under O. The Dutch van is normally included in the name in the text but when the name is listed it is put under the part printed with a capital, thus van Rooyen is listed under R. In US such a name is usually spelt with a capital V and in many American journals (and in *Bergey's Manual*) an author such as C. B. van Niel is listed under V. The German von is usually not included in the name.

Titles present problems, particularly in those countries which do not use such titles. The best advice is to use in the text the name as printed on the byline (e.g. Lord Stamp, in the text), but in the references to reverse the name and title (e.g. Stamp, Lord, in the references). Knights are known by one of their forenames, which should be used in the text (e.g. Sir Graham Wilson) but which should be abbreviated to an initial in the references (e.g. Wilson, Sir G.).

[1, BAC.64. 2, BOT.73. 3, ZOO.27; 32(c)(i). 4, BOT.73Fc. 5, ZOO.ApE3. 6, BOT.R73C. 7, ZOO.ApD21. 8, BAC.Ap9B. 9, BOT.R73B. 10, BAC.R10a(2). 11, BOT.R20A. 12, BAC. (1958) Annot. p. 110. 13, ZOO.ApD15. 14, ZOO.ApD37. 15, BAC.63. 16, BOT.R23A.

17, BOT.(1975)R73C. 18, BAC.59. 19, BOT.R73F. 20, ZOO.ApD16. 21, ZOO.ApD17. 22, ZOO.ApD18. 23, ZOO.R31A.]

ph or f? Names differing in spelling only by *ph* or *f* are homonyms.[1]
(See **f or ph?**)
[1, ZOO.58(7).]

phage type, phagotype. A subdivision (usually of a species) determined by the susceptibility of a bacterium to one or more of a series of bacteriophages which are themselves called **typing phages**. The type is often seen as a pattern of lytic zones in the places where phage has been spotted on a plate sown with the strain under test. Because it includes the word type, this is one of many such names frowned upon by nomenclators, who prefer **phagovar**. However, it is not a mutant or a variety but a subdivision of the species, and phagovar seems an inappropriate term.

phagomosaic. The pattern of phage sensitivities. Used by Leonova, Hatenever & Chistenkov (1967, *IJSB.* **17**, 303–5) in connexion with streptococcal phages, in which they pointed out that like staphylococcal phages, the phage divisions are patterns of sensitivity rather than sensitivity to one specific phage.

phagovar, *syn.* **phage type, lysotype.** Term used in BAC.[1] for **phage type** (q.v.).
[1, BAC.Ap10BT.]

phanerogram (phanérogramme). Visible characters: defined by A. Lwoff as 'What can be seen, such as the shape of a virus.'

pharmacognosist. A taxonomist who has an equal interest in the chemistry of plants and in their identification. Term used by Bate-Smith to describe Hegnauer, a pioneer of chemotaxonomy.

phasic variation. The antigenic nature of flagella may not be constant and in some bacteria (e.g. salmonellas) may regularly occur in two or more different forms. The original discovery by Andrewes was of two forms (hence diphasic variation) but later Edwards showed that three forms may occur and occasionally more phases have been found. Phasic variation is associated with salmonellas and similar organisms (e.g. the Arizona group), but outside a few groups such as the enterobacteria, antigenic analysis of flagellated bacteria is still in its infancy.

phenetic, *adj.* **1.** Applied to a classification based on overall activity, as determined by an unprejudiced assessment of all known characters. Since it uses all observable characters, a phenetic classification will make use of genetic data if these are available. In contrast to **phyletic** and **phylogenetic**, the word phenetic does not have any evolutionary implications, except in the sense of showing the end-product of evolution.
2. Empirical; applied to a classification of the non-overlapping, hierarchical type, based on observed facts, and excluding all *a posteriori* inferences.
3. Existential; since characters are taken at face value and without weighting, phenetic taxonomy has been compared with existentialism; both reject divine relevation or 'intuition' as a means of finding *operational truth* (Smith, 1965, *Syst. Zoo.* **14**, 148–9).

phenetic gap. A discontinuity in the phenetic space, assumed by the absence of **OTUs**. The task of phenetic classification is to discover the limits of clusters and the existence of phenetic gaps between clusters and to further ascertain that such gaps are not due to poor sampling of OTUs. LRH.

pheneticist. Word used by Mayr for those who practice numerical taxonomy (a term that he dislikes), which he regards as being based on nominalistic principles, the fallacy of which lies in the misinterpretation of the causal relation between similarity and relationship.

phenogram. Dendrogram intended to show the phenetic relations of the organisms.

phenome. The observed constitution of an organism; the difference between it and the **genome** becomes indistinct at the molecular level.

phenon. **1.** Taxonomic group in which the degree of similarity is established by numerical methods. Thus instead of using categories such as species and genera, numerical taxonomists work with groups that have, say, 85% and 75% similarities which are called the 85% phenon and the 75% phenon respectively.

2. Zoologists use phenon for a sample or group of phenotypically similar organisms; in a sexually dimorphic organism there will be one phenon for the male and another for the female. The difference between the phenon of the zoologist and of the numerical taxonomist is seen in Mayr's statement (1969) that *There is hardly a species that does not contain several if not dozens of phena.*

3. Users of the word phenon are divided on how the plural should be formed; numerical taxonomists, using the word in sense **1** have phenons; zoologists, using it in sense **2**, have phena.

phenospecies. **1.** A word suggested by Stanier for a group of bacteria whose phenotypic description is known and which is judged to have a rank corresponding to species. Such a phenospecies can be more precisely defined than, say, a **nomenspecies**, which is a category designated more by name than by content or the characters of the included strains.

2. Phenospecies is used by Michener in the sense of a species recognized and separated from others by degrees of difference; when they become better known and are more thoroughly studied, such species may be upgraded to **biospecies**.

phenotype. The observable characters of a strain constitute the phenotype, and are the product of the interaction between the hereditary potential (**genotype**) and the internal and external surroundings. Organisms with the same characters may, however, be descendants of different genotypes.

phenotypic. Pertaining to the phenotype. Phenotypic characters are those that are detected when a strain is submitted to a series of tests; they characterize or describe the organism as it exists, and not as it might have existed in some earlier form (the **genotype**), which is a hypothetical state and the subject of much unprofitable speculation.

phenotypic plasticity. The expression of a range of variation from one genotype in response to differences in environment. The variants themselves are known as phenotypic modifications.

phenotypical, -ly. Variation influenced by the conditions under which the organism grew.

-philic. Suffix indicating a liking for a substance or state, e.g. halophilic, thermophilic.

phobotaxis. Reversal of direction of movement by bacteria under the influence of a stimulating agent.

phosphorescent. Self-luminous; luminous without heat or incandescence. Sometimes used to describe the photobacteria, but a better term is **luminescent**.

photobacterium. **1.** In roman, with lower case initial letter, the common name for bacteria that are able to generate and emit light. Also called the luminescent bacteria. In the plural, photobacteria, used as a common name for phototrophic procaryotes in *Bergey's Manual* (1974).

2. In italic with initial capital, a generic name.

photochromogen, -ic. An organism that produces a carotenoid pigment when grown in the light. The term is applied to some mycobacteria (cf. **scotochromogen**).

photocopy. Copy of a document made directly from a document or book on light-sensitive or heat-sensitive paper. A special form of photocopy can be made on non-sensitized paper, and is normally known by the trade name (Xeroxing) of the process.

Photocopying would not normally be regarded as effective publication of a scientific name, but Xeroxing is in rather a different category, and probably would be acceptable.

photolithography, photo-offset. Printing method in which the block is made from a photograph of the original material. Often used for second (and subsequent) impressions of a book. (See also **lithography**.)

photosynthesis, -thetic. Production of organic compounds, particularly carbohydrates, from CO_2 and water, using sunlight as a source of energy. Algae and green plants are the main photosynthesizers. Photosynthetic bacteria, which contain pigments (bacteriopurpurin, various carotenoids, or green chlorophyll-like pigment), are divided into two main groups: (1) photolithotrophs which obtain their carbon from inorganic sources, and (2) photo-organotrophs which obtain carbon from organic sources.

phototaxis. Attraction to light; repulsion by light is negative phototaxis, see **taxis, 2.**

phototroph, -ic. General term for the adjective applied to bacteria that make use of light as a source of energy.

Phototrophic procaryotes is the term used in *Bergey's Manual* (1974) for the photobacteria; they include the blue-green algae, the red photobacteria, and the green photobacteria.

-phyceae. Suffix which indicates the name of a taxon of the rank of class in the algae.

[BOT.R16A.]

-phycidae. Suffix indicating the name of a subclass in the algae.

[BOT.R16A.]

phyletic. Said of an arrangement *which aims to show the course of evolution* (Cain & Harrison, 1960). A vague term unless qualified as it may refer to cladistic, patristic, or chronistic components of evolution, or to any combination of them. An unambiguous alternative is evolutionary.

phylogenetic. Adjective which, relating to a classification, requires that taxa should reflect, as far as possible, their evolutionary history. By implication, it assumes that a phenetic approach to classification does not do this (see Colless, *Syst. Zool.*, 1969, **18**, 123).

phylogenetic noise. Used by Colless (*Syst. Zool.*, 1969, **18**, 123) for *a false grouping suggested by convergence*.

phylogenetic speculation. A favourite expression of those authors (including this one) who regard most bacterial phylogenies as the products of vivid imaginations.

phylogeny. *The expression of evolutionary relationships between organisms.* Constance (*Taxon*, 1964, **13**, 257–73) found differences between the views of British and US botanists on the meaning of phylogeny or phylogenetic classification. In English it means genealogy whereas in American it means self-reinforcing classification based on a maximum correlation of characters which, as Heywood points out, is how **phenetic** classifications (as opposed to phylogenetic) are defined.

phylogram. A tree-like figure intended to express graphically the evolution of an organism and its relationships. See **cladogram**.

phylum, *pl.* **-a.** The category or level below kingdom in a hierarchical classification of living things. Just as, at a lower level, species is often regarded as something rather special, so at a higher level, much is made of the phyla, particularly by zoologists.

Phylum is not used in bot. or bact., in which the primary division of a kingdom is a **division**.

physiologic(al) race. In bot. preferred to physiological form for a non-morphological subdivision of a species; their names are not subject to BOT. In mycology may be further subdivided into biotypes, sometimes designated by letters and numbers.

-phyta. Suffix indicating the rank of Division of plants (except fungi). The name should be taken either (1) from distinctive characters of the division (when it will be a descriptive name), or (2) from the name of an included genus, which will be automatically typified.
[BOT.1975, R16A.]

-phytina. Suffix indicating the rank of subdivision in the plants (except fungi).
[BOT.R16A.]

PI. See **peculiarity index**.

pilus, *pl.* **-i.** Later synonym of **fimbria**. Pilus is used by American workers, who seldom use fimbria; it is also used by most workers in discussing the genetics of bacteria.

It is suggested that pilus be reserved for the genetic element, and fimbria for the morphological entity.

placement. Allocation to a position (level) in a hierarchy, most often to a genus and species.

place names can be used to form specific epithets (specific names in zoo.). See **geographical name**.

plate out, *v.* **1.** In bact. the act of spreading a culture or **specimen 2** over the surface of a solidified nutrient medium in a Petri dish with the object of obtaining isolated colonies to next obtain a pure culture.

2. In gel-diffusion precipitation, the act of putting soluble antigens into one set of cups or cavities, and antisera into other cups or troughs to detect antigenic similarities where precipitates form.

pleiotropy. State in which a single gene is concerned in the formation or control of several characters.

plenary power. Authority to make decisions conferred on a committee by the members of an organization. ZOO.[1] authorizes the Zoological Commission to suspend the application of rules under plenary powers, and given guidance on how these powers should be exercised.
[1, ZOO.79.]

pleomorphic, -ism. 1. The state in which individual organisms of a strain show variation in shape and size. It is a characteristic that may have value in the identification of the micro-organism, e.g. *Corynebacterium diphtheriae*.
2. A stricter interpretation is the ability to change shape or form, all the individual elements of a strain at any one time being the same shape. This is an extension of an older (discredited) view held by Zopf, Metchinikoff, and others that the morphology of bacteria was inconstant, and that they could easily change from spheres to rods or spirals, especially with changes of environment. Cf. **monomorphic, -ism**.

pleonasm. In nomenclature, almost, but not quite, the same as a tautonym; it is equivalent to the zoo. *virtual tautonym.*[1] Literally it means using more words than are necessary. In biological nomenclature, it is the use of generic and specific names that refer to the same subject, though the words used may be in different cases, e.g. *Arizona arizonae*. A **tautonym**, on the other hand, is the exact repetition of the generic name in the species name, e.g. *Fusiformis fusiformis*, which is not permissible under BOT.[2]
[1, ZOO.R69B(2). 2, BOT.R23B.]

plesiomorph. A primitive character; one that has not changed in form from that found in ancestors.

plural forms of generic names should not be underlined in manuscript or printed in italic. As a scientific and Latin word the generic name is always used in the singular, when the plural is needed it becomes a common word in the language being used and the method of forming the plural in the modern language should be followed. In English the plural is formed by adding *-s* or *-es* to the singular, which gives shigellas, salmonellas, and so on. However, words ending in *-um* or *-us* make unpleasant-sounding plurals when treated in this way, and *-a* is used for *-um* words (medium becomes media) and *-i* for *-us* words (Bacillus → bacilli).

Bact. and bot. names for ranks above genus are latinized plurals, substantives or adjectives used as substantives, and do not have (as scientific names) a singular form. As common names some, such as Enterobacteriaceae, are used as collective nouns, as enterobacteria, which itself is plural.

pole, polar, *n.* and *adj.* The end of a rod-shaped bacterium; the adjective used mainly to describe the apparent site of origin of a **flagellum** or flagella. Appearances may be deceptive and peritrichate flagella, arising from the long side of the elongated bacterial cell, may twist together and separate only near their extremities, to give the appearance of a whip attached to the pole.

polychotomous. Division into more than two branches, or pathways in an identification key. Cf. **dichotomous**. LRH.

polyclave. A **multiple-entry identification system** (q.v.).

polygenic, *adj.* Determined or controlled by many genes.

polygeny, *n.* State in which several genes are involved in the production of a single character.

polymorphic, polymorphous. Having many shapes; extended to the meaning of many variations or phases.

polymorphism. 1. Genetic variation occurring within a population, for which Davis & Heywood prefer Huxley's term, morphism, because polymorphism has become debased and may refer to any kind of variability.

2. Developmental polymorphism is seen in the different stages of a life cycle, as in the malaria parasites (*Plasmodium* spp.).

polynominal (polynomial). A name made up of many words, often descriptive. In common use to describe plants and animals before **binominals** (or binomials) became the accepted form.

polyphasic taxonomy. Term used by Colwell, but not come into widespread use, to signify successive or simultaneous taxonomic studies of one and the same set of organisms by radically different techniques, e.g. by numerical taxonomy, amino-acid sequence determination of homologous proteins, molecular hybridization of nucleic acids *in vitro*, etc. LRH.

polyphenic. Experimental procedures which yield many characters, e.g. pyrolysis/gas–liquid chromatography. LRH.

polyphyletic. Taxa not developed from a common ancestor, but classified together for some other reason, e.g. by convergent evolution they have developed some gross similarities.

polythetic. Said of an arrangement in which organisms with the most shared features are grouped together. A member of a polythetic group does not have any essential character, neither does the possession of a unique character suffice to place an organism in a particular taxon. Polythetic groups are shown in **diagnostic tables** (see Table D. 1, p. 96), from which it is possible to form a **dichotomous key** (Table D. 2, p. 97) to yield **monothetic groups**.

polytrichate, -ous. Having several flagella, looking like a tuft, at one pole of a rod-shaped bacterium. This compound of two Greek words is preferred to **multitrichate**.

polytypic. Species based on more than one type; a species containing two or more subspecies each based on a different type. Used in zoo. for species composed of races separated geographically, and, according to Mayr, only important when geographical separation is accompanied by variation.

In bact. *Pasteurella multocida*, an organism made up of races of bacteria that cause septicaemic diseases in many different kinds of animal and named *P. aviseptica*, *P. lepiseptica*, *P. bubiloseptica* and so on, might be considered to be polytypic, in contrast to *Pasteurella pestis*, which is monotypic. See also **superspecies**.

population. Operational unit used in the process of classification. They are not named as such in a classification, but they must be classes of something, or have members – which are the individuals. It had been said *Individuals are identified. Taxa are classified.* Populations can be considered as being made up of individuals (identifiable), which, in turn form taxa (classifiable).

population concept. Difficulty in defining the different ranks of micro-organisms has given rise to the development of the population concept of **species**. A single bacterium implanted on a suitable medium will soon grow into a colony of millions of (more or less) similar bacteria, and these in turn will, when sub-cultured to fresh medium, produce further similar progeny. We known that variations of one sort and another are constantly taking place so that the progeny will not all be exactly alike and a few will show a change in their characteristics. The population concept is built round a group such as this, and several individuals (or strains) may be needed to describe the whole population; in this it differs from the **type concept** in which one strain is chosen, and may determine the characterization of the organism. (See also **representative strains**).

position. The place of an organism or taxon in a hierarchical classification, usually indicated by giving the name of the next higher taxon to which it belongs.

positive result. In numerical taxonomy, the Jaccard coefficient (see **S**) counts only those characters scored positive for both **OTU**s as similarities (as opposed to the matching coefficient, **M** (q.v.) which counts as similarities all matches, whether positive or negative). Care must therefore be exercised when using the Jaccard coefficient that the coding of results as positive does have a greater meaning than a mere convention or tradition. Generally, in bact. positive result means changed in some way from an uninoculated control. For certain tests, the allocation of a positive result is arbitrary, e.g. a bacterium sensitive to penicillin can be coded as Sensitive, positive, or Resistant, negative. LRH.

post-. Sometimes a fatuous prefix, analogous to the fatuous *pre-*. Occasionally seen in scientific writing, and more common in English (postgraduate) than American (graduate).

PPLO. Laboratory jargon for a member of the Mycoplasmatales. Started as a shortened form of pleuropneumonia-like organism, which was an illiterate way of saying an organism similar to those isolated from animals suffering from pleuropneumonia. Now archaic.

pre-. Described by Woodford (1968) as the fatuous *pre-*, this is commonly an unnecessary prefix. Examples given by Woodford are pre-cooled and pre-sterilized, and he asks what advantage these words have over the words without the prefix.

Authors should pause whenever they want to write pre... (pre-requisite, pre-determine, pre-reduced and so on) and ask themselves 'Is this prefix really necessary?' Sometimes, as in the question, it is! In general, when a hyphen would be used, the *pre-* is unnecessary; when a hyphen would not be used (as in prepare and prefix) the *pre* is an essential part of the word. The same test cannot be applied to US words, where the hyphen is much less used.

preamble. An introductory section which explains the aims and purposes of the nomenclatural codes. BOT. and ZOO. have preambles, but BAC. has **general considerations** that fulfil the same purpose.

predicate, *n*. Term used in logic and grammar, of which only the grammatical uses concern us. According to *SOED* these are:

1. Statement made about a subject, e.g. 'Time flies', time is the subject and flies the predicate.

2. An appellation that asserts something.

3. A quality or attribute.

In taxonomy descriptions of organisms may include predicates of kinds **1** and 3 in lists of characters such as 'gelatin liquefied', 'urea hydrolyzed' and 'Gram negative'. The grammatical importance of recognizing the predicate is that two words such as Gram and negative are, as a predicate, unhyphenated but as a composite adjective are hyphenated (a Gram-negative bacterium). See **hyphenated qualities**.

predictive value is a quality of a classification; prediction is interpreted in different ways by workers.

The use of a classification to predict future evolution is not only speculative but a form of scientific dishonesty. In another sense, a new set of characters may be tested against an existing classification and, when this is not unduly upset, the new tests may be regarded as having a high predictive value. Attempts may also be made to predict the characters of an unknown organism because of a supposed relationship to a known. Jeffrey thinks that the better a system of classification *reflects the various degrees of overall similarity... the greater... its predictive value*. DuPraw has yet another use for the term; writing of non-Linnaean taxonomy, he says that such classifications *attain a maximum in predictive value (= repeatability)*...

pre-empted name. An occupied or appropriate name. A name applied to a taxon and not therefore available for application to another taxon of the same rank.

preferable name. A name preferred by one worker may be anathema to another, and the codes of nomenclature do not recognize *preferable names*. A **legitimate name** may not be rejected because another is preferable.[1]

[1, BAC.55(3), BOT.62.]

preferred spelling. 1. A name in which the transliteration conforms to accepted – not always the same in different countries – or classical usage.

2. Apart from scientific names, the spelling of the same word in different countries may vary, as, for example, many English words that end in *-in* have *-ine* endings in French, as penicillin, penicilline. Even in English-speaking countries the preferred spelling may vary, as in spelt (English) and spelled (US), and editors of different journals are known to have preferences; authors are advised to consult a recent issue of the journal of their choice, and to observe the trend of **editorial whim**. A useful list of preferred spellings will be found in the guide to the preparation of manuscripts by Sir Graham Wilson (1965).

preoccupied name. A zoo. term for a junior or later homonym; a name identical with one given previously to a different taxon.

[ZOO.G1.]

preprint. A **separate**. A printed or duplicated typescript, which has been distributed before publication in a scientific journal or, rarely, book. Not a procedure often used in microbiology, but a summary of the classification to be used in the 5th edition of *Bergey's Manual* was issued as a preprint. Preprints should be indicated as such and should be dated.[1] Subsequent authors should not quote preprints as sources of information and they should not be indicated in References or Literature Cited. This restriction arises because preprints are usually of very limited distribution and will be generally unavailable in libraries.

[1, ZOO.R21D.]

prevalid, *adj*. Applied to the name proposed or published before the **starting date for nomenclature** relating to the organism. BOT.[1] states that the name of the original author of such a name should be followed by the word *ex* and then the name of the author who validly published the name after the starting date for bot. nomenclature; this replaces an earlier use of square brackets around the name of the original author.

The term *prevalid* has not been used in bact. literature, but may come into use after 1 January 1980, the new starting date for bact. nomenclature. BAC. includes **Provisional Rules**[2] that deal with the citation of names included in **Approved Lists**, and names that are reused or revived.

[1, BOT.R46E. 2, BAC.Prov.A1; B1, 2.]

primitive. Ill-developed, early, simple. In relation to characters of micro-organisms it is speculative to describe some as primitive and others as advanced or derived, because we do not have any information about characters of organisms described before the end of the nineteenth century. For example, we do not know whether the ability to use organic substrates is a primitive character or the result of adaptation and specialization.

principal component, coordinate analysis. A technique of numerical taxonomy whereby the original *n* dimensions (one for each character) are reduced to a smaller number, or even only two or three for display purposes. The object is to extract synthetic dimensions (towards which each character makes a different contribution) such that the first accounts for the greatest possible amount of the total variance, the second the next greatest amount of variance in the residual variance (after subtraction of that due to the first dimension), and so on, until there is little or no variance left. By this procedure, the differences between clusters is maximized and the differences within clusters minimized. If the analysis begins with correlations (or covariances) between characters, it is principle component analysis; if between **OTU**s, it is principal coordinate analysis (Gower). LRH.

Principle. A general law that applies (or is assumed to apply) to nomenclatural codes. The Principles are said to *form the basis* of botanical nomenclature.[1] BAC. and BOT. have Principles which outline the scope of the codes. The content of the Principles of the two codes do not coincide and some Principles of BAC. are found in the **Preamble** of BOT. The Principles and General Considerations, taken together, of BAC. approximately correspond to the Principles and Preamble of BOT. and with the Preamble of ZOO.

[1, BOT.Pre.]

principle of connectivity. Two different entities should be distinguishable by at least two pairs of differing character-states.

printing of author citations and qualifying phrases. Author citations and qualifying phrases should stand out in print from the scientific names to which they refer and to do this a different typeface is used. When the scientific name is printed in italic the author citation and phrases should be in roman or boldface type; when the name is in boldface (as in *Bergey's Manual*) the citation and phrases should be in roman or italic.[1]

In this dictionary qualifying phrases are sometimes printed in italic; this is to distinguish them from other parts of the text and should not be taken as

examples of the printing practice to be adopted when they appear after scientific names.
[1, BAC.33aN2.]

printing of suprageneric names. The choice of printer's type to be used for names above the category of genus is not laid down in any of the Codes, and different countries adopt different **conventions**. The Royal Society recommends that scientific names above genus should be printed in roman with an initial capital, and most journals in Great Britain follow this advice. In journals of other countries it is common to see these suprageneric names in italic type, but this usage is not recommended by the *CBE Style Manual* (p. 183).

printing practice. This varies from one country to another and each printer (or publisher) has a house style. The only accepted practices, reinforced by the codes of nomenclature, are (1) that the first letter of the name of a taxon of genus rank or above should be printed as a capital (upper case) letter,[1,2] and (2) the first letter of the specific epithet (specific name) should not be capitalized, even when the word is derived from the name of a person or place,[3] but BOT. allows capitals in certain circumstances.[4] BAC. recommends the use of a different fount for scientific names.[5] See also **typeface**.
[1, BAC.10a. 2, ZOO.28. 3, BAC.59. 4, BOT.73F. 5, BAC.Ad.A.]

priority is established in nomenclature by the date of valid publication; the precedence thereby gained constitutes the rights of a name. Priority is regarded as the first principle of nomenclature[1] and although the intention (stability of nomenclature) is excellent, the practice can be vexatious. In attempts to achieve stability all the codes (except VIR.) accept the principle of priority, by which the name for a taxon that is first published in accordance with the rules becomes the **correct** (**valid** in zoo.) **name**. Names published before certain dates need not be considered in determining priority; these dates are: for algae, 1753; bacteria, 1980 (see below); fungi, 1801, 1821 (for details see BOT.[2]); protozoa, 1758; viruses, not yet decided). In **combinations** the priority of each element must be determined separately.[3]

Provision is made by each code for exceptions to be made to the rules governing priority. ZOO. formerly excluded names that had not been used as senior synonyms for 50 years and which might be regarded as forgotten names (*nomina oblita*). This rule involved zoologists working in some fields to make unnecessary literature searches (which the rule was intended to prevent) for the non-use of the name in the preceding 50 years, and so the rule was changed. The 50-year **Statute of Limitation** has been abolished, and the Zoological Commission has been given guidance on dealing with cases referred to it.[4] Stress has shifted from priority to a more realistic stability and acceptance of names that have gained widespread usage. When the Commission refuses to exercise its plenary powers in a particular case it is required to specify, in the Opinion rendered, the name that is to be used and the action (if any) to be taken.

Names in regular use may be conserved by the bodies governing the nomenclature of all the biological disciplines, and such conserved names are conserved against all earlier synonyms based on the same type and against earlier homonyms.[5,6] In bot.[7] priority applies to names of taxa of families and below but is not mandatory for names above families, though authors are advised to follow the principle. In zoo.[8] priority applies to names from families to names of subspecies, and in bact.[9] from order to subspecies.

Priority applies (1) to legitimate names[9, 10] and epithets; (2) only to the stem of specific epithets (names) and not to the whole word; this means that a change in ending (to agree in gender with the generic name) does not affect priority. The ending of an epithet containing a so-called **orthographic error** may be changed by a subsequent author (or editor) without affecting priority, and the genitive of a patronymic incorrectly latinized may (and should) be changed by a later worker. Priority does not extend to a name (epithet) outside its own rank, and when a taxon is changed in rank the priority of its name must be established anew.[11]

Revision of BAC. completely alters the rules on the priority of names of bacteria. The starting date for bacterial nomenclature – and the priority of its names – changes from 1 May 1753 to 1 January 1980, on which date all names in the Approved Lists of Bacterial Names will be deemed to be validly published for the first time.[12]

Names of bacteria validly published (by the old or new rules) before 31 December 1977 will be considered by the Judicial Commission with the help of members of taxonomic subcommittee, and those accepted will, together with the names of the authors who originally proposed them, be listed; when approved by ICSB the **Approved Lists** will be published in *IJSB* before 1 January 1980. Names published according to the rules of BAC.[13] between 1 January 1978 and 1 January 1980 will be added to those in the Approved Lists, but after 1 January 1980 further names will not be added.

All other names of bacteria, including names validly published under the rules of earlier codes but not included in an Approved List, will not have any standing in nomenclature, but they can be re-used as they will not be added to lists of rejected names; they will be treated as if they had never been used or published. See **reuse of names**.

The purpose of the new rules in BAC. is to make it unnecessary, after 1 January 1980, to search the older literature for names and descriptions validly published before that date.[14]

[1, BOT.PRIII.ZOO.Pre. 2, BOT.13. 3, BAC.23aN1. 4, ZOO.79(a), (b). 5, BAC.56bN1. 6, BOT.14N5. 7, BOT.1975, 11; 16; R16B.8, ZOO.23(a–b), (c). 9, BAC.23a, b. 10, BOT.45. 11, BOT.60. 12, BAC.23aN3. 13, BAC.24aN1. 14, BAC.28aN3.]

probabilistic identification matrix. In numerical identification, the computer-stored data table in which the responses of each taxon to each test is recorded as a probability figure for a positive result. The values range from 0.01 to 0.99. LRH.

probabilistic similarity index. An index proposed by Goodall (*Nature*, 1964, **203**, 198), in which the rarity or commonness of a feature in the population being studied determines the weight to be put on it.

probability approach to identification mentioned by Cowan & Steel is based on subjective clinical pathological assessment or practical laboratory experience. It is quite different from the *probabilistic identification* of numerical taxonomists in which objective probabilities are calculated by statistical methods.

procaryote, -ic, *n.* and *adj.* Cell (or pertaining to cell) in which the nuclear material (DNA, as in the bacteria, blue-green algae and many viruses, or RNA in some viruses) is not enclosed by a nuclear membrane. This is in contrast to the eucaryote cell, which has a separate nuclear body.

Procaryotae is the plural feminine of the Latin noun, procaryota, and is used by R. G. E. Murray in *Bergey's Manual* (1974) as the name of a kingdom of microbes (bacteria and blue-green algae).

progressive (step-by-step) method of identification aims at making a rapid identification by a logical approach. It is essentially a method for bact.; a few characters of the unknown are determined (often within 24 h of purifying a culture), these are compared with the characters shown in a *first-stage diagnostic table* and usually indicate the likely genus or small group of genera. Further tests are made (based on the *second-stage table* of the probable genus) and the characters thereby determined are compared with those shown in the second-stage table. Often the identification in then evident, but sometimes additional tests are required and a third-stage table has to be consulted. Occasionally, very specific confirmatory tests will finally be carried out (e.g. serological or animal pathogenicity tests). Examples of the **diagnostic tables** and diagrammatic representation of the groups of genera will be found in Cowan & Steel (1974).

projection routine. A computer routine in which, by **principal component analysis**, information about the characters of an organism is compressed into *n* numbers (usually a small number, 2 or 3); these are the co-ordinates that indicate the location of the organism in an *n*-dimensional character space. The same routine is used to define the subspaces occupied by different taxa in a multidimensional space. An unknown can be identified by computing its co-ordinates in the same multidimensinal space, and it is identified when its distance from the centre of a reference taxon is less than the radius of that taxon.

prokaryote, protokaryote. Other, but less-frequently used, spellings of **procaryote** (q.v.).

pro parte, *abbr.* ***p.p.*** In part. In author citations indicates that only a part of the taxon or circumscription of a previous author (who is named) is being considered by the author who attaches the qualifying phrase to the citation.

prophage. A bacteriophage that is carried by a lysogenic bacterium in a form that multiplies at the same time and in association with the genetic material of the bacterial cell. It can be transmitted to other cells by conjugation and by transduction.

propose, -al, *v.* and *n.* **1.** Advocating that a stated action should take place: the proposal is the statement of the thing to be done. In nomenclature a new name may be proposed by any author, but many authors are hesitant and may doubt their competence to propose a name, believing (wrongly) that such proposals should only be put forward by committees or workers supposedly more expert than themselves. Half-hearted and indefinite proposals will lead to difficulties in the future, for both BAC.[1] and BOT.[2] state that a name is not validly published when it (1) is not accepted by the author, (2) is proposed in anticipation of the future acceptance of the taxon to which it is applied, or (3) is merely mentioned incidentally.

An author should avoid vague phrases such as 'the name *X-us albus* seems to be appropriate', 'I suggest the name *Y-us blankus* for this organism', or 'The name *Z-a livida* might be acceptable', for it could easily be argued in each case that he did not accept the name; instead he should be purposeful and **designate** or state categorically that 'I name this organism *R-us rubrus*'.
2. General usage allows the verb to mean 'put forward for discussion' and some authors of scientific papers use it as a synonym of suggest; but in taxonomy both the verb and the noun have strict and more limited meanings.
[1, BAC.28b. 2, BOT.34.]

proposed International Code of Nomenclature of Viruses. A code put forward by ICNV after discussion at the Moscow Congress of Microbiology in 1966. In many respects it followed BAC. of that time but in details such as the endings of names it followed ZOO. In this dictionary reference to the rules of the proposed code is indicated by the abbreviation VIR.

The proposed code recognized categories from phylum to subspecies; categories of ranks below subspecies were not to be subject to the rules of the code. To avoid changing names then in regular use, the Law of Priority was not to be applied to virus names published before the code. There were rules on the endings of names to be applied to ranks above genus, recommendations on the formation of names (anagrams and siglas were forbidden), orthography (references to this section (6) is made in the set of rules, VR, approved in 1975), and the designation of nomenclatural types. This code was never accepted by virologists and the requirement that a species name should be a binary combination did not find much support.

ICNV approved a set of Rules to serve as a guide and as a substitute for the proposed code but the numbers of the Rules do not correspond with the numbered Rules in the proposed code. Some of these Rules, which are not as detailed as the rules of the proposed code, were modified in 1975. References to the Rules in this dictionary are shown by the abbreviation VR.

propositum, *abbr.* ***prop.*** Latin, proposed. A name proposed for conservation or rejection may have the abbreviation prop. added to the qualifying phrase, as in nom. cons. prop.

pro synonymo, *abbr.* ***pro syn.*** As a synonym. In an **author citation** this qualifying phrase indicates that the name was published as a synonym and not as a new name.[1] When, in the synonymy of a name, a publishing author C quotes an unpublished name of another worker D, this can be indicated in the form D ex C pro syn.

(See also **ex 1(2)**.)
[1, BOT.R50A.]

protista. An obsolete name for organisms, often unicellular, that were not regarded as being either animals or plants; they included algae, fungi, protozoa, and bacteria.

protologue. All the information given by an author when a name is first published;[1] this includes the description, diagnosis, synonymy, list of specimens, source material, place of isolation of cultures, figures, micrographs, discussion, references, and any comments of the original author.
[1, BOT. 1975, 7.]

protonym. An effectively but not validly published name, or a **devalid** name that is reused and validly published by a latter author.

protoplast. The bacterial cell without the rigid cell wall, and consequently without the characteristic shape of the normal bacterial cell; the protoplasm is intact and is surrounded by the cell membrane. When the original cell is Gram positive, protoplasts will be Gram negative, since an intact cell wall is necessary to retain the Gram-positive state. Thirteen users of the word (Brenner *et al.*, 1958, *Nature*, **181**, 1773) issued an agreed statement that *protoplast* should be strictly limited to structures in which cell walls are known to be absent. McQuillen says that **sphaeroplast** is used for the less perfect stage, i.e. when the cell wall is only modified or remains in fragmentary form.

prototroph. An organism that is nutritionally independent and able to synthesize growth factors that it needs from simpler substances (cf. **auxotroph**).

protype. Synonym of neotype (q.v.).

Provisional Committee for Virus Nomenclature, *abbr.* **PVNC.** A provisional committee set up by the Executive Committee of IAMS to consider proposals for a code of nomenclature for viruses. At the 1966 Congress it became the **International Committee for the Nomenclature of Viruses.**

provisional name. A name proposed in anticipation of the future acceptance of a group of organisms. Also known as a ***nomen provisorium***. Such a name is not validly published and will not have any standing in nomenclature.[1,2]
[1, BAC.28b(2). 2, BOT.34.]

Provisional Rules. The revision of BAC. published in 1975 contains six Provisional Rules numbered A1, A2, B1 to B4, placed between Rule 36 and Rule 37a; these are to come into effect on 1 January 1980. The Provisional Rules deal with the author citation of names in Approved Lists, the transfer of taxa whose names are in an Approved List, and the reuse of names published before 1980 but not included in any Approved List. See **citation of a name**.

pseudocatalase activity is shown by some leuconostocs and pediococci when grown on media of low glucose content. It is not inhibited by azide or cyanide, and is sensitive to acid (cf. **catalase**).

pseudocompound. A compound word in which the case endings of all parts are retained, as in *Myos-otis*[1] and *choleraesuis*. In a true compound only the last word retains its case ending, other words forming the compound are represented by their stems, as in *chlororaphis*.
[1, BOT. 1975. R73G.]

pseudogeneric name. A name given to a genus for which species were not named. Term used by Buchanan (1925) for names used by Trécul, *Urocephalum*, *Amylobacter*, and *Clostridium*.

pseudoisonym. See **isonym**.

pseudomonad. A general term used for oxidative Gram-negative rods that have the general characters of members of the genus *Pseudomonas*. Some pseudomonads may for one or other reason be excluded from that genus, thus certain plant pathogens that produce yellow pigment were removed by Dowson, who created a new genus *Xanthomonas* for them, but they are still within the bounds of the more general group known as pseudomonads.

pseudomutation. Used by Drake (1974, *Symp. Soc. gen. Microbiol.*, **24**, 41) for adaptations not necessarily due to changes in the nucleic acid. **Phasic variation** (q.v.) in *Salmonella* is given as an example. LRH.

psychrophil(e), *n.*, **ic,** *adj.* Literally, liking cold; used for organisms that can grow at low temperatures and generally unable to grow at the body temperatures of warm-blooded animals, but there is little agreement on optimal temperature for growth and temperature limits. Some workers define psychrophils as those organisms with an optimal temperature for growth below 20 °C. Eddy, on the other hand, suggested that psychrophil should be applied and restricted to those whose maximum growth temperature was below 35 °C whatever the minimum or optimum growth temperature might be. The optimal temperature is usually above 20 °C and is always near the maximum temperature for growth and on this basis many psychrophilic bacteria could be regarded as mesophils. The main feature that distinguishes psychrophils from mesophils is their ability to withstand and grow in the cold. The use of psychrophil was discussed by Ingraham & Stokes (1959, *Bact. Rev.* **23**, 97). Cf. **psychrotroph.**

psychrotroph, *n.*, **-ic,** *adj.* Applied to organisms that are able to grow at 5 °C or below. The term describes only the lower range of growth, and is not affected by the temperature range or by the optimal temperature for growth. It should not be confused with **psychrophil** (q.v.), which is generally more descriptive of the maximum growth temperature, and indicates merely a tolerance of low temperatures. See Eddy (1960, *J. appl. Bact.* **23**, 189). In the dairy industry, an organism able to grow at 7 °C or less is described as psychrotrophic.

publication. In nomenclature this term has a special meaning and is important in establishing the priority of a name. Publication means the first significant appearance of a new name, and each of the codes specifies what it regards as the essentials for publication of a name; BAC. and BOT. distinguish between **effective** and **valid publication**. ZOO. does not make these distinctions and has simple rules which determine whether a name has been published.[1] Publication in zoo. means reproduction of multiple copies in ink on paper, distributed by sale or gift for scientific and public information. A recommendation advises against mimeographing and lists methods that do not constitute publication; these include microfilms, microcards, students' notes (even when printed), proof sheets, mention at a scientific meeting, labelling of a specimen, deposition of a document in a library, and, after 1950, anonymous publication.

See also: **acceptance for publication**; **available**; **date of publication**; **editors and publication**.

[1, ZOO.7, 8, 9; R8A.]

publicity. Taxonomists seldom seek the limelight but taxonomic change and the creation of new taxa should be made widely known if they are to be effective and if they are to be noticed by other workers. ZOO. comments on this and, so that they will not be overlooked, recommends to authors that they should send copies of their published works to appropriate abstract journals.[1]

[1, ZOO.ApE24.]

punched-card identification. Punched cards in various forms can be adapted to identification schemes; in some the holes are already punched, in others the user removes parts of each card or notches specified holes to the edge of the card.

In the notched systems sorting is made by inserting a needle through the hole representing the character to be tested, the pack (US deck) is shaken gently and the cards notched at that hole will fall out of the pack. Which cards fall out will depend on whether the notch represents a positive (as is usual) or negative character state. In other systems cards are punched at areas representing particular characters and the sorting is made by holding the whole pack up to the light (the **peek-a-boo system**).

Punched-card systems can be used in step-by-step identification; the first minimal difference set of cards will identify genera (or small groups of genera) and a more specialized set, using more characters and two rows of holes to deal with variable characters, will be needed to identify the species. The value of the identification system lies in the quality of the information used in the preparation of the master cards (i.e. the characters recorded for the known taxa).

punctuation of scientific names and author citation. Only ZOO. mentions punctuation; in general, author citation is not separated from the scientific names by any punctuation,[1] but when more than one citation or notation follows the name each should be separated by a comma.[2] In zoo. (but not in bact. and bot.) citations, a comma should separate the author's name from the date of publication when this is cited.[3]

Breaks in the scientific name–author citation sequence other than by commas are made by round brackets when the author has transferred a species from one genus to another. The name of the author of the specific epithet (name) is put in brackets before the name of the author who made the change.[4] See **name change**.

[1, ZOO.51(b). 2, ZOO.ApE13. 3, ZOO.22. 4, BOT.49; ZOO.51(d); R51B.]

pure culture. Culture is a convenient term for microbial growth and a pure culture is ideally the progeny from one microbial cell, in which case it is more accurately described as a single-cell culture or clone. More often the pure culture is obtained by plating and replating single well-isolated colonies, when it is *assumed* that the discrete colony has developed from the multiplication of a single bacterium or (in the case of bacteria that grow in chains or clumps) from one unit of the organism. The assumption is false when the colony is on a **selective** or **inhibitory medium** containing a selective bacteriostatic substance which prevents multiplication but does not kill some organisms that may have been implanted; colonies on such media are often a mixture of two or more different kinds of organism. Colonies from such media must be replated on non-inhibitory nutrient media and well-isolated single colonies picked, perhaps replated, and tested for purity by examining the characters of several colonies, which, in a pure culture, should be similar. Phasic, and other, variation may cause differences in the appearances of colonies in pure cultures and only by practical experience can this be distinguished from mixed cultures.

purpose of a name. The essential purpose of a scientific name is to label it unambiguously; *it is not intended to describe the characters or the history of the taxon:*[1,2] the point needs to be brought home to bacteriologists because some believe, and others expect that, in Nelsonian terms, each name should do its duty and tell us something. Nothing is further from the requirements of the nomenclatural codes, with the possible exception of the virus cryptograms which are attempts at descriptive labels.

Another purpose of a name is to indicate the taxonomic rank,[2] which the name does by the suffixes attached to the stem.

A biological scientific name is purely a means of referring to the taxon to which it is attached; with that requirement the name should be unique to the taxon.

[1, BAC.Pr4. 2, BOT.Pre.]

putative synonym. R. E. Buchanan used this term but as far as is known did not define it and we can only speculate on the precise meaning he attached to it; he was an active churchman and possibly based the term on Canon law. A putative marriage, while invalid in law, is one made in good faith by at least one of the contracting parties. From this we might assume that by a putative synonym Buchanan meant a name proposed in good faith as a name for a new taxon but it was invalid because, when published, the taxon to which it was applied included the type of an earlier name which should have been chosen. J. G. Holt (personal communication) thinks that Buchanan used putative to mean subjective (see **subjective synonym**).

pyrogram. Chromatogram of pyrolysis products; pyrolysis/gas–liquid chromatography can be used for the identification of micro-organisms and has been found useful in distinguishing between some of the mycobacteria.

Q

Q technique. Techniques of multivariate analysis which are based on direct comparisons between the objects, not the features. Conventionally, in numerical taxonomy the data table is arranged with the organisms (or **OTUs**) down the side and the features along the top. Q techniques start then by comparing the rows of such tables with each other (e.g. by calculating similarity or matching coefficients). Some numerical taxonomy methods begin by comparing the columns (e.g. by calculating correlation coefficients between features); these are called **R techniques** (q.v.).

qualifying phrase, *syn.* **indicator, notation.** A tag appended to a scientific name, usually after the author citation, which qualifies or describes a quality either of the name or of the taxon. We can divide these tags into (1) nomenclatural qualifying phrases such as ***nomen ambiguum***, ***nomen rejiciendum***, ***epitheticum specificum conservandum***, and (2) taxonomic qualifying phrases, ***species incertae sedis***, ***mutatis characteribus emendavit*** (when the description of the taxon, not the name, has been amended).

BAC.[1] gives guidance on the typeface to be used for the qualifying phrase. To draw attention to the phrase it should be printed in a typeface different from that used for the scientific name it qualifies; thus a binomial printed in italic should have the qualifying phrase in roman or boldface; the name of an order, printed in roman, should have a qualifying phrase in italics or boldface. Some

US journals would print the ordinal name in italic; these should print the qualifying phrase in roman or boldface.

In this dictionary these phrases are sometimes in italic, not as examples of how they should be printed when serving as qualifying phrases, but to make them stand out from other parts of the text.

[1, BAC.33aN2.]

qualitative character. A character or feature that cannot be measured precisely and is described in general terms such as strong (weak), much (little), large (small), positive (negative), produced (not produced), or by a series of + and − signs. Bacteriologists are fond of recording data as a series of degrees of positiveness (−, ∓, ±, +, ++, +++ etc.); characters recorded this way are called semi-quantitative characters.

qualitative notation. Characters that are either present or not present (absent) can be indicated in tables and lists by plus (+) and minus (−) signs and for computers usually by 1 and 0. In such a case these symbols are being used in a qualitative notation. Letters and other signs may be used to express characters qualitatively, such as A for acid production, and AC for acid and clot production. For a fuller discussion see **symbol**.

quantitative character. A character that can be measured precisely and expressed in numerical terms. When used in **diagnostic tables** such characters may be indicated as being above or below a particular value, e.g. pH >7 or >8.

quantitative notation. Expression of characters by numbers, signs, or symbols to which some quantitative measure is attached. For example, agglutination tests may be recorded by a series of plus and minus signs in which, say, ++++ corresponds to complete agglutination of the particulate suspension with a clear supernatant fluid, through +++, ++, + to indicate smaller proportions of agglutinated particles, and ±, in which agglutination is not visible to the naked eye and is seen only by observation through a hand lens which gives low degrees of magnification; − would indicate absence of agglutination (it is incorrect to say '— shows no agglutination', for you cannot demonstrate a negative).

Other examples of quantitative notations are seen in tables of characters of micro-organisms but few authors have made any serious attempt to standardize the quantitative aspects of **symbolic notations** (q.v.), and every table (or set of tables) must be accompanied by an explanation of the symbols.

quantitative systematics. The application of statistical methods to problems of systematics, and especially to taxonomy.

quantitative taxonomy. 1. Methods of multivariate statistics applied to taxonomic problems.

2. Expression of taxonomic characters in a quantitative manner.

Quellung reaction. A swelling of the capsule of pneumococci when they are mixed with homologous (i.e. of the same serotype) antiserum.

question mark after a name or any other suggestion of doubt about the name or the taxonomic position of a bacterium will invalidate the publication of the name.[1] This is not so in bot. provided that the name is accepted by the author.[2] (See also **doubt**, **taxonomic doubt**.)

[1, BAC.28b(1). 2, BOT.34N1.]

quoad. As to. Qualifying phrase used in an author citation to indicate the part of a taxon named by an earlier author that is being referred to by the present author.

quotation. Extract from a book, paper, letter, or, occasionally, speech. When the quotation is from a letter or conversation, insert the words 'personal communication' within brackets after the name of the author; if from some unpublished lecture give the name of the lecture or the occasion in brackets after the author's name. References to personal communications and unpublished lectures should not be included in Lists of References, or Literature Cited at the end of a paper or book. It is tactful to gain permission from the maker of the remarks or statement before committing them to print and quoting him or her as the source. (In this dictionary there are some 'personal communications'; may I request the indulgence of any persons quoted in this manner for instances where I have been unable to ascertain whether permission was given – LRH.)

quotation marks, or quotes. When made from a written source, an exact quotation should be inserted between single quotation marks 'one turned comma at the beginning and one apostrophe at the end' (Collins, 1962). A quotation within a quotation has double quotation marks at the beginning and end. American usage is the reverse of this; a simple quotation has double quotation marks, a quote within a quote has single marks. Words left out are indicated by three points (ellipsis) thus . . . ; material inserted to improve the clarity should be enclosed within square brackets. A paraphrase should not be inserted within quotes.

R

race. 1. A vague term used by nineteenth-century botanists and bacteriologists (such as Cohn) for subdivisions for **form species** separated by chemical differences in their physiology. Obsolete in bact.

2. *Physiological race* is a subdivision of Form (Forma); the term is used in mycology in preference to physiological form (Ainsworth, 1973) and, since confusion is not introduced, appears to comply with BOT.[1]

3. Biological races are recognized in zoo. for morphologically similar organisms that can be distinguished only by their ecological requirements. However, examples among protozoa have been given specific rank, e.g. *Trypanosoma equiperdum* and *T. evansi*.

[1, BOT.4.]

raise, *v.* Used by some (usually non-medical) immunologists for the older and clumsier verbal phrase 'the stimulate the production of antibody'; this longer form was often shortened to produced or production in sentences such as 'antibodies were produced against X antigen', or 'X antibody production' at the expense of truth, for they suggested that the act of inoculating an antigen produced antibody.

rank. In taxonomy rank indicates the relative position in an orderly sequence and is almost synonymous with **category**. The various ranks can be likened to the shelves of a book stack, each shelf representing a different rank and all the books

Table R. 1. *Ranks above genus recognized by the different codes*

	BAC.[1]	BOT.[2]	VIR.[3]VR[5]	ZOO.[4]
Kingdom	.	+	.	.
Division	.	+	.	.
Subdivision	.	+	.	.
Class	+	+	(+)	+
Subclass	+	+	.	.
Order	+	+	(+)	+
Suborder	+	+	.	.
Superfamily	.	.	.	+
Family	+	+	+	+
Subfamily	+	+	(+)	+
Tribe	+	+	.	+
Subtribe	+	+	.	.
Genus	+	+	+	+

Bold type indicates ranks most used (as distinct from available) in bact. (+): in provisional code but not in the Rules (Fenner, 1976) [1, BAC.5b. 2, BOT.3; 4. 3, VIR. 14; 16.VIR.Pr6; 3. 4, ZOO.35; 42; ApE6. 5, VR14, 16.]

Table R. 2. *Ranks below genus recognized by the different codes*

	BAC.[1]	BOT.[2]	VIR.[3]VR[5]	ZOO.[4]
Genus	+	+	+	+
Subgenus	+	+	(+)	+
Section	.	+	.	.
Subsection	.	+	.	.
Series	.	+	.	.
Subseries	.	+	.	.
Species	+	+	+	+
Subspecies	+	+	.	+
Variety	.	+	.	.
Subvariety	.	+	.	.
Form	.	+	.	.
Subform	.	+	.	.

Bold type indicates ranks most used (as distinct from available) in bact. (+): in provisional code but not in the Rules (Fenner, 1976) [1, BAC.5c; 11. 2, BOT.3; 4. 3, VIR.11; 14. 4, ZOO.42; 45. 5, VR11, 14.]

(taxonomic groups) on that shelf are of the same rank, level, or category.[1] But ranks in different orders are not necessarily equivalent, nor are the criteria that determine rank standardized in any way; thus the family Enterobacteriaceae is approximately equivalent to the genus *Streptococcus*. Rank is merely a taxonomic convenience and does not, as some would imply, convey phylogenetic information. Organisms are moved from one rank to another as easily as a librarian moves a book from one shelf to another; an example is the pneumococcus, 'ranked' by different authors as a genus and as a species.

Each organism belongs to a number of taxa of consecutive rank, and of these the species is basic.[2] This is the hierarchical system. Hierarchical systems are built up to suggest – but do not prove – relatedness between taxa of microorganisms placed at different ranks. Two or more individual strains may have so many characters in common (i.e. they are so much alike) that they may be combined to form a taxon with the rank of **species**; several species may, because of the sharing of (fewer) common characters, be combined into a taxon with the rank of **genus** and so on, each successive higher rank having fewer common characters. But the creation of a hierarchical system is not merely collecting together units that share characters; it is a way of life of taxonomists and taxonomic thought requires every taxonomic unit, however different from all its fellows, to be found a place in each rank (from species upwards) of the system. *Listeria monocytogenes* was monotypic when created; it was at the same time a genus and a species and it had also to be fitted into taxa of higher ranks, and it was included in the family Corynebacteriaceae. The requirement to place an organism in a series of consecutive ranks has worried thoughtful bacteriologists for many years and it was only in the eighth edition of *Bergey's Manual* (1974) that a break was made with this tradition by a major group of bacterial taxonomists. And that break has not gone unchallenged.

The various codes of nomenclature do not all recognize the same ranks;[3,4,5,6] family, genus, and species are recognized by all codes, but tribe is not a rank accepted in virology (see Tables R. 1 and R. 2). Because the provisional virus code has not been accepted, the many ranks shown in it that are not included in the guiding Rules of ICTV are shown within brackets in Tables R. 1 and R. 2.

The relative order (sequence) of the ranks must not be changed.[7]

[1, BAC.Pr8N2(iii). 2, BOT.2. 3, BAC.5b. 4, BOT.3. 5, ZOO.35; 42; 45. 6, VIR.11, 14, 16. 7, BOT.5; BAC.5a.]

react, *v.* **1.** In chemistry indicates that one substance acts upon or with, another.

2. In serology used loosely to indicate that a reaction may take place when antigen and antibody are mixed, or meet in a gel-electrophoresis system.

recent. A relative term, often better avoided or the time event more precisely stated since a paper or book may be read and quoted many years after it was written (cf. **country**). In biology sometimes used as the opposite of fossil, or even to indicate that the material could be living. Sometimes used in this dictionary in dating the recent revision of BAC., due to the conflicting statements on its title page and Rule 1a.

recognize, recognition. 1. To accept (acceptance of) a name as valid, correct, or legitimate; this is an objective (legalistic) recognition.

2. To accept (acceptance of) a taxon as a distinct entity; this subjective recognition of a taxon may not be acceptable to others, as, for example, the recognition of *Arizona* as a genus separate from *Salmonella.*

3. To identify (identification of) an organism as a member of a named taxon.

recombination is a term used by microbial geneticists to indicate the end result of the transfer of genetic material from one organism to another. There are three main mechanisms: transformation, transduction, and conjugation.

recommendation. Advice given in the codes on the formation and use of scientific names. Recommendations have less force than rules; together they constitute a guide to good behaviour in nomenclature. A name contrary to a recommendation cannot, on that account, be rejected,[1] but the example set should not be followed by later authors.

[1, BOT.Pre.]

recover, *v.* Bact. term for the act of re-isolating an organism from a site (or host) artificially contaminated (infected) with the bacterium. One of the techniques used to obtain a pure culture of a pathogen from a mixed culture of a pathogen and a non-pathogen is to inoculate an animal and recover the pathogen from the site of inoculation or the bloodstream of the infected animal.

Apart from animal inoculation a site may be contaminated, exposed to some agent or treatment, and an attempt made to recover the organism from the site; if the organism has survived the treatment it may, in fact, be recovered.

Often used incorrectly for the original attempt to isolate an organism from a **specimen 2** (Wilson, 1965, *Mon.Bull.PHLS*, **24**, 280).

redundant, -cy. Of characters, the inclusion of more than one character dependent on the same reaction, e.g. acid production from glucose/methyl-red test; acid production from lactose/acid reaction in litmus milk.

reference marks. See **note indicators**.

references. 1. Heading used for List of references cited in a book or paper (see **literature cited** and **name–date system**). These are usually arranged alphabetically by the surnames of the first-named (US senior) author.

2. There are two reasons for putting references in scientific papers; (1) to give the authority for statements based on the work or theories of others, and (2) as a space-saving device; e.g. a reference to a paper describing a technical method, or the results of tests, saves space in the journal and the time that would be needed to read the detail.

3. In nomenclature, references are generally to papers in which an author has described or named a taxon (**author citation**).

reference strain. A culture of a bacterium that is neither a type nor neotype strain, but is used as a standard of reference in a comparative study. It does not have any nomenclatural status,[1] but since it is chosen because it is typical (characteristic, or conforms with the original description) of a taxon, it may well be more useful to the taxonomist than a nomenclatural type strain.

[1, BAC.19.]

register (registration) of names. From time to time taxonomists have despaired of bringing order to microbial nomenclature. There is so much synonymy, or in Churchillian terms, there are so many names chasing so (relatively) few defined organisms for so long a time that a halt should be called, and the acceptance of new names should be dependent on compliance with a few new rules or regulations. Ainsworth & Ciferri (1955, *Taxon*, **4**, 3) suggested that new names should be registered; from this followed the idea of a list of accepted or conserved names which, when approved by the international nomenclature committee appropriate to the discipline, would be entered in registers to be kept by each committee. A liaison committee would keep an eye on the possible duplication of names (particularly at generic level) in the different disciplines. Authors of new names would be required (*a*) to consult the appropriate register to make sure that the name had not been used before, either as a valid (available) name, or as a synonym; (*b*) to publish the name in one or more officially designated journals (possibly in different languages), together with a full description of the organism, or reference to a description in another journal, and (*c*) to request the addition of the name to the appropriate register.

In practice, the registration of names of cultivars is one of the methods used in the regulation of names of cultivated plants.

In bact. the **Approved Lists** will automatically fulfill the purpose of registers, but only of names proposed before and approved by 1 January 1980. BAC. also officially designates the journal in which the Approved lists, and (after 1 January 1980) future new names, shall be published, thus meeting requirements (*a*) and (*b*) above.

rejected name. A name that has been rejected by the appropriate Commission or Committee authorized to do so. An individual worker may decide to reject a name for his own use, but official rejection of a name (to which the term *nomen rejiciendum* is applied) can only be made by the **Commissions.**[1] Rejected names are published as Appendixes to BAC. and BOT.[2] and rejected zoo. names are placed in the appropriate Official Index of Rejected Names.[3]

Rejection of a name is often the consequence of the conservation of another name; the basis of rejection or the grounds which suggest that rejection is desirable are considered in **rejection of a name** (q.v.).

[1, BAC.56a. 2, BAC.Ap4; BOT.ApIII. 3, ZOO.78(f).]

rejected synonym. A taxon has only one **correct name**, which is one of the synonyms (different names for the same unit); all other names applied to the taxon are rejected synonyms. These rejected names are important because if for any reason the correct name has to be changed, one of the rejected names will become the new correct name.

rejection of a name. 1. A procedure by which the Judicial Commission (bact.), General Committee (bot.), or International Commission (zoo.) dispose of names that are unwanted, unused and, although **legitimate** and **available**, have a nuisance value in nomenclature as they prevent the legitimate use of later synonyms that are in fact used regularly, often in the belief that they are the right names to use for the taxa concerned. The rejection of an unwanted name may be made at the same time as the **conservation** of the wanted name, and BOT.[1] shows the rejected and conserved names in parallel columns; BAC.[2] includes lists of rejected names, and in zoo. they are published in Official Indexes.[3]

2. When a worker (or better a group of workers) thinks that a name should be rejected he should state the case for rejection, present it to the secretary of the appropriate commission and ask for the name to be rejected, usually in favour of some other name. There is no need for the sponsors of rejection to think of reasons why the name should be retained; there will always be opponents who will present the case for the other side.

3. A name may not be rejected merely because it is inappropriate,[4] but in bact. and bot. a name must be rejected by an author, and not used if it is **illegitimate,**[5] which means that the name was not published in accordance with the rules of nomenclature. Apart from the technicalities of **publication**, a name may be illegitimate because it is **superfluous,**[6] or a junior **homonym.**[7] Names are not legitimate (available in zoo.) when the author did not consistently follow a binominal system of nomenclature in the work in which the name was published.[8] In bot., but not in bact. and zoo., **tautonyms** must be rejected.[9]

Other reasons for the rejection of a name on the grounds of illegitimacy are:

(1) if a bacterial name is a later homonym of the name of a protozoon, alga, or plant;[10] (2) if the name is used with different meanings (i.e. based on different types) and has become a persistent source of error[11] (a good example is *Aerobacter aerogenes* which before being rejected (Opinion 46) had been applied to two different bacteria: *Klebsiella aerogenes* and *Enterobacter aerogenes*), or has persistently been used for a taxon that does not include the type of the name;[12] (3) if its application is uncertain (a **nomen dubium**);[11] (4) when the characterization of the taxon was based on a culture that consisted of two or more kinds of bacteria.[11]
[1, BOT.ApIII. 2, BAC.Ap4. 3, ZOO.78(f). 4, BAC.55; BOT.62; ZOO.18(a). 5, BAC.51a. 6, BAC.51b(1); BOT.63. 7, BOT.64; ZOO.53. 8, BOT.23(3); ZOO.11(c). 9, BOT.23. 10, BAC.51b(4). 11, BAC.56a. 12, BOT. (1975) 69.]

relatedness value. A measure of the relation between two strains of the same or different species as determined by in-vitro nucleic acid homology experiments. The value can be calculated by dividing the product of the logarithm of radioactivity counts in the two heterologous nucleic acid reactions by the product of the logarithm of the counts in the two homologous reactions. Values of 0.10–0.20 (or greater) are said by Weissman & Cole to be characteristic of a species.

relation. An association that may be assumed when two organisms are compared and are found to have some or all characters in common. With micro-organisms the relations between two or more strains are determined by this association of characters (i.e. multivariate association), but only in a limited number of cases can the relatedness or **relationship** (q.v.) be determined.

relationship. The suffix -ship implies a genetic connexion, and the word is often wrongly used, and should be replaced by **relation**. In microbiology most relationships are only assumed, few (such as that between *Escherichia*, *Shigella* and *Salmonella*) are known. There are different degrees of relationship or relatedness; when genetic material can be transferred to produce forms that bear and reproduce the characters of both parents the degree of relatedness is close; there are other cases where transfer of perhaps non-nuclear material shows a less close relatedness. In-vitro nuclei acid homology experiments indicate the degrees to which single strands of DNA from different bacteria can form stable duplexes (double-strand DNA) and this is assumed to indicate degrees of relationship between the bacteria.

American usage does not distinguish between relation and relationship.

relative affinity. See **affinity**.

relative order. Term used for the order of sequences of categories (ranks) fixed by the Codes[1,2] and shown in Tables R. 1 & R. 2, p. 220, which also show the suffixes attached to names that indicate the ranks above genus. The sequence should not be changed, in spite of the apparent waywardness of biologists in placing tribe at a lower level than family.
[1, BAC.5a. 2, BOT.5.]

relevance coefficient. Sokal & Sneath proposed a coefficient of relevance as the ratio of the number of characters applicable to two taxa to the total number of characters used. LRH.

remodel, *v.* To change in form. A taxon is remodelled when it is divided or united with another taxon of the same rank, when its circumscription and its diagnostic characters will be altered. A change in name occurs only when (1) the type is excluded, (2) on union with a taxon of longer lineage, (3) by a change in rank. [BAC. 37–40; BOT. 51–6.]

repeatable, -ility. 1. Of a test (usually to determine a characteristic), the quality that ensures that the same result is produced when the test is carried out on more than one occasion or by different individuals.

2. Of an identification system, one that, given the same set of individuals, will always produce the same sorting (or identification) when the same procedures are applied by different workers.

3. Of a classification, syn. of **predictive value** (DuPraw, 1965, *Syst. Zoo.* **14**, 22).

repent. 1. Term used by Lewin & Lounsbery (1969) to describe the motion of bacterial cells which lack flagella, but show motility by gliding, as in the flexibacteria which are *motile, though they do not swim.*

2. In bot. used for creeping growth on or just below the soil surface.

3. Zoologists use repent for creeping or crawling.

replacement culture. See **substituted culture**.

replacement name. Zoo. term for a new name or a synonym adopted to take the place of a name found to be invalid. When a rejected junior homonym has (an) available synonym(s) the oldest must become the replacement name.[1] In bot. an illegitimate name (or epithet) is replaced by the oldest legitimate name. If none exists, a new name based on the same type is the first choice, or a new taxon may be described and named.[2]

[1, ZOO.60. 2, BOT.72.]

replica, replicate. Literally a copy, facsimile, or duplicate. In microbiology used for subcultures made at the same time from the same inoculum (colony, suspension); when the inoculum was a type culture these subcultures may be distributed to and maintained in different laboratories as authentic descendants of the type culture. All the ampoules prepared from one suspension, freeze-dried at the same time and under identical physical conditions, form the most stable and uniform replicates. In genetics the words replicate and replication have other (and specialized) meanings.

replica plating. A technique developed by the Lederbergs by which the same colony or colonies can be transferred to several different plates without change in the spatial relations between the colonies.

Reports of committees raise difficulties in **author citation**. It is usual to ascribe authorship of new names or new combinations to the committee as a whole, but there are not any rules as to how this should be done in practice. Author citation seems to have four variations: (1) the names of all the committee members, (2) the name of the chairman followed by *et al.* (this is the form used in *Index Bergeyana*), (3) the name of the Committee, or (4) the words 'Report of' followed by the name of the Committee and date.

representative strains. The type concept deals with names, and the type of a name is not necessarily representative or characteristic of the taxon named. Some taxonomists who take a broader view find the type concept restrictive and regard it as an example of the tunnel vision of nomenclators; they see the species as a population of similar units, with a need for several strains to indicate the range of variability among members of the species. Gordon (1967, in *The Ecology of Soil Bacteria*, eds. T. R. G. Gray & D. Parkinson), who holds this view, believes that a species can be defined only as *a group of newly isolated strains, old stock strains, and their variants, that have in common a set of reliable characteristics separating them from other groups of strains.* In 1975 (*ASM News*, **41**, 715) she made the point that as long as strains survive they belong to the same species and bear the same name, even although a strain may lose a particular characteristic such as pathogenicity.

This approach shows up one of the main deficiencies of the nomenclatural type strain as a taxonomic unit and the need for several strains to represent (rather than typify) a taxon. Even if the strains are picked at random they must, taken together, be more representative than any single strain.

reprint. **1.** Reproduction of a portion of a journal, usually those pages that include one **paper** or article. Sometimes called an **offprint**; see also **separate**. **2.** The printing of additional copies of a book from the same type setting, with only minor changes or corrections of printer's errors. More often called a new **impression**.

republication of new name. An author should not publish a name *as new* in more than one publication (paper in a journal or book), neither should he publish the same paper in more than one journal or language without stating clearly that the new name has already appeared elsewhere, giving the full bibliographic reference to the earliest publication of the name. This sensible recommendation of ZOO.[1] should be followed by biologists of all disciplines. [1, ZOO.ApE22.]

request for an Opinion. Opinions are not given by Judicial Commissions unless they are asked for, but any worker (who need not be a member of any committee or represent any group) can ask for one. The method of doing so is described in BAC.[1] The request is considered first by the Judicial Commission and if approved by 10 or more members is then submitted to ICSB for final approval.[2] [1, BAC.Ap8. 2, BAC.58a; c.]

resemblance. The result of the comparison of characters of two organisms. The more characters that are alike (or similar) the greater the resemblance, so that the comparison can be expressed mathematically as a Similarity Coefficient, **S**.

resistogram. The resistance of an organism to antibiotics is often tested and reported as an **antibiogram**. Apart from antibiotics some antibacterial substances tested at critical concentrations may show differences between individual strains of a bacterial species, and the method may be used to subdivide a species into types, possibly of epidemiological significance. Resistance to 8 substances has 256 theoretical combinations; a strain resistant to all is designated ABCDEFGH, while one sensitive to A, B, and E is CDFGH. Elek found that among strains of *E. coli*, the resistogram type showed a correlation with serological typing.

respiration, respiratory metabolism. Unlike respiration in animals, which concerns the transport of oxygen, in bacteria respiration involves the transfer of hydrogen to acceptors. The hydrogen transfer is mediated by dehydrogenases; when the acceptor is oxygen, the process is known as aerobic respiration or

oxidation; when it is inorganic (e.g. nitrate) because gaseous oxygen does not take part, the process becomes anaerobic respiration; when the acceptor is organic, the process is **fermentation**.

retention of a name (epithet). When a taxon is divided, the part that contains the type retains its name (or epithet) unchanged.[1] When taxa of equal rank are united, the one with the oldest legitimate name (epithet) retains that name.[2] A species transferred to another genus retains its epithet unless the new combination is a later **homonym**[3] or (bot.) a **tautonym.**[4]

When a taxon of rank between genus and family (bot.)[5] or between subgenus and order (bact.)[6] is changed in rank, the stem of the name is retained; the ending is altered unless the name so formed must be rejected for another reason.

[1, BAC.39b; 40b; BOT.52, 53. 2, BAC.38; BOT.57. 3, BAC.41a; BOT.55. 4, BOT.23. 5, BOT.61. 6, BAC.48.]

reticulate evolution. Evolutionary hypotheses which permit the fusion, as well as separation, of lineages. In bact. evolution, there is growing evidence that considerable reticulate evolution has occurred. LRH.

retroactive. Applying to actions (i.e. giving a name) that occurred before the rule or recommendation was made. BAC.[1] and BOT.[2] specificially state that the rules are retroactive.

The Zoological Congress 1972 rescinded the Statute of Limitation which restricted the application of priority rules to names used within the last fifty years, and matters affecting the application of rules to unused senior synonyms should be referred to the Zoological Commission, to be dealt with under its plenary powers.[3]

[1, BAC.2. 2, BOT.PrVI. 3, ZOO.23(a–b); 79(b).]

retroculture. Re-isolation of an organism previously inoculated into an animal (or other host) to determine its pathogenicity. An unnecessary word for which recover or reisolation is preferable.

reused name. After 1 January 1980, an author in proposing a new bacterial name may use a name validly published before 1980 provided it was not included in one of the **Approved Lists.**[1] The organism newly named need not be related in any way to the organism originally so named. However, if the organism has the same circumscription, occupies the same position, and has the same rank as the organism to which the name was applied before 1980 it is to be regarded as a **revived name**, and the abbreviation nom. rev. added to the name.[2]

A reused name (and a revived name) will be regarded as a new name, should be published in *IJSB*, and authority claimed for it by the new author. The word **non** may be used to indicate that the taxon differs from that to which the name was applied by an earlier (pre-1980) author.

Provisional rules for author citation of reused names have been drafted and will come into use in 1980.[3] (See **citation of a reused name**.)

[1, BAC.N1. 2, BAC.24a, 28a. 3, BAC.Prov.B1–B4.]

revalidate. A name proposed before the starting date of the nomenclature of a discipline may be revalidated, or made valid, by being accepted and used by a later author, to whom the name would be attributed.

reviser. 1. Agent-noun for one who revises or makes a revision. In biology implies the person who carries out the taxonomic work, subjects his results to a critical appraisal, and before committing himself, thinks out the implications of a revision of a taxon. It is not applied by biologists to the person who deals with a manuscript and prepares it for the printer.

2. In zoo. *the term first reviser is to be rigidly construed*[1] and in the case of names published simultaneously (synonyms) he must make clear that he regards the names as applying to the same taxonomic unit; he must also state which of them he chooses as the name of the unit. In making this choice he should base it on the name that serves best the interests of nomenclatural stability.[2]

[1, ZOO.24. 2, ZOO.R24A.]

revision. In taxonomy, the end result of a critical re-appraisal of work done on a group of organisms, usually supplemented by new work, or re-examinations by more modern techniques. The revision may be a relatively short piece of work dealing with only a few specific aspects, or, when the amount of material to be investigated, evaluated, and correlated is large, it may take up most of the working life of the reviser; such revision may be delayed in the hope (usually unwarranted) that the reviser can complete his work so well as to make his revision definitive.

A good example of a lifetime's devotion to the revision of one group is **Kauffmann's** revision of Bruce **White's** serological classification of the genus *Salmonella*, which extended over about 25 years.

revived name. The abbreviation nom. rev. is used in BAC[1] for a name proposed after 1 January 1980 that perpetuates a name for a taxon of the same circumscription, position, and rank as one used by an earlier author, which name was not included in one of the **Approved Lists** (cf. **reused name**).

The earlier author of the name may be indicated by adding the word ex, the name of the original author and the year of original valid publication,[2] and the bracketed words are followed by nom. rev.

[1, BAC.28a; Prov.B2. 2, BAC.Prov.B3.]

rhizobi-um, *pl.* **-a. 1.** A bacterium living in the root nodules of leguminous plants. **2.** Name of a genus of root nodule, nitrogen-fixing bacteria.

rhizosphere. The microenvironment surrounding the roots of a plant. The microbial flora of the rhizosphere has an important bearing on the health and nutrition of the plant.

rhapidosomes. Submicroscopic rod-shaped particles; detected in lysates of old cultures of some procaryotic microbes such as *Saprospira gracilis*, *Proteus mirabilis*.

RNA. Ribonucleic acid; a nucleic acid made up of nucleotides in which the sugar is ribose, and the bases are adenine, guanine, cytosine, and uracil.

robust. Strong, reliable. Used by Edwards & Cavalli-Sforza (*Biometrics*, 1965, **21**, 362–75) about the subjects of a classification. When the subjects were stable and reliable their separation was useful for many different purposes. Only robust classes of objects were thought worthy of bearing names. Also used about the statistical methods (of numerical taxonomy) themselves which are said to be robust if they each give the same general result with the same set of **OTU**s.

room temperature. Although popular, this is probably the worst-defined value in microbiology. In England, the temperature of a laboratory may vary between 15 and 20 °C, only rarely is it above 22 °C. In US laboratory temperatures are seldom below 25 °C, and often reach 28 °C.

The expression, room temperature, may be used in the definition of **thermotolerant**, and in the descriptions of so-called free-living bacteria such as some mycobacteria (but not the pathogenic species), and micrococci.

R technique. Techniques of multivariate analysis used in numerical taxonomy which are based on direct comparisons between the features, not the objects (cf. **Q technique**).

rule. Regulations designed to make effective the principles of nomenclature. BAC.[1] further adds to their function the putting of the nomenclature of the past in order and to provide for the nomenclature of the future. Cf. **Principle, Recommendation**. LRH.
[1, BAC.GC6(2).]

Rules of nomenclature have been drawn up for plants (wild and cultivated), animals, bacteria, and viruses. The varius **codes** have many features in common but each shows differences that are either idiosyncratic or are modifications to suit the particular biological discipline. The microbiologist may be concerned with all the codes because the mycologist uses the botanical code, and the protozoologist the zoological code. The bacteriological code was based largely on the botanical code; the virologists hoped to combine the best from both bacteriological and zoological codes in their rules.

Although suggestions have been made for the compulsory use and registration of names, the rules cannot be enforced and they serve merely as a code of good behaviour.

There may be confusion about what is covered by these rules; they relate only to nomenclature, i.e. to the labelling or naming of different taxa. Classification is not governed by any rules except those of good sense, but virologists intended to insist on good descriptions before the grouping and classification is approved by their taxonomic committee (see **Rules of virus nomenclature**).

Rules of virus nomenclature. Virologists used common names for the viruses until the middle of the twentieth century when some plant pathologists toyed with the classical binomial nomenclature of botany, but this experiment was received with horror and scorn by almost all other virologists. By the efforts of a few, and notably by C. H. Andrewes, systems of nomenclature for viruses have been developed and tried out but virologists have been more interested in describing their viruses succinctly than in naming them.

With the development of a draft code, a compromise between the bacteriological and zoological codes and with relatively simple rules, it seemed as if virus workers were beginning to accept a binomial system of nomenclature and simultaneously a coded system that combined a name and description in the form of a **cryptogram**. By the seventies some agreement seemed to be in sight and eighteen rules were accepted by ICNV; of these, Rule 4 said that an *effort will be made towards a latinized binomial nomenclature*. Progress was slow but the idea of arranging viruses into genera and families was welcomed; latinized names for virus groups (genera) seemed to be acceptable but the use of epithets for the members (species) of the groups received little support, and in 1975 Rule 4 was amended by deleting the word binomial.

S

S, similarity coefficient. Sneath's original measure of overall similarity, in which he ignored characters negative in both strains being compared:

$$S = \frac{a}{a+b+c},$$

where a = number of positives shared by both strains being compared, and b and c = number positive in one strain and negative in the other. The similarity index is usually expressed as a percentage by multiplying by 100; 100% S equals identity, 0% S no resemblance at all. This percentage may be called the *S value*.

Other workers, e.g. Tsukamura (*J. gen. Microbiol*, 1969, **56**, 265–87), use the symbol S, and speak of the S value for the result of a formula which includes characters that are negative in one or both of the strains being compared. Cf. **M**, the matching coefficient, in which 'negative features' are included.

saccate, like a sack, describes the appearance of the liquefied area in gelatin stab culture produced by some bacteria. May be difficult to distinguish from **infundibuliform** (funnel-shaped) liquefaction.

saccharoclastic, saccharolytic. Capable of breaking down carbohydrates. These terms can be used to describe the action of *Brucella* spp. on carbohydrates which, although not producing detectable acid in the usual peptone-containing media, attack glucose (and other sugars) until it is no longer detectable in the medium.

St John-Brooks, Ralph. Curator of the National Collection of Type Cultures from its foundation in 1920 until his retirement in 1946. He was the first Secretary representing medical bacteriology of the International Committee on Bacteriological Nomenclature, set up by the International Society for Microbiology (later IAMS) at the first congress in Paris (1930), and later became the Secretary General of the International Association. With R. E. Buchanan and R. S. Breed was one of the authors of the first Bacteriological Code.

Born in 1884, St John-Brooks died in his native Ireland in 1963.

saltation. Used in mycology to describe a permanent variation or mutation in the fungal mycelium.

saprobe. Term introduced (by Martin, G. W., 1932, *Bot. Gaz.* **93**, 421) for a fungus that uses dead organic material and commonly causes decay.

saprophyte. Strictly an organism that obtains its food supply from dead and decaying material, but in practice it is used for one that lives a parasitic existence on the tissues of a living animal without doing any harm to the host. It does not cause an infection or penetrate the tissues. In man bacteria which live in situations such as the nasopharynx may more properly be called **commensals**.

Plant pathologists distinguish saprophyte (a plant using dead organic material as food and producing decay) from a *necrophyte* which lives on dead material that is *not part of a living host*.

satellite, -ism. In microbiology, growth around another (usually larger) colony. Satellitism is seen when a blood agar plate uniformly seeded with *Haemophilus influenzae* is spot inoculated with *Staphylococcus aureus*; a factor (V) essential for the growth of the influenza bacillus is produced by the staphylococcus and diffuses from the colony into the medium. Close to the staphylococcal colony the colonies of *H. influenzae* are large, but become smaller as the distance from the staphylococcal colony increases and the concentration of the V factor decreases. The other essential growth factor (X) required by *H. influenzae* is supplied by the blood in the medium.

An unusual US variant of satellitism is *satellitosis* (Russell, 1965).

schema. A word used by Kauffmann (and copied by others) writing both in German and English for the English word scheme, e.g. Kauffmann–White schema. There is no reason why this usage should continue when writing in English.

schizotype, schizotypification. A nomenclatural type proposed by Korf & Rogers (1967, *Taxon*, **16**, 19–23). It is what they describe as an *implicit lectotype*, formed when an author treats in one publication all the *syntypes* of a taxon but retains only one of the eligible syntypes, excluding all the others; finally he fails to designate the one retained syntype as a holotype.

Schizotypification is the term used to describe this implicit typification by omission.

scientific name. The Latin or latinized name of an organism. For the ranks of genus and above the name is uninominal, and has an initial capital letter. For species and subspecies the binominal and trinominal combinations are printed in italic. A subgeneric name (epithet in bot.) has an initial capital letter; when present it is not counted in the two word (binary) binominal combination of a species, or in the three word (ternary) trinominal combination of a subspecies.[1] A generic name is printed in italic, ranks above genus are printed in roman. American journals often use italic for all scientific names.

[1, ZOO.6.]

score, *v.* The action of summing **similarities** and **differences**. Often used as synonymous with the verb code, but Lockhart advises the verb encode for the action of recording the appropriate symbol in the data column of the character.

scotobacteria. A vernacular name for bacteria, rickettsias and mycoplasms, which are described as procaryotes indifferent to light; in contrast, the photobacteria which consist of procaryotes that are **phototrophic**.

These terms, scotobacteria and photobacteria, are preferred by the members of the Bergey's Manual Trust to latinized scientific (or pseudoscientific) names for the divisions of the Procaryotae.

scotochromogen, -ic. An organism able to produce pigment in the dark. Applied to some mycobacteria.

screening test. A test used by diagnosticians to detect or to eliminate a particular organism or group of organisms. Screening tests may be morphological, tinctorial, biochemical, serological, sensitivity or resistance to phages or antibiotics. A well-known example is the use of a urea-containing medium to detect and exclude urease-producing organisms when looking for salmonellas or shigellas among intestinal bacteria.

section, subsection. Subdivisions of a genus. The name of a subdivision of a genus is a combination made up of the generic name, a term indicating the rank (subgenus, section, etc.) and the subdivisional epithet (with an initial capital letter).[1] The subdivisional epithet is either the same as the generic name or a plural adjective agreeing in gender with the generic name.

Subdivisional names are not taken into account in determining the priority of homonymity of a generic name. In zoo. a section has the status of a subgeneric name.[2] The ranks themselves do not seem to have been used by bacteriologists, but they formed temporarily an unnecessary elaboration of the hierarchy during the operation of the 1966 Code. They are now regarded as **informal categories**.[3]

[1, BOT.21. 2, ZOO.42(d). 3, BAC.11.]

segregate, segregation. Separation, splitting into two or more parts; when this involves the nomenclatural type, the part containing the type or name-bearer must retain its original name and a new name must be given to the other part(s)[1] – often called the excluded part, or a segregate that excludes the type.

[1, BAC.37a.]

selective medium. 1. To isolate a particular organism or group of organisms from a specimen containing a mixed microbial flora, the diagnostician often uses media containing substances that prevent the development of colonies by organisms it is not wished to isolate. The unwanted organisms are not killed by the inhibitory agent but remain dormant and multiply when transferred to non-inhibitory media. The unwanted organisms may lurk in the base of a colony that has developed on the selective medium, and, on subculture, give the lie to the assumption (too often made) that a single colony is the result of growth from a single organism, and consequently yields a pure culture. Colonies from selective and **inhibitory media** should *always* be subcultured to a plate of a non-inhibitory medium. Neglect of this precaution is the commonest source of failure to isolate a pathogen present in faeces, and also the commonest source of misidentification and **exotic organisms**.

2. Geneticists use the term selective medium for one capable of supporting growth only of bacteria adapted to certain of its constituents (e.g. antibiotics).

semispecies. 1. Term suggested by Mayr to indicate the intermediate nature of geographically isolated groups (populations) for which the exact ranking (species or subspecies) is difficult to decide on the evidence available.

2. Also used by Mayr for the component species of a superspecies, for which Amadon later suggested the name **allospecies** (*Condor*, 1949, **51**, 250–8).

senior. 1. Of synonyms, a term for the earliest published name in a list of synonyms.

2. Of authors, applied to the more experienced of two or more authors of a scientific paper, whose name often appears last; this usage is becoming less common, and perhaps should now be described as rare. In US the word senior is used for the first-named author.

***sensu*, *abbr.* sens.** In the sense of . . . , and followed by the name of an author who applied the scientific name to which it refers.

sensu amplo*, *abbr.* sens. ampl.** In the broad sense. One of the qualifying phrases recommended by BOT.[1] for use when the diagnostic characters of a taxon have been considerably altered, but without excluding the type. A more common phrase to indicate amendment of circumscription is ***emendavit.

[1, BOT.R47A.]

***sensu lato*, *abbr.* sens. lat.** In the general sense. In bact. used most often in connexion with pasteurellas. When writing of all species included in the genus

Pasteurella before 1944 (when van Loghem proposed that certain species should be excluded to form a new genus *Yersinia*), the appellation *sensu lato* may be used, but when limiting the discussion to species retained in the genus, after *Yersinia* had been removed, *Pasteurella sensu stricto* is appropriate.

sensu novo. Suffix applied to taxa that have been reclassified; to distinguish from bacteria with similar name but with another and probably wider circumscription. For example, used by Mordarska, Mordarski & Goodfellow (*J. gen. Microbiol.*, 1972, **71**, 77–86) for nocardias of the spp. *N. asteroides*, *N. brasiliensis*, and *N. caviae*.

sensu stricto, *abbr.*, **s. str.** In the strict sense. Used for example in speaking of the bacterium that causes haemorrhagic septicaemia in various animals, as a pasteurella (or *Pasteurella*) s. str., in contrast to the plague bacillus, a pasteurella *sensu lato* (now generally transferred to a newer genus, *Yersinia*).

To limit the application of a specific epithet in *Klebsiella pneumoniae*, Bascomb *et al.* (*J. gen. Microbiol.*, 1971, **66**, 279) added *sensu stricto* when they referred to the species as defined by Cowan *et al.* (*J. gen. Microbiol.*, 1960, **23**, 601) and *sensu lato* when the name was applied to the more broadly defined species of I. Ørskov and many other workers.

separate. Part of a journal or book that contains a single paper (or group of related papers), bound apart (separately) from the remainder of the journal (book) and sold to the author for distribution to colleagues. Often referred to as a reprint. In nomenclature, separates will be important when they are issued or sold to the public before the publication day of the journal; in that event a proposed new name will date from the date printed on the separate *unless there is evidence that it is erroneous.*[1] ZOO.[2] indicates the responsibility of editors and publishers to ensure that the date of issue of the parts of a serial publication is correctly shown; reprints of parts should contain information that will enable accurate citation to be made to the original pagination and date of publication. A preprint should be identified as such, and should carry its own publishing date. As corollary to the duties laid on editors and publishers, an author must not distribute any separates before the date of publication printed on them.[3]

[1, BOT.30. 2, ZOO.R21B, D. 3, ZOO.R21A.]

sequence of ranks. The relative order of the taxonomic ranks is given in the codes and this should not be changed; the order is shown in Tables R. 1 and R. 2, p. 220.

[BAC.5a, b. BOT.3; 5]

sequential method. Identification by proceeding in stages or steps.

1. Characters are considered one at a time and taxa that do not share a character with the unknown are eliminated. The method is used in punched-card systems of identification, and in multiple-entry keys.

2. Several characters are compared simultaneously; in bacteria eight to ten carefully chosen characters are often sufficient to permit identification down to the level of genus or at least to a small group of genera. The next step consists in comparing other characters of the unknown with known characters of the probable genus. This is the step-by-step (progressive) method of Cowan & Steel.

serial. A publication that appears at regular (occasionally irregular) intervals. Often has the word Journal in its title, but may be Proceedings or Transactions of a society.

series, subseries. Ranks between section (subsection) and species; the names used are not competitive with generic names in establishing priority. Although these ranks appeared in BAC., they seem to have been used in bact. only in the genus *Streptomyces*. In BAC. they are regarded as **informal categories.**[1] In bot., series is used in the *Flora URSS for* **species aggregate**.

[1, BAC.11.]

sero-fermentative phage-type. A term used by Kauffmann in his redefinition of species as *a group of related sero-fermentative phage-types*. This statement, a complicating elaboration of his earlier definition of species, is not easy to understand, and no enterobacteriologist of my acquaintance has been able to explain it to me. See **species**.

sero-fermentative type. The original Salmonella Subcommittee of the International Committee on Bacteriological Nomenclature published a list of *Salmonella* species that were defined serologically. Later Kauffmann thought that these should not be described as sero-types but as sero-fermentative types, and he gave as examples *Salmonella typhi* (which differs in some biochemical reactions from all other salmonellas) and *S. choleraesuis*, the type species, which is unusual in failing to ferment both arabinose and trehalose. Later still, he modified his ideas and introduced the term **sero-fermentative phage-type** (q.v.).

serological type. A subdivision made on the basis of antigenic difference. In the genus *Salmonella* the subdivisions have been treated as species, but in other genera, e.g. *Staphylococcus*, as subspecific groups. Among the streptococci a distinction is made between serological group (roughly equivalent to species) and serological type (a subdivision of serological group and therefore of subspecific level). See also **serotype**.

serology. 1. Literally, the science of serum reactions. Now replaced by **immunology** in microbiology.

2. In microbiology, the study of the reactions between micro-organisms and antibodies in the serum of infected or immunized animals (including man). Serological reactions are reactions between antigen(s) and antibody(ies), and the different manifestations or kinds of reaction (agglutination, precipitation, complement fixation, mouse protection, and so on) depend on the physical conditions in which the tests are carried out. On the basis of serological tests, differences between apparently similar organisms may be found, and the subdivisions formed are called serological types or serotypes.

serotaxonomy. Application of serological techniques to taxonomy, particularly to the characterization of micro-organisms at the subspecific level, e.g. *Salmonella* serotypes. Diffusion methods of antigenic analysis are seldom used outside the field of bact.

serotype. A shortened form of serological type. These are subdivisions made on differences in antigenic structure of culturally indistinguishable bacteria; occasionally the serotypes may also be distinguished by their biochemical characters. The Salmonella Subcommittee (1934, *J. Hyg.* **34**, 33) suggested that *new serological types should receive specific names*, and for many years enterobacteriologists equated serotypes with species. With the rapid multiplication of salmonella serotypes the absurdity of the situation became obvious, and, with Kauffmann a notable dissenter, there has been a general movement away from this equation of serotype with species. See **sero-fermentative type, species**.

BAC.[1] is probably fighting a losing battle when it tries to eliminate the use of the suffix -type and recommends serovar as a substitute for serotype, which is very well established.

[1, BAC.Ap10.]

set. The ensemble of characters. In earlier numerical taxonony writings, the term battery was frequently used. With the initial hostile controversies somewhat now abated, the less militaristic word set is preferred. LRH.

set theory. A discipline between elementary logic and elementary mathematics which aims at introducing some system into general ideas that deal with 'sets' (classes or collections). Set theory is applied to 'methodological taxonomy', or to the resolution of philosophical questions posed by taxonomic thought and work; Simpson (1961) in *Principles of Animal Taxonomy*, pp. 19–23, discusses this application.

sexduction. Transfer of genetic characters by conjugation of sex factors; e.g. the R factor of *E. coli* has been transferred by the sex factor F, to *Proteus mirabilis.*

sex hair. Term used for F **fimbriae** (or pili) found on the surface of male cells of *Escherichia coli*; they differ from other fimbriae found on bacteria in their apparent involvement in conjugation and in the transfer of antibiotic resistance.

SI. Abbreviation of Système International d'Unités (SI unit), which derives all measurements from seven basic units, (SI symbols in brackets): length, metre (m); mass, kilogram (kg); time, second (s); electric current, ampere (A); thermodynamic temperature, kelvin (K); luminous intensity, candela (cd); amount of substance, mole (mol). Symbols for units do not have a plural form and, except at the end of a sentence, are not followed by a full stop (period).

Multiples and fractions of SI units are shown by prefixes (e.g. 10^6, mega (M); 10^3, kilo (k); 10^{-3}, milli (m); 10^{-6}, micro (μ); 10^{-9}, nano (n).

In microbiology the adoption of SI units has required the rejection of older units and symbols such as micron, μ for 10^{-6} metre and millimicron, mμ for 10^{-9} metre; the equivalent SI units for these are micrometre (μm) and nanometre (nm) respectively.

sibling species. Term used by zoologists for 'good' species (in the genetical sense) that morphogically can be separated only with difficulty, or not at all.

sic. Thus, so. Enclosed in brackets *sic* is used in quotations to show that the spelling, which may look erroneous or absurd, exactly reproduces that used by the author of the quotation. In a paper by Baumann *et al.* (1968) the title was shown as 'A study of the *Moraxella* group. II. Oxidative-negative species (genus *Acinetobacter*).' The text of the paper showed that the word oxidative was an error for oxidase, and in a reference to this paper it should be shown as '... II. Oxidative-[*sic*] negative species...'. Square brackets are used to indicate that the word enclosed by them was not used by the original author.

Unlike the nomenclatural duty of an author (and editor) to correct an incorrectly spelt scientific name (orthographic error or variant), in giving a reference an author should quote the title exactly (errors included).

sigla. An alien and unnaturalized word, possibly an abbreviation of sigilla (*SOED*). A word made up of initials or other characters; an abbreviation. Used by virologists in words such as Echo (made up from EnteroCytopathogenic Human Orphan) virus. See **acronym**.

sign. 1. A conventional symbol (such as a plus sign, +) used as an indication of a particular reaction; the indication given by the sign should be explained, but need not be repeated whenever the sign is used.

Signs may be used in the text of a manuscript, but have their greatest usefulness in tables, and, in taxonomy, in **diagnostic tables** or tables of characters.

2. Signs used with special meanings in taxonomy: ≡, between scientific names, indicates **nomenclatural synonyms**; = indicates **taxonomic synonyms**.

signature, *abbr.* **sig.** Letters (or letter and figure), usually in small capitals at the foot of the first page of a sheet of printed matter. ZOO.[1] recommends the use of such a sign or signature to identify a sheet when the pages are not numbered. [1, ZOO.ApE15.]

significant. In statistics, a significant difference is one that is unlikely to occur by chance, but in taxonomy significant differences are *taxonomically useful attributes*.

similarity coefficient. See **S**.

simultaneous publication of names. When two or more names for a single taxon or the same name for different taxa, are published in the same paper or book the action of the first revising author determines the priority of one of the names.[1,2] The first reviser of simultaneously published synonyms should state clearly that he believes the names apply to the same taxon and he must choose one of them as the name of the taxon. If none of the names has an apparent advantage he should select the name that appears first in the paper (excluding the summary) or book.[3]

[1, ZOO.24. 2, BOT.64. 3, ZOO.R24A.]

Skerman, Victor B. D. Australian bacteriologist who made his mark by developing the first successful dichotomous key for bacteria, and for his advocacy of the standardization of methods used in the characterization of bacteria. His first published key was revised in the form published in the 7th edition of *Bergey's Manual* and greatly improved in his *Guide* (1959, 1967), which also contained descriptions of recommended methods.

skyline diagram. An alternative to the **Sneath diagram** as a method of arranging % **S** values to show grouping and divisions between major groups, which are represented as dips between solid columns of the histograms which gives the diagram its name. An alternative name is **bar diagram**.

smear. Film of microbial growth spread on a microscope slide which, after fixation, will be stained.

Sneath diagram. Term used by Hutchinson, Johnstone & White (*J. gen. Microbiol.*, 1965, **41**, 357–66) for the arrangement of S (similarity) values in the form of a shaded triangle. By suitable shading (to indicate different S values as percentages) and rearrangement the relations between different organisms can be appreciated at a glance.

Sneath, Peter, H. A. Pioneer numerical taxonomist. Sneath applies statistical principles to the relations of microbes as shown by multivariate analysis, or analysis of relations in all possible directions or combinations. He applied his ideas to results obtained by a systematic study of *Chromobacterium* spp., and later collaborated with a zoologist, R. R. Sokal, in writing up the theory and practice of what they called **numerical taxonomy**. The shorter version (published in *Nature*, 1962, **193**, 855) is much the clearer, and more suitable for the non-expert. An early application of numerical taxonomy to a wide range of bacterial genera confirmed the soundness of classifications made by older taxonomists which were based on subjective assessments of relatedness, but Sneath's survey also pointed out relations (e.g. between *Vibrio* spp. and Enterobacteriaceae) that had been overlooked, relations that have since been confirmed and extended by other workers using different methods.

solidus. A sloping line (/, US, slant line), used in typography for fractions, and in scientific work in place of the word *per* in rev/min, a unit symbol for revolutions per minute. The use of the solidus in mathematical expressions may lead to ambiguities which can generally be resolved by using brackets.

solubilize. **1.** To alter the nature of a substance to render it subsequently dissolvable.

2. Frequently, just an ugly word for dissolve, make a solution.

sorb, desorb. Sorb is an ambiguous word for removing something (e.g. antibody) when the mechanics of the removal are unknown to the author, who consequently is uncertain whether to use **adsorb** or **absorb** (q.v.). Desorb is the reverse process, releasing the removed substance.

sorption. A term which expresses the author's uncertainty as to whether a process is an ad- or ab-sorption.

sorter. A device described by Olds to make easier the identification of a micro-organism. Characters similar to those used in diagnostic tables are recorded on punched metal plates which are superimposed on each other. A light behind the plates (as in the **peek-a-boo system** for cards), show a hole or holes for organisms that have the characters tested. The sorter is a practical machine but must not be expected to make an unassailable identification.

sp., *pl.* **spp.** **1.** Abbr. of species. Should be used only in taxonomic works. Useful to avoid repetition of specific epithets when these are unnecessary for the understanding of a passage, e.g. *Shigella* spp. for several species of *Shigella*.

2. Used to indicate a strain of an unidentified species in a known genus, again used only in taxonomic works. In this case, *Shigella* sp. (note: singular) means a species not identified of the genus *Shigella*.

special classification. Classification derived for a special purpose in mind from the outset. Used to contrast with general classification, special classifications necessitate a deliberate selection of characters pertinent to the selected purpose. LRH.

speciation. **1.** Evolutionary differentiation leading to the creation or the origin of species; the process involves the development by organisms of different characters, and the finding by the taxonomist of the gaps (discontinuities) between the diverging forms. Though orginally not referring to biological species, John Locke's aphorism is applicable, *The boundaries of the species, whereby men sort them, are made by men.*

2. Zoologists use the word speciation for the multiplication of species at any one time (cladogenesis) and also for the transformation of species in time (phyletic evolution, anagenesis).

3. As a synonym for identification in a title such as 'Presumptive speciation of group D streptococci...'. This is a misuse of the word and is to be avoided.

species (abbreviated to **sp.** or **spp.** (plural)). **1.** A category, definable only in terms of position (below genus) in a hierarchical system.

2. A taxonomic group, definable in terms of the characters of the constituent members.

3. A concept; that it is useful cannot be denied, but the user must realize that the species does not exist and is not an entity.

In the pre-Darwinian era the species was thought to be of divine creation and to be unchangeable, but as theories of evolution became accepted, the idea of immutable species had to be abandoned. In bot. and zoo. the terms 'biological' and 'evolutionary' species are used but apply only to a minority of organisms. In organisms that reproduce sexually we can say that those units that cannot interbreed cannot be of the same species, but since we are not able to test out all possible cross-fertilizations, we cannot explore all the interbreeding possibilities. In microbiology we are not concerned with sexual reproduction as this term is generally understood. However, microbial genetics has shown that although binary fission is the rule, a form of sexual reproduction occurs with some organisms, and an exchange of genetic material takes place; after this process (recombination) the organism has characters of both parents.

So much has been written on species and the species concept that it is almost true to say that there are as many ideas on species as there are biologists, and many a biologist has changed his ideas during the course of his working life. Ravin defined three very practical kinds of species, each with different sets of properties in which the first (**genospecies**) consists of mutually interfertile forms and corresponds most closely to the species concept of botanists and zoologist; the **nomenspecies** made up of individuals resembling the nomenclatural type, and **taxospecies** those units based on overall similarity and determined (as **OTU**s) by numerical methods.

To summarize: a species is a group of organisms defined more or less subjectively by the criteria chosen by the taxonomist to show to best advantage and as far as possible put into practice his individual concept of what a species is.

species aggregate. A device used by botanists to group together species that are morphologically similar and difficult to distinguish; it is similar to the zoologist's collective group. The components of a species aggregate are sometimes known as **microspecies**. Species aggregates came into being as *species collectiva*, described by Davis & Heywood as *often a confession of ignorance*; species aggregate may be used when further work is likely to give a good chance of making an identification or placing the unit in a better taxonomic position, but it should never be the ultimate category of a biological unit.

Hawksworth advocates avoidance of the species aggregate concept as it is *merely a method of avoiding the issue...or making more critical observations.*

species group. Zoo. term for the ranks of **species** and **subspecies**; it does not include **infrasubspecific forms** to which the provisions of ZOO. do not apply.[1] The name of a species consists of two words (binomen: generic+specific names) and that of a subspecies three words (trinomen: generic+specific+subspecific names). Each taxon is defined by reference to its type specimen. The subspecies that contains the type specimen of the species must have the same subspecific name as the species, and it is the *nominate subspecies* (cf. **nominate subordinate taxon**).

[1, ZOO.45, 46, 47.]

species-group name. A zoo. term for that part of the name of a species or subspecies that is peculiar to the taxon; it must be published in combination with a genus-group name.[1] It is the equivalent either of a specific name or a subspecific name.

A species-group name may be treated either as a noun or as an adjective. As a noun it should be in the nominative singular and stand in apposition to the generic name, or it may be in the genitive. When an adjective[2] it can be in the nominative singular and should agree in gender with the generic name, its ending changed when needed by transfer to a genus with a generic name of different gender.

[1, ZOO.11(g). 2, ZOO.30.]

species incertae sedis, *abbr.*, **sp. incert. sed.** Species of uncertain taxonomic position. The specific status of the group should be reasonably certain, but the allocation to a larger group (of generic rank) may be debatable. A bacteriological example is *Haemophilus vaginalis*; it is likely that at least two different organisms have borne this label, and in the 1974 edition of *Bergey's Manual* one was placed among a *species incertae sedis* appendix to the genus *Haemophilus* Winslow *et al.* because that would be *the place in which readers will most probably look for it.*

A *species incertae sedis* cannot be validly designated as the type species of a genus.[1]

[1, ZOO.67(R).]

species indeterminata. A species that cannot be identified from the original description. This taxonomic qualifying phrase could be applied to many bacterial species (or names of species) that were first named and described before the era of pure cultures.

species inquirenda. An incompletely identified species, one that requires further characterization and comparison with other species.[1] A qualifying phrase also used for species of doubtful taxonomic position, but a better Latin tag to attach to them would be *species incertae sedis.* A *species inquirenda* should not be designated as the type species of a genus.[2]

[1, ZOO.Gl. 2, ZOO.67(h)].]

species name. The name of a species. In bact. and bot.[1] it is a binary or **generic name** and a word for the species or **specific epithet**. It is synonymous with **specific name**.

Not used in zoo., for which the correct word is **binomen**.

[1, BOT.23.]

species nova, *abbr.* **sp. nov., sp. n.,** or **n. sp.** New species. Authors of new names for micro-organisms use the convention sp. nov., etc. as the authority, instead of appending their own names, when proposing a new species name (but not in subsequent publications). This is not usual in bot. or when thought desirable as it may be ambiguous when there is more than one author. BOT. deprecates the use of *nobis* (abbr. nob., of us),[1] and recommends that authors should always append their names (similarly, *mitri*, of me).

If used, the abbreviation sp. nov., etc., must not be repeated by subsequent authors, who should replace it by the normal **author citation**.

CBE Style Manual (p. 49) says the Latin form (sp. n.) is preferred by botanical journals, the English form (n. sp.) by most other journals.

[1, BOT.R46F.]

specific epithet. Used in bact. and bot. for the second element of the name of a species; thus *coli* in *Escherichia coli* is the specific epithet. If the specific epithet consists of two or more words, they must refer to a single concept and should be united; if they were originally hyphenated, they should be joined by subsequent authors.[1,2] In writing about a species the specific epithet should not be used alone as a noun but should always be preceded by the generic name or the first letter of the generic name. It is slipshod to write 'strains of coli were...' for 'strains of *Escherichia coli* (or *E. coli*) were...'; the better form takes up little more time to write and few more pieces of printer's type. An epithet can be useful as an adjective, e.g. 'abortus fever', or attached to another word, as coliform, coliphage. See **literacy**.

BAC.[3] and BOT.[2] give similar advice about the choice and formation of specific epithets, and all codes state that the epithet may be taken from any source whatever and may even be composed arbitrarily; in bot. it must not be a **tautonym**.[2]

Specific epithets may be (1) adjectives, which must agree grammatically with the generic name (*Staphylococcus aureus*, *Sarcina flava*), (2) nouns in the nominative, in apposition to the generic name (*Spirillum rubrum*), or nouns in the genetive (*Klebsiella edwardsii*, *Staphylococcus lactis*).[3] Recommendations made by the codes for the formation of specific epithets from the names of individuals are described in detail under **personal name**, and in Tables P. 1–P. 3, pp. 198, 199.

A specific epithet combined with an illegitimate generic name is not thereby rendered illegitimate.[4]

The priority of a specific epithet is decided independently of the priority of the generic name to which it is attached.[5]

[1, BAC.12a. 2, BOT.23. 3, BAC.12c. 4, BAC.32b; BOT.68. 5, BAC.23aN1.]

specific name. 1. In bact. and bot. is the combination formed by the generic name+the specific epithet.

2. In zoo. the specific name is the second element of the combination of generic+specific names that form the **binomen** or name of the species. The specific name must be published in combination with a genus name; when cited alone (i.e. without the generic name), the specific name does not have any standing in nomenclature. ZOO.[1] requires a specific name to be, or be treated as (*i*) an adjective in the nominative singular which agrees in gender with the generic name; or (*ii*) a noun in the nominative singular in apposition to the generic name; or (*iii*) a noun in the genitive, or (*iv*) an adjective used as a noun in the genitive, derived from the specific name of an organism with which the animal is associated. A name published before 1931 must have been accompanied by a description, definition, or **indication**.[2] A name published after 1930 must be accompanied by a statement of distinguishing characters, or by a reference to such a statement, or be deliberately proposed as a replacement for an **available name.**[3]

When a species is transferred to another genus it may be necessary to change the ending of the specific name to agree in gender with the new generic name.

[1, ZOO.11(g). 2, ZOO.12. 3, ZOO.13.]

specimen. 1. Individual taken as a representative of a larger group; e.g. single plant or animal. *Syn.* strain (isolate) representative of a microbial taxon.

2. A portion or sample of material, e.g. specimen of sputum. From such a specimen a culture may be made and one or more strains (isolates) isolated.

spectrum. Apart from it usual meaning, of light split by passage through a prism according to wavelengths, the word spectrum is used in taxonomy in the form of an analogy. An antibiotic spectrum describes the sensitivity or resistance of an organism to different antibiotics. Another use is to the resemblance between the intergrading of closely similar species and the merging of colours in the visible spectrum. Cowan & Steel (1965, 1974) who are critical of hierarchical schemes, arrange bacterial genera in two spectra, one consisting of Gram-positive, the other of Gram-negative bacteria.

spelling of names. In general, the spelling of the original author should be followed. BAC.[1] and BOT.[2] authorize the correction of **orthographic errors**. ZOO., while it allows correction of *incorrect original spelling*,[3] is less permissive and does not accept incorrect transliteration, improper latinization, and the use of an inappropriate connecting vowel,[4] as inadvertant errors and therefore does not permit their correction by later authors. The only corrections allowed by ZOO. are those made by an author's agents (his secretary or printer) and holds the author responsible for linguistic and grammatical errors.

As names can be formed in an arbitrary manner[5] and even an anagram[6] can be used as a scientific name, the detection of an unintentional error can be difficult.

An incorrect spelling does not have a separate status in nomenclature and it does not enter into homonymy; it cannot be used as a **replacement name.**[3] See also **variable spelling**.

[1, BAC.57cN1. 2, BOT.73. 3, ZOO.32(c). 4, ZOO.32(a)(ii). 5, ZOO.11(b). 6, ZOO.Ap.D26, 41.]

sphaeroplast is a term used for a bacterial cell in which the cell wall has been modified; this state may be brought about by lytic enzymes such as lysozyme, or by growth in the presence of an antibiotic such as penicillin. Thus a sphaeroplast is a structure less well-defined than a **protoplast** q.v. US spelling is spheroplast.

sphere. In numerical taxonomy, term sometimes used for clusters derived from the concept of a centrally placed **OTU** (see **centrotype**) and a fixed radius of taxonomic distance around it. In some methods such clusters are more correctly called ellipsoids, and, to be pedantic, when defined in taxonomic spaces of more than three dimensions, hyperspheres and hyperellipsoids. LRH.

splitter. Descriptive term for a taxonomist who enjoys dividing taxa into several units of the same rank or category. Heller was a good example of a bad splitter: she raised the genus *Clostridium* to a family, which she then divided into two subfamilies and twenty-six genera. Taxonomists of contrasting philosophy are known as **lumpers** or **clumpers**.

Mayr believes that splitters classify characters rather than groups of organisms, and in doing so forget Linnaeus's dictum that it is the genus that gives the characters, and not characters that make the genus.

It is generally true that a man is a splitter of those organisms on which he is expert, and a clumper of those about which he knows – or cares – little.

spore. More correctly **endospore**, a resting phase, a body within the main part of the vegetative microbial cell. The spore is more resistant to drying and moderate temperatures than the vegetative cell. To be sure of killing bacterial spores a temperature–time combination of 126 °C–20 min is essential, and this combination should be used for sterilizing cultures and contaminated apparatus. Spores are resistant to disinfectants and only the oxidizing agents have significant sporicidal action.

Spores are found in relatively few bacteria; therefore, the ability to produce spores is a useful (and indeed valuable) taxonomic character. Some spores are spherical, others are ellipsoidal; they may or may not be wider than the body of the vegetative bacterium. These differences – together with the position of the spore (terminal, subterminal, or central) – have been used as diagnostic characters. The difference between subterminal and central is a matter of the utmost nicety and is so subjective that the only significant differentiation is between the terminal and subterminal positions.

The spores of fungi are not usually as resistant as those of bacteria; the spores of aquatic fungi may be motile. Fungi may have two sporulating states, the imperfect in which asexual spores are produced, and the perfect, characterized by sexual spores (e.g. ascospores, basidiospores).

square brackets. 1. Used to surround additional words inserted into a quotation, usually to clarify the meaning.

2. In the antigenic formulae of salmonellas, square brackets are used to enclose antigens or antigenic factors not present in all strains of the serotype.

3. In ZOO. square brackets (*a*) surround the name of an author who published anonymously and whose identity was only subsequently discovered;[1] and (*b*) enclose the date of publication when found only from external evidence.[2]

4. In an *allospecies* name, square brackets enclose the name of the superspecies placed between the generic name and the allospecific epithet.

5. Formerly used in bot. to indicate, in an author citation, that the name was proposed before the starting date of the nomenclature of the group. Now obsolete and replaced by **ex.**[3]

[1, ZOO.R51A. 2, ZOO.R22A(3). 3, BOT.R46E.]

stable, stability. 1. A state in which change is uncommon ; the antonym of variability. The **Approved Lists** authorized under BAC. have, as one of their objects, greater stability in the nomenclature of bacteria.

2. Not easily moved or changed. The first aim of the codes of nomenclature is the attainment of stability in the scientific names of organisms, an aim doomed to failure for, as knowledged increases, the taxa named are themselves subject to change. Because of the small size of the subjects, the criteria used in the classification of micro-organisms differ from those used for plants and animals;

with the development of bacterial genetics and molecular biology, advances in our knowledge of small organisms are now much greater, so that the revision of classifications of microbes goes on at a great pace. Each revision makes some changes in the definition of taxa concerned, and may introduce problems of nomenclature, some of which involve **change of name**. Thus stability in nomenclature, whilst wholly desirable, is not likely to be attained within the foreseeable future.

One approach to stability is the greater use of **conservation**, whereby names are conserved against earlier synonyms, so that once it has been fixed (conserved), a name remains permanently attached to at least one culture. The problem is discussed more fully under **change of name**.

stabilate. Living material in which the biological characters have been stabilized, as by **freeze-drying** or very low temperature storage. Lumsden & Hardy (1965, *Nature*, **205**, 1032) defined it as *a population of an organism preserved in viable condition on a unique occasion*. The amount of such a stabilate will be limited, as will the number of samples laid down on the unique occasion.

stage in a life-cycle. In mycology the term perfect stage (sometimes state) implies the production of sexual spores; the imperfect stage is that in which only asexual spores are produced.

The complexities of nomenclature brought about by the life-cycles of some fungi are not made any easier by BOT.,[1] but fortunately more lucid explanations are available, and those in difficulty should consult Ainsworth (1973) or Hawksworth's *Mycologist's Handbook* (1972).

[1, BOT.59.]

standing in nomenclature is a phrase frequently used by the older taxonomists. It refers to the status of a name in relation to the rules of nomenclature, and 'no standing in nomenclature' means either that the name does not conform to the rules (e.g. a name proposed provisionally), or that the rules do not apply to them (e.g. vernacular names, or names of infrasubspecific taxa).

BOT. says that a name *has no status under this Code* unless it has been validly published.[1]

[1, BOT.12.]

starting date for nomenclature. Bact. nomenclature originally dated from 1753 when Linnaeus published the first edition of his *Species Plantarum*; this unrealistic date was chosen when bact. nomenclature was regulated by BOT.[1] and was one of the reasons why many bacteriologists did not treat it seriously. The absurdity of the situation was well illustrated by the name *Serratia marcescens* given by Bizio in 1823 to the red emenation that appeared on bleeding polenta. With the crude microscopes available at that time, spherules were seen and these were almost certainly yeast cells (*Rhodotorula*), but by persistent use the name has become attached to a bacterium that can also produce a red pigment on polenta.

After discussion for several years, bacteriologists have agreed on a new starting date, 1 January 1980, and in anticipation a completely revised version of BAC.[2] was published in 1975 and became effective from 1 January 1976. There was a short interregnum, but names published after 1 January 1977 should conform to the new Code.

Botanists (and hence mycologists) chose different starting dates for different groups; some fungi started at 1801, others at 1821, algae start at 1753 with exceptions. The nomenclature of fossil plants begins in 1820. In general, plant nomenclature started in 1753[1] (Linnaeus's *Species Plantarum*).

Zoological nomenclature starts with the publication in 1758 of the 10th edition of Linnaeus's *Systema Naturae*.[3]

[1, BOT.13. 2, BAC.1a. 3, ZOO.3.]

state. **1.** Given in BAC.[1] (p. 126) as a preferred term for colonial variants that may be defined by the surface antigens of bacterial cell walls, e.g. R, S, and M states (letter designations are preferable to the words rough, smooth, and mucoid, which in some bacteria may be misleading).

2. In numerical taxonomy, *state* refers to the nature of a character as simple or complex; in a two-state character (e.g. motility) there are only two responses, positive (present) or negative (absent); in a multistate character (e.g. colony surface) there may be more than two responses (e.g. rough, smooth, mucoid, papillated, etc.). However, many apparently two-state characters (e.g. indole production) can be measured quantitively and may then become multistate characters.

3. Sometimes used as a synonym of **stage**.

[1, BAC.Ap10B.]

status. See **standing in nomenclature**.

status novus, *abbr.* **stat. nov.** A qualifying phrase used to indicate that a taxon has been changed in rank (e.g. from subspecies to species) without a change of name or epithet. As with other phrases indicating something (status, name, combination) that is new, stat. nov. is used only at the time of publication, and should not be repeated by later authors.

status of names under consideration. When a name has been referred to a Commission for an opinion existing **usage** should be maintained until the decision of the Commission is published,[1] or confirmed by the next congress.

[1, ZOO.80; BOT.15.]

statute of limitation of the Law of Priority. Zoologists tried to increase the stability of nomenclature by excluding names that had not been used as senior synonyms for 50 years, thereby treating them as forgotten names (ZOO.[1]). But it was found that the Statute of Limitation was unsatisfactory when put to the test, and it was repealed in 1972. In its place, ZOO.[2] says that each case in which a zoologist thinks that the application of the law of Priority will cause confusion or disturb stability must be referred to the Zoological Commission, to be dealt with under a revised article (ZOO.[3]), which defines the plenary powers of the Commission.

[1, ZOO.1964, 23(b). 2, ZOO.1972, 23(a–b). 3, ZOO.79.]

Statutes of the International Committee on Systematic Bacteriology replace Provisions 4 and 5 of earlier editions of BAC. They deal with membership of the Committee and of the Judicial Commission, the duties of the Chairmen and the Secretaries, the functions of the publications committee and of the many taxonomic subcommittees.

The Statutes also define the procedure for handling proposals for amending the Code, interpreting its Rules, and requests for Opinions.

stem, *n.* **1.** Part of a name (word) to which case endings are added. In nomenclature, that part of a name to which the **suffix** appropriate to a category is added. ZOO.,[1] which includes a useful table of stem, points out that the grammatical stem is not always employed. BOT.[2] requires the stem to be retained when a taxon between genus and family is changed in rank.
2. In numerical taxonomy, the single, starting section of a dendrogram. LRH.
[1, ZOO.ApDT2; Gl. 2, BOT.61.]

stem augmentation. Addition of a letter to the stem of a word, made before adding a suffix. In bot.[1] an adjectival epithet can be made from a personal name ending in a consonant by adding *-i* (the stem augmentation) before the suffix. Thus, *Desmodium griffithianum* (instead of '*D. griffithanum*').
[1, BOT. (1975) R73c(e).]

step-by-step system. Informal and descriptive term for the sequential method for the identification of bacteria put forward by Cowan & Steel. See **sequential method 2**.

sterile. 1. Incapable of reproduction; refers to the reproductive defects of an organism.
2. Free from living (reproducible) micro-organisms; refers to the state of culture media and other material that has been submitted to treatment to render any micro-organisms contained in it incapable of reproduction. The nonsensical phrases 'partial sterility' or 'almost sterile' are sometimes used, and are the justification for this entry.

strain, *n.* A term used more in bact. than in the other microbiological disciplines, and is used most commonly in medical bact., where the alternative **isolate** is virtually unknown. A strain is a living culture and the term cannot be applied to an illustration or to an organism that has not been grown in culture. Strain can have at least two meanings.
1. Descendants of single colony (sometimes a single cell, in which case **clone** is appropriate) isolated from a **specimen 2**. More than one strain may be isolated from a specimen, and the number of strains is often determined by the number of different colony forms that develop when the specimen is plated on a particular medium. The same specimen may be plated on several different media, and those from each medium may be considered to be separate strains. Further and more detailed examination may show that all are alike and that the specimen yielded, in retrospect, only one strain.
2. Several separate isolations from one or more patients (animals) in an epidemic (epizootic); these *epidemic* (*epizootic*) *strains* are assumed to have derived from one source of infection and can be expected to have similar characters. There is a belief that mutational change occurs more often among epidemic strains than in strains isolated repeatedly from the same individual. It is doubtful if this is so, for there is evidence that a strain can change (mutate) during sojourn in one individual.

A strain may be labelled by a number or a name (often that of the patient or other source). Several specimens may be examined from the same source (e.g. patient) and the problem to be decided may be whether the cultures from the specimens are to be regarded as separate or the same strain. When all ascertainable characters are identical, the cultures are best regarded as being of one strain.

stratiform. In layers. Term used in bact. for the appearance of **gelatin liquefaction** that starts at the top of a gelatin stab culture and, as growth continues, extends downwards. Organisms that do this are fairly strict aerobes.

style. Editorial and printing **conventions**. The editors of journals usually issue instructions (or advice) to contributors in which the journal style is set out, and authors should study and follow these requirements. A most comprehensive guide to style and good writing was published in the *Journal of Hygiene* in 1940, by its then editor, G. H. F. Nuttall, and the Royal Society has produced *General Notes on the Preparation of Scientific Papers*. Editors of American journals dealing with biology have issued the *CBE Style Manual* (3rd edn.) which can be obtained from the American Institute of Biological Sciences, 3900 Wisconsin Avenue, N.W., Washington, D.C. 20016.

European Life Science Editors (ELSE) and the Ciba Foundation have sponsored a guide under the title *Writing Scientific Papers in English* by M. O'Connor & F. P. Woodford, published by Associated Scientific Publishers, Amsterdam.

sub-. Prefix attached to the name of a category or rank to indicate a lower category, e.g. family and subfamily. To indicate the subrank the codes of nomenclature specify particular endings; in this dictionary these are listed collectively under **suffix**, and individually under the name of the rank above, except for **subgenus** and **subspecies** which, like their parent categories, do not have distinctive endings.

subgeneric epithet. The word that, in bot., is the name of a subgenus. When the subgenus contains the type species of the correct name of the genus to which it belongs, the subgeneric epithet must repeat the generic name.[1] Other subgenera must have different epithets and within one genus all the subgeneric epithets must be different.[2] A subgeneric epithet that does not repeat the generic name may be a plural adjective which agrees in gender with the generic name.[3]

The subgeneric epithet is written and printed with an initial capital[3] and is enclosed in brackets between the generic name and the specific epithet.[4] Alternatively its rank may be indicated by the abbreviation subg.

A subgeneric epithet should not be formed by adding *Eu-*, *-oides*, or *-opsis* to the name of the genus to which it belongs,[3] but the suffixes may be added to the name of another genus it resembles.[5]

[1, BOT.22. 2, BOT.64. 3, BOT.21. 4, BOT.R21A. 5, BOT.R21B.]

subgeneric name. Name of a subgenus in bact.[1] and zoo.,[2] nomenclaturally co-ordinate with the generic name. The subgeneric name is printed with an initial capital[3] and is enclosed in brackets between the name of the genus and the specific epithet (name).[4]

When a genus is split into subgenera the one that contains the type species of the correct name of the genus must, as its subgeneric name, repeat the name of the genus. Thus if *Bacillus* is divided into subgenera, the subgenus that contains *Bacillus subtilis* must be named *Bacillus*, and the species name would be written *Bacillus* (*Bacillus*) *subtilis*.[5] The subgenus is known as the *nominate subgenus* because it bears (repeats) the name of the genus.

[1, BAC.10b. 2, ZOO.43. 3, BAC.10a; ZOO.28. 4, BAC.10c; ZOO.6. 5, BAC.39a, b; ZOO.44(a).]

subgenus, *abbr.* **subg.,**[1] **subgen.** **1.** The only category between genus and species recognized by BAC. and ZOO., but one of several in BOT. (see Table R. 2, p. 220).

2. Individual taxon of the rank subgenus.

[1, BOT.21Ex.; R22Ex.]

subjective synonyms. Different names that have been attached to what are believed to be similar organisms, or to different descriptions of the same organism. They are distinguished from **objective synonyms** which are based on the same type. See Table S. 3, p. 253, for the synonymy of adjectives qualifying the word **synonym**.

subordinate taxon (taxa). 1. Taxa of lower rank that are included in the taxon of higher rank.

2. Ranks whose names start with *sub-* and are followed by the name of another rank, are subordinate (of lower rank) but are **nomenclaturally co-ordinate.**[1] It follows that names of subordinate taxa are governed by the same rules and recommendations as the taxon of immediately higher rank.[2]

3. The subordinate taxon that contains the type of the higher taxon and is known as the **nominate subordinate taxon**; the name must repeat the name of the type.[3]

4. Mention of subordinate taxa within a newly created taxon does not constitute valid publication of the name of the new taxon.[4]

[1, ZOO.36. 2, BAC.10b. 3, BAC.39a; BOT.22; ZOO.37, 44, 47. 4, BOT.34(5).]

subordination of characters. A belief that certain characters define certain ranks, and other characters define other ranks; e.g. morphological characters define genera and biochemical ones define species.

This attractive theory is not supported by evidence from bact., where morphological shapes and tinctorial appearances are much more limited than in mycology and the macrobiologies.

subspecies, *abbr.* **subsp., ssp. 1.** In a hierarchical system the rank immediately below **species**. It is the lowest rank to which the rules of BAC.[1] and ZOO.[2] apply. An abbreviation for subspecies is used only in works on taxonomy.

2. Individual taxon of subspecies rank.

3. The scientific name of a subspecies must be written as a ternary combination[3] or trinomen;[4] the use of a binomen (generic name+subspecific epithet) is not permitted.[5]

[1, BAC.5d. 2, ZOO.1, 45(c). 3, BAC.13a. 4, ZOO.5. 5, BOT.24.]

subspecific epithet or **name.** The third element of the trinominal that forms the scientific (Latin) name of a subspecies. The same epithet may not be applied to more than one subspecies within a species or even within the same genus;[1,2] this rule is intended to avoid confusion should the subspecies ever be raised to species within the same genus.

BAC.[3] and BOT.[4] require that a term denoting the subspecific rank should be placed between the specific and subspecific epithets. A binary combination (omitting the specific epithet) is not allowed for the name of a subspecies, and a name of a subspecies published in that form should be altered without change in the citation of the author.

[1, BAC.13c. 2, BOT.R24B; BOT.64. 3, BAC.13a, 32a(2). 4, BOT.24.]

substituted culture. It was not unknown for a strain maintained by serial subculture at regular intervals to become contaminated and eventually replaced by another micro-organism. This has happened in every culture collection (and in many smaller private collections) and accounts for some of the anomalies of named cultures.

Since the introduction of freeze-drying techniques for the maintenance of cultures, substitution or replacement by a contaminant is much less likely to occur, and should not go undetected.

substitute name. a new name (epithet) published as an *avowed substitute* (*nomen novum*) for an older name must take as its type the type of the old name it replaces.[1]

[1, BOT.7.]

substrate. In connexion with the growth of micro-organisms it refers to the nutritional substances forming the growth medium; in enzymology it means the substance(s) attacked by an enzyme.

substratum. Term used by mycologists for the material on, or in, which a saprophyte is living. Preferred to **substrate** for which enzymologists have a prior claim to specialized usage.

suffix. Letter or syllable added to the stem of a word to complete or form a new word or name. In biological nomenclature different suffixes are used to indicate the rank of the taxon named; these suffixes are shown in Tables S. 1 and S. 2 below. The suffixes used in bact. and bot. are similar; those used in vir. and zoo. are different but have a basic similarity to each other.

Table S. 1. *Suffixes added to the stem of a name to indicate rank*

Rank	BAC.[1] *Bacteria*	BOT.[2] *Algae, Fungi* and *Plants*	VIR.,[3] VR[4] *Viruses*	ZOO.[5] *Protozoa*
Order	-ales	-ales	—	—
Suborder	-ineae	-ineae	—	—
Superfamily	—	—	—	-oidea
Family	-aceae	-aceae	-viridae[4]	-idae
Subfamily	-oideae	-oideae	-virinae[3]	-inae
Tribe	-eae	-eae	—	-ini
Subtribe	-inae	-inae	—	—
Genus	—	—	-virus[4]	—

[1, BAC.9T. 2, BOT.17, 18, 19. 3, VIR.16. 4, VR.15, 16. 5, ZOO.29, R29A.]

Table S. 2. *Suffixes indicative of higher ranks of Algae, Fungi, and Cormophyta*

Rank	*Algae*	*Fungi*	*Cormophyta*
Division	-phyta	-mycota	-phyta
Subdivision	-phytina	-mycotina	-phytina
Class	-phyceae	-mycetes	-opsida
Subclass	-phycidae	-mycetidae	-idae

[BOT.R16A.]

Note. Two of the suffixes, *-idae* and *-inae* indicate different ranks in bact./bot. and vir./zoo.

Other suffixes are added to personal and geographical names to form **generic names** (q.v.), and specific names (epithets) (Tables P. 1–P. 3, pp. 198, 199).

Some botanists seem to prefer the word *termination* to suffix.

summary. The name of a new taxon or a new combination applied to an old (-established) taxon should be written out in full in the summary of a paper, whether the summary appears at the beginning or end of the paper. The generic name may be abbreviated after its first mention and abbreviations are normally used for qualifying phrases such as sp. n., comb. nov., etc.

Even when the summary (or abstract) appears at the beginning of a paper the appearance of a new name or combination in the summary is not, for purposes of priority, its earliest publication.[1]

[1, BAC.33bN2.]

superfamily. 1. The highest category subject to ZOO.;[1] a category of the family-group above family. The name of a taxon of this rank is formed by adding the suffix *-oidea*[2] to the stem of the type genus. A category not used in bact. or bot.
2. Individual taxon of superfamily rank.

[1, ZOO.35. 2, ZOO.R29A.]

superfluous name. 1. A name applied to a taxon that included the **type** of a name (or epithet) which should have been adopted if the rules of nomenclature had been followed. Such a name is illegitimate and is to be rejected.[1]
2. A name superfluous on publication may become correct later if it derives from a **basionym** which is legitimate.[1] The epithet of an illegitimate name should not be adopted in a new combination applied to the same taxon.[2]

[1, BOT.63. 2, BOT.TR72A.]

superspecies. A term used by Mayr for a monophyletic group of allopatric species morphologically too different to be included in one species. The derivation is explained as follows: *super* (beyond) is the counterpart of *sub* (below or within); *supra* (above) the counterpart of *infra* (below). Mayr meant his superspecies to be the counterpart of the subspecies. Although the superspecies is highly controversial in zoo., the principle behind it would be usefully applied in microbiology to *Pasteurella multocida*, different strains of which infect different animals, and at one time were given names such as *P. aviseptica*, *P. boviseptica*, *P. suiseptica*, and so on.

suppress. Used in zoo. literature. Equivalent to reject in bact. and bot.

suppressed generic name. When it is necessary to suppress a well-known generic name and an author wishes to show that the name has been replaced by another, the suppressed name can be shown very simply by putting an equals sign (=) before the suppressed name and enclosing both sign and name within brackets, e.g. *Escherichia* (= *Bacterium*) *coli*. But this method of indicating a suppressed name must be used with caution and proof-reading must be vigilant; omission of the equals sign would indicate *Bacterium* as a subgenus of *Escherichia*. The usage of the equals sign does not please all zoologists and ZOO.[1] says that a synonym should never be cited between the generic and specific elements of a binomen.

[1, ZOO.R44A.]

suppression of a name. Rejection of a name, prohibition of its use in a certain position, or attached to a particular taxon. Application can be made to the Zoological Commission to suppress a name on either of two grounds:[1] (1) to allow the same name published at a later date (i.e. a junior homonym) to be applied or continue to be applied to another taxon; official sanction is needed because suppression overrides both the Law of Priority and the Law of Homonymy; (2) to allow a later name applied to the same taxon (i.e. a younger objective synonym) to be retained for use (see **suppression of unused senior synonym**); this suppression overrides the Law of Priority but not the Law of Homonymy.

Suppression of a name should not be made by an individual, but he should ask the Commission of his discipline[2] to suppress the name and state his reasons for making the request. BAC.[3] does not deal with these problems in quite the same way; the bact. Judicial Commission would (or could) act by the **rejection** of one name or names and the **conservation** of another.

[1, ZOO.79(a). 2, ZOO.23(a–b). 3, BAC.56a.]

suppression of unused senior synonym. When the correctness of a zoo. name in current use is questioned and the stability of nomenclature is threatened by the existence of an unused senior synonym for the same taxon, application may be made to the Zoological Commission for the suppression of the unused name.[1] A prima facie case is made when it can be shown that the senior synonym has not been used for the last 50 years, and that the name it would displace has been applied to the taxon as its presumed valid name by at least five different authors and in at least ten publications during the same 50 years. The Commission in considering the application will treat each citation on its merits, and listing of the name in synonymy, in an abstract journal, in a nomenclator or index of names will not normally be accepted as evidence of usage.[2]

The qualifying phrase ***nomen oblitum*** may be added to the author citation of a name suppressed by the Zoological Commission.[2]

[1, ZOO.23(a–b). 2, ZOO.79(b).]

S value. See **S, similarity coefficient**.

swarm. Some motile bacteria spread as a thin film of growth over the moist surface of a solid medium. This swarming type of growth is shown especially by *Proteus* and *Clostridium* spp.

symbiosis. A term to be treated with care in microbiology; it is often incorrectly used for **synergism**. Symbiosis means physical contact between two organisms, each of which is a symbiont. There are many examples of symbiosis between bacteria and insects, plants, and animals, and each of these associations is usually specific (as between *Rickettsia prowazekii* and *Pediculus humanus*, and *R. tsutsugamushi* and *Leptotrombidium akamushi*). Symbiosis between different bacteria (an example is parasitism of bacteria by *Bdellovibrio* species) seems to be much less frequent.

symbol. 1. Typographical or written character to which a meaning has been assigned. Many of these symbols (or signs) are well-known and readily understood, as plus (+) and minus (−) signs, but others are applied with special meanings, and since the same character may be used to indicate different things, the particular usage should, in all cases, be stated. *Symbols are international – though not universal* (Ellis, 1971) and should replace abbreviations which may vary from one language to another (e.g. DNA, ADN). Symbols for SI (Système International d'Unités) units consist of seven basic units, with prefixes for multiples or fractions. Some of the basic units differ from units of similar name (e.g. litre) in the metric system as used before 1960. Some old units such as the micron have disappeared and been replaced by new units (micrometre, μm). Unlike abbreviations, symbols for units stand alone and, except at the end of a sentence, are not followed by a full stop; they do not take a plural form.

2. * † ‡ § || ¶, see **note indicators**.

3. In tables showing characters of organisms different symbols are used; there are also may different systems, some qualitative, others quantitative; some of the best known, or more useful, are described under **symbolic notation**.

symbolic notation. System of recording characters by symbols, usually with the object of recording the characteristics of organisms in **tables**. There are many such systems, but the only symbols with universal meanings are plus (+) = character present and minus (−) = character not present. Since about 1950 there has been an attempt to standardize the use and meanings of symbols for characters of the Enterobacteriaceae, and, since 1965, to give them a quantitative significance. A summary of recent usage follows.

(*a*) Kauffmann (1954): + = positive; ± = weakly positive; (+) = weakly and slowly positive; X = late and irregularly positive or negative; − = negative; −+ = gelatin liquefied slowly; (for sugar reactions) ++ = acid and gas formed; +− = acid without gas; −− = neither acid nor gas produced. With this system there was some difficulty in distinguishing between (+) and X reactions.

(*b*) Edwards & Ewing (1955): + = promptly positive; X = delayed positive; − = negative; v = some cultures positive, some cultures negative.

(*c*) ICSB Enterobacteriaceae Subcommittee Report (1958): + = positive in 1–2 days; X = late and irregularly positive; − = negative, or no reaction; d = different biochemical types; (+) gelatin liquefied slowly, urea hydrolysed slowly, etc. The X and d symbols continued to cause difficulty, and an attempt to clarify them is shown next.

(*d*) Cowan (1956): + = positive 1–2 days; (+) = delayed positive (3–4 days); X = late and irregularly positive or negative (mutative); − = negative; d = different strains, each consistent in itself, gave different reactions.

(*e*) Cowan & Steel (1965): + = 80–100% of strains positive; d = 21–79% positive; − = 0–20% positive (i.e. 80–100% negative); ( ) delayed reaction in biochemical tests.

(*f*) Sedlák (1959) introduced a symbol / for not tested, but later used this slant line in the form +/− to indicate that the reaction (or character) might be either positive or negative. His usage in 1969 was + = > 90% positive; +/− = > 50% (i.e. 50–89%) positive; −/+ = < 50% (i.e. 11–49%) positive; − = 90% or more negative (10% or less positive); (+) = delayed or weak reaction.

(*g*) Lautrop (1969, personal communication); + = > 95% positive; d + = 76–94% positive; d = 25–75% positive; d − = 5–24% positive; − = less than 5% positive.

(*h*) Scottish authors often used other symbols; the following are from Gillies & Dodds (1965) for sugar reactions; 1 = acid without gas; + = acid and gas; − = acid not produced; ( ) = delayed reaction.

sympatric. Where two forms or species occur together, i.e. have the same area of distribution.

symplesiomorphy, -ic. State in which primitive or **ancestral characters** are shared by different species.

synapomorphy, -ic. The state when derived characters are shared by species.

synchronic. Occurring at the same time; contemporaneous. The word may be used in discussing so-called evolutionary classifications but we should remember the caution issued by Breed, Conn & Baker (1918), namely that the organisms we study are descendants and not representatives of primordial micro-organisms, and that one present-day bacterium must not be considered as the ancestor of another.

syneresis (water). Condensation water.

synergism. **1.** A state in which two (or more) organisms establish an organization of mutual benefit (cf. **commensal**), often described as **symbiosis**. A simple demonstration *in vitro* of one-sided synergism is shown when *Haemophilus influenzae* is grown on blood agar (the blood of which provides one essential growth factor, X or haemin) with a culture of *Staphylococcus aureus* (which provides V or co-enzyme I); the association is not specific, and any other organism that provides co-enzyme I would do equally well.

2. Antibacterial substances can act synergistically, although the individual may not have any apparent inhibitory action. A good example is the combined activity of colistin and sulphonamide on most strains of *Proteus mirabilis* and *P. vulgaris* (Möller & Holmgren).

synergy, -ic. Two or more activities applied to the same object or with the same purpose, which have greater force than the sum of the independent activities. In biology synergic action may be either advantageous or disadvantageous to the organism, see **synergism**.

synisonym. An **objective synonym**. Names based on the same name-bearing epithet (basionym).

synonym, *abbr.* **syn.** A word with the same meaning as another word; often applied to words with slightly different shades of meaning and a special use. In biology the words synonym and synonymy are applied to different names given to the same organism (animal, plant). Nomenclaturists recognize two main kinds of synonym, and in the various disciplines many adjectives are applied to them (see Table S. 1, p. 248): *objective* (nomenclatural) synonyms are different names attached to the same organism, and means that they are based on the same type; *subjective* (taxonomic) synonyms are different names attached to what are believed to be similar organisms, but this is a matter of opinion as the names are based on different types. At the genus level objective synonyms are names based on the same type species; at the species level objective synonyms are based on the same type strain; in bact. this is an unlikely event and most bact. synonyms will be subjective.

A point worth making is that only names can be synonyms or synonymous; taxa cannot be synonyms. The following is an example of incorrect usage: '*Streptococcus mutans* is a clearly defined, distinct species and not synonymous with *S. bovis* or *S. sangius*.'

In bact.[1] and zoo.[2] the adjective *junior* is applied to all later published synonyms, and *senior* to the one first published, usually the correct (valid) name. Blackwelder thinks that the many terms applied to synonyms can be confusing and misleading and he prefers **correct name** and its **rejected synonyms**.
[1, BAC.24a. 2, ZOO.Gl.]

Table S. 3. *Synonymy of, and adjectives qualifying, synonyms*

Synonyms based on the same type	Synonyms based on different types	Other adjectives applied to synonyms
absolute	facultative	junior
homotypic	heterotypic	older
isonym	metonym	putative
objective	**subjective**	rejected
nomenclatural	**taxonomic**	senior
obligate	—	younger
synisonym	—	—
typonym	—	—
exact*	qualified*	—

* Proposed, but not accepted, at XIIth Int. Bot. Congress, 1975.

synonymy. 1. A list of the different names that have been applied to a particular taxon.
2. The relations between the different names associated with the taxon.[1]
[1, ZOO.Gl.]

synthetic. Loosely used for artificial, something simulating (or a substitute for) a natural product. In English, often an illiterate usage for defined, as in describing media made of constituents whose chemical nature is known and defined. In American, an acceptable adjective for **defined** in describing these media. Strictly, synthetic media implies media made up (i.e. synthesized) from chemical elements, and this is never the case in microbiology.

syntype. When a new taxon is based on more than one specimen (strain, culture) and the author of the name does not designate one of them as **holotype**, each of the specimens listed in the original paper becomes a syntype.[1] This term is preferred to **cotype**.[2]
[1, BOT.7. 2, ZOO.R73E.]

system. A system of classification, according to Heywood, is the arrangement of groups relative to one another, which agrees with the definition of Meeuse, the method or pattern of classification. The system is important in that it may determine how the information is filed, whether alphabetically by name or numerically by a code number. With an alphabetical arrangement the system will be upset by a change (perhaps due to priority) in the name of a taxon.

systematics. A plural adjective used as a noun to embrace the many approaches to the study of orgnisms with the ultimate object of characterizing them, sorting them, and arranging them in an orderly manner. Systematics includes everything that contributes to the organization of organisms, both intrinsic and extrinsic, and man's attempts to arrange them neatly and to label them so that the knowledge available can be transmitted to others succinctly. The advantage of the hierarchical system of arrangement is that the characters of the higher taxa are (or should be) shared by all the subordinate taxa; if this is not so there is no justification for having ranks higher than the practical ones of genus and species. But if the ideal hierarchical system has any logic behind it, it should lead to logical approaches to classification and the really practical aspects of systematics, namely the identification of the unknown organism. In the task of identification (which Mayr thought was the lowest task a systematist had to do) the bacteriologist at least can be systematic and approach the unmasking of the unknown by a series of steps, whether he follows a dichotomous key or a series of diagnostic tables. Thus, among the medical bacteria, about ten well-chosen tests reveal sufficient characters of the organisms to identify most of them to the level of genus (Cowan & Steel, 1974). Classification, too, can be made systematic and the last twenty years has seen many advances in that direction; these advances are made by the newer arrivals in the bact. field; by the biochemists, the geneticists and the molecular biologists, who have all contributed detailed information upon which classification can be built and explained. They have been helped by mathematically inclined biologists and their computers, which have taken the drudgery out of the analyses and sorting procedures.

Of the three parts of taxonomy as defined here, only the communications system lags behind; we are still burdened with archaic nomenclature and with all the attendant complexities of the nomenclatural codes; people may not like those codes and the type of names they force us to use, but they dislike coding schemes (which often have a numerical basis) even more.

systematist. One who practises systematics. Mayr, a zoo. systematist, defined his field as (1) identification, (2) classification, (3) the study of species formation and evolution. Thorne thought that the classification element should indicate phylogenetic relationships as well as the evidence (fossil series or 'educated inferences') allowed.

Microbial systematists are seldom worried by species formation and evolutionary trends, though some exercise their minds in that direction. To many, systematics is a field they enter reluctantly and only because their work leads them there, but they seldom regard themselves as systematists. Many start as identifiers of unknown organisms and are then gradually drawn into the systematist's net. Mayr, who takes a poor view of the identifier's work, regarding it as a lowly task, overlooks the fact that, in microbiology at least, there is little money available for microbial systematists to indulge exclusively in theoretical studies and most systematic work is a side-line of a very practical investigation.

systematology. *Syn.* of **systematics**. Not in *SOED*, but occurs several times in William Bulloch's *The History of Bacteriology* published in 1938 by Oxford University Press.

T

table of characters. Tables are useful for the characterization of both large and small groups of bacteria; and particularly for showing differences between members of small groups, e.g. for distinguishing between the species of a genus or a small number of related genera as in the Enterobacteriaceae.

Tables of the characters of taxa may be constructed in two ways: (*a*) arrange taxa in columns, and show characters (or tests) on lines; (*b*) arrange taxa on lines, and show characters (or tests) in columns (see Table T. 1).

Table T. 1. *Arrangements of tables of characters*

	(*a*) Species					(*b*) Test				
	A	B	C	D		1	2	3	4	5
Test 1	+	+	+	−	Species A	+	+	+	−	−
Test 2	+	+	−	−	Species B	+	+	−	−	+
Test 3	+	−	−	−	Species C	+	−	−	−	+
Test 4	−	−	−	+	Species D	−	−	−	+	+
Test 5	−	+	+	+						

Usually the number of characters (or tests) that appears in a table exceeds the number of taxa, and then construction (*a*) is to be preferred. This arrangement is used by workers on the enteric bacteria, and appears in Edwards & Ewing's *Identification of Enterobacteriaceae*, in publications from C.D.C. Atlanta (US Public Health Service), and in Cowan & Steel (1965, 1974).

The content of the tables should consist of symbols (+, −), letters (A for acid, etc.), figures (pH values, percentages) and in all cases should be explained in footnotes to the table (see **symbolic notation**).

Generally, descriptive matter should be avoided in the body of tables; characters should be stated in the left-hand column and their presence or absence in the taxon indicated by symbols, as in Table T. 2.

Table T. 2

	Taxon		
	1	2	3
Shape			
Spherical	+	+	+
Straight rod	−	−	−
Curved rod	−	−	−
Arrangement			
Singly	+	−	+
Pairs	+	+	+
Chains	−	+	−
Clusters	−	−	+
Gram reaction	+	+	+

tactic(al) response is the movement towards (positive) or away from (negative) a physical or chemical stimulus. Higher organisms respond to a stimulus directly by moving to or from the stimulus, but bacteria appear to be unable to do this, and they show their response by reversing the direction of their movements (**phobotaxis**).

tag. In bact. collections, a label other than the accession number associated with a culture. The tag may sometimes be better known than the number, e.g. Oxford strain of *Staphylococcus aureus* rather than NCTC 6571. However, the referral to a culture by a tag, rather than collection number, may be a dangerous practice: not all BCG strains of *Mycobacterium tuberculosis* are derived from the same strain. LRH.

take up a name, *v.* The act of validly publishing a name first used before the starting date of nomenclature of the discipline; by this action the name, if in the appropriate form, can be made legitimate (bact., bot.) or available (zoo.). Such a name dates from the date of publication by the author who takes it up, and the author citation should read F (name of original, pre-starting date author) ex E (name of publishing author).[1]
[1, BOT.R46E.]

tautonym, tautonymy. 1. A tautonym is the exact repetition of the generic name in the specific epithet (name in zoo.), e.g. *Rattus rattus, Fusiformis fusiformis.* Tautonyms are allowed in zoo.[1] but not in bot.;[2] they are not mentioned in BAC. and so are not forbidden.

2. *A virtual tautonym* is created when the name of a species is *virtually the same as the generic name*;[3] a ZOO. example is *Scomber scombrus*, and a bact. example is *Arizona arizonae* (see also **pleonasm**).

3. *Absolute tautonym.* If a new zoo. genus contains among its species one having the generic name as its specific or subspecific name, either as a valid name or a synonym, that species automatically becomes the type species of the genus.[4]

4. *Linnaean tautonym.* In a zoo. genus established before 1931, a species name in the synonym dating from before 1758 will become the name of the type species if that name is one word identical with the new generic name.[4]

5. *Automatic tautonymy.* The name of a taxon above the rank of family that contains the type of the higher rank must have a name based on the stem of the same generic name as the higher rank, modified only by the ending appropriate to the rank.[5] Example: Pseudomonadineae is the name of the suborder that contains the type (the genus *Pseudomonas*) of the order Pseudomonadales.[6]
[1, ZOO.18(b). 2, BOT.23. 3, ZOO.R69B(2). 4, ZOO.68(d). 5, BOT. (1975) 16. 6, BAC.Ap9T.]

taximetrics, taxometrics. Taximetrics is etymologically more correct than taxometrics and taxonometrics. Silvestri and his colleagues derive their taxometrics from the word taxon (a taxonomic group) rather than from taxis (order or arrangement). Taximetrics is a word used to describe numerical taxonomy, and taximetrist could be used for numerical taxonomist. Rogers describes taximetrics as a new word for an old concept by grouping specimens by the number of characters they share, but this sense is not quite that of Sneath & Sokal, which is the use of statistical methods in calculating similarities.

Mayr (1966, *Syst. Zoo.* **15**, 88) thinks that a disadvantage of taximetrics may be that, in the words of a mischievous friend, it suggests 'the science of taximeters'.

taxis. From the Greek ταξις, arrangement. In biology used in compound words with quite different connotations.

1. As the stem of words such as taxonomy, taxinomy, taximetrics.

2. As the second part of words to indicate the attraction to or the repulsion of an organism from some physical or chemical stimulus, e.g. phototaxis, chemotaxis, thermotaxis. These tactic responses are known as positive (towards) or negative (repelled).

taxon, ***pl.*** **-a.** A taxonomic group of any rank;[1] to be distinguished from **category** which is a group of organisms of a particular rank or level. A genus is a taxon and so is a species, and because it is of a lower rank the species of a genus is a *subordinate taxon* of that genus. The term *nominal taxon* means a named taxon. Each taxon is a group of organisms sharing several characteristics, and the more characters the members of the group have in common the lower will be the rank of the taxon; however, there are not any standards or agreed lists of characters as to what makes a species, a genus, or any other rank.

In zoo. a taxon has been defined as a formal unit at any level in a hierarchical system (Simpson) sufficiently distinct to be worthy of bearing a name (Mayr).

Each taxon above species with a given circumscription, position, and rank can have only one **correct name**[2] (**valid** name in zoo.[3]).

[1, BAC.GC7; BOT.1. 2, BAC.23a; BOT.11. 3, ZOO.23.]

Taxon. Official bulletin of the International Association for Plant Taxonomy, edited and published by the International Bureau for Plant Taxonomy and Nomenclature, Utrecht, The Netherlands.

taxonomic category. A subdivision in a hierarchical system. Each category is a subjective concept, whether it is a species, genus, or family (for only the individual is real), and each has a place (rank) in the hierarchy of bacteria, fungi, protozoa, animals, or plants.

taxonomic character. A character of a taxon. Each character contributes to the qualities of a taxon; a character exclusive to the group distinguishes the taxon from other taxa and at different times will be labelled a diagnostic, differential, distinguishing, pathognomonic, peculiar, specific (without implying an attachment to the category species), or taxonomic character. Of these adjectives, distinguishing is the most precise. Each character is a taxonomic character as it is one of the sum that makes up the description of the whole taxon.

taxonomic distance. Term used by numerical taxonomists to express the relations between specimens in terms of a multidimensional space, each character having one dimension. This concept cannot be shown graphically but in a simple form can be represented by a model, as was shown by Lysenko & Sneath. Identical OTUs occupy the same space and the taxonomic distance is zero.

taxonomic doubt. A bact. name published with a question mark or other suggestion that the author is uncertain or doubtful about it (even although he accepts the name) is not to be regarded as validly published.[1] In other words, an author who is uncertain about his facts, or the wisdom of his ideas, or the correctness of a name should leave well alone, and not publish muddled thoughts. There is no place in taxonomy for unnecessary doubt.

Botanists, on the other hand, regard a question mark as an indication that the taxonomic position only is questioned, and that the name is accepted by the author. Such a name is, in bot., taken as validly published.[2]

[1, BAC.28b(1). 2, BOT.34N1.]

taxonomic group. A group of organisms (plants, animals) of any rank.[1] BAC.[2] enlarges on this in *a named group in a formal taxonomy*, and adds that *it may or may not correspond to a category.*

[1, BOT.1. 2, BAC.GC7.]

taxonomic position. 1. The place of a taxon in a **hierarchy**, particularly the level of the category or rank.

2. The taxon to which an organism is assigned after it has been identified. In the case of a new taxon, indicates the taxa it most closely resembles, and the taxon to which it is subordinate.

taxonomic synonym. Synonyms are different words (names) for the same genus, species, and so on; taxonomic synonyms are different names based on different types but which are considered to be of the same taxon. (The types are: species for a genus; strain or specimen for a species.) Since these synonyms are not based on the same type, taxonomic synonymy depends on a personal judgement and is subjective; they are sometimes referred to as **subjective synonyms**. In bot. also known as heterotypic synonyms (see Table S. 3, p. 253).

Indicated by an equals sign (=) in lists of synonyms.

taxonomist. A person who practises or studies taxonomy. As much of taxonomy is highly subjective, taxonomists are individualistic to a degree, and as a group they can be described by the aphorism as 'all out of step but our Jock'. Taxonomy is as much an art as a science and its interpreters are artists. Most are argumentative; many are aggressive; some are good and are able to make a useful contribution to science. Taxonomists are jealous of their preserve and make things difficult for the newcomer by the use of a complicated (and often unnecessary) terminology. Few taxonomists will agree on a definition of taxonomy itself, and each is a merry sharp-shooter of his colleagues.

taxonomy means all things to all men but never the same thing to all men. *Taxonomic questions cannot be decided by the mechanical application of a set of rules because they are matters of judgement* (MRC Memo. no. 23, 4th edn, 1977); here taxonomic refers to classification. In the title *The identification, taxonomy and classification of*. . . taxonomy means nomenclature. To others, taxonomy is equivalent to *systematics* as defined in this dictionary, and there are many other fine differences in meaning. Disagreement on what taxonomy is or does is not confined to bact., it exists also in bot. and zoo., so that it cannot be considered a science (though it has scientific elements and aspirations) but rather as a scientific art intended to bring order to what otherwise would be untidy or even disorderly. Migula thought that the task of the taxonomist was to bring order from chaos, and he probably summarizes best what each taxonomist thinks he does. Nevertheless, taxonomy is the happy hunting ground of the semasiologist (semasiology = the meaning of words and their sense-development, *SOED*); hence the need for this dictionary.

Simpson defined taxonony as the *theoretical study of classification, including its bases, principles, procedures, and rules.* He went on to say that the subjects of taxonomy were classifications, and the subjects of classifications were organisms. Davis & Heywood give a similar definition but attribute it both to de Candolle, 1813, and to Simpson. Cowan (1965) dealing with the principles of taxonomy, regarded it as almost synonymous with systematics, and divided it into three parts: (1) *classification*, the orderly arrangement of units into groups; (2) *nomenclature*, the labelling of the units defined in (1); and (3) *identification* of unknowns with the units defined and labelled in (1) and (2), which collectively formed *the trinity that is taxonomy*.

SOED gives the derivation of the word taxonomy as adapted from the French taxonomie, itself derived from the Greek *taxis* = arrangement, and *nomia* = distribution. Mayr (1966, *Syst. Zoo.* **15**, 88) gives a different derivation, *taxis*+*nomos* (arrangement+law) but says that if it was from a Greek word it should be taxinomy; he excused the spelling taxonomy because it was introduced by de Candolle in 1813 *as a French word*. But to a mere bacteriologist it seems likely that de Candolle, to whom the word is attributed, would not be averse to **hybrid names**, and that he might well have formed *taxonomie* from the Greek *taxis* and the Latin *nomen*, the hybrid meaning the arrangement of names, or arrangement and naming.

taxospecies. A term applied by Ravin (*American Naturalist*, 1963, **97**, 307–18) to a group of organisms which have a high coefficient of similarity.

taxotron. The name given by Véron (*Bull. Inst. Pasteur*, 1969, **67**, 2739–66) to the punched cards and facilities for their sorting (either mechanically or by hand) that constitute *an easily utilizable 'machine for taxonomy making'*. One of the simple uses of the system is the superposition of punched cards to detect similarities and differences among the characters of strains. An alternative name suggested was *bactériotron*.

technique (US **technic**). **1.** Special skill applied to doing something.

2. Description of a **method**, often with details of the apparatus and reagents used.

temperate phage. Bacteriophage that reproduces by passing through a prophage state in a lysogenic bacterial cell.

term. In technical writing *term* is used for words, phrases, and concepts peculiar to the discipline. In bact. and bot. term designates the rank of a taxonomic category (species, genus, and so on) but not the name (epithet) given to a particular taxon of given category.[1]

[1, BAC.GC7; *Reg. Veget.* 1968, 56.]

terminal. 1. Ends, or tips, of identification keys.

2. Computer hardware; an elaborate electric typewriter directly connected to a telephone, which enables a user to submit (and receive) data to (from) a computer. By so-called time-sharing systems, computers are now easily accessible and only a terminal is needed to carry out computer work. LRH.

termination. End, ending (of a name). Preferred by some botanists to the alternatives *ending* or *suffix* (1976, *Taxon*, 25, 170).

terminology. The correct use of terms; the use of special terms in an art or science. It differs from nomenclature, which is the choice and application of names to particular entities or concepts, and in biology follows rules (articles) in the various codes of nomenclature.

ternary combination. Combination of the binominal of the species name, a term denoting the rank, and the subspecific epithet, e.g. *Corynebacterium diphtheriae* var. *gravis*. (Note that after 1 January 1976 the use of the rank-name variety (abbr. var.) is not allowed in a *new name*.[1] but presumably the older usage shown above remains acceptable.)
[1, BAC.5c; 13a.]

tessellation. The verb tessellate means the making of a mosaic or chequer-board, and tessellation has been suggested as the name for a diagnostic table or **table of characters** (q.v.).

test, *v.* and *n.* In microbiology the act and result of applying some procedure, chemical or physical, to a culture of a micro-organism, to determine the effect of the one on the other, thereby ascertaining a character(istic) of the organism. Examples of (1) chemical tests are (*a*) the suspension of oxalic acid papers above a culture to detect indole production; (*b*) addition of indicator to a medium containing carbohydrate to detect acid production by the observance of a colour change; (2) a physical test, exposing a culture to 60 °C for 30 min followed by subculture to detect survival.

Not all tests are independent, and the number of tests carried out does not necessarily indicate the number of characters determined. (See **linked characters**.)

test kits. In bact., commercially made sets of dispensed media for various tests which require the user to add simply a suspension of the test organism. Of considerable use in routine identification, they have also been used in classification studies. It is hoped that by large-scale industrial manufacture, greater standardization in media may be achieved. LRH.

thermobiology. The study of life at high temperatures and the effect of heat on biological processes and living organisms.

thermophil(e), *n.*, **-ic,** *adj.* Terms applied to organisms capable of growing at temperatures much greater than that of the animal body, and usually not growing at 40 °C or below. A distinction may be made between obligate and facultative thermophils; strict or obligate thermophils grow best at 65–70 °C, but not below 40–42 °C; facultative thermophils have maxima about 50–65 °C, but are capable of growth at much lower temperatures (about 25 °C).

See also **thermotolerant**.

thermotolerant organisms are those able to survive and multiply at temperatures between 45 and 50 °C, and are also able to grow at **room temperature** (q.v.); as this definition is of US origin, it probably means that growth will occur down to 25 °C.

title of paper. The title of a paper is important in that it will either attract or repel a reader; a long title looks uninteresting and O'Connor & Woodford recommend that it should be limited to not more than 12 words and 100 characters and spaces. Abbreviations (except of qualifying phrases in author citations) should be avoided as they may mislead.

The title is the basis of indexing the subject matter; it should be brief but informative. When a paper deals with the rearrangement of taxa, name the main group but do not list the lesser groups. If it describes and names a new taxon include the scientific name in the title but if, as often happens, there is taxonomic doubt do not hide it. A title such as *Some strains in search of a genus* is brief, informative, and invites further reading, whereas '*Ordnung in das Chaos* Migula' does not, and is quite uninformative.

T_m is the abbreviation used for the melting temperature of double-stranded nucleic acids, indicated by an increase in absorbance at 260 nm when dilute solutions of, say, DNA are heated. One of the ways by which the base composition of DNA can be estimated.

token word. Term used in BOT. for a word (not intended as a name) used by an author to indicate a taxon.[1]
[1, BOT.20(1).]

topology. Term used by Colwell for the connecting links between taxa; as examples she gives similarity in DNA base ratios between *Pseudomonas* and certain *Vibrio* spp.

topotype. Material collected from the same place as the type specimen is said to be topotypical. It may be collected by the original author or by a later worker.

transcribe, transcript, transcription, *v.* and *nn.* **1.** The verb has two meanings, (i) to make a copy from an original, and (ii) to write in other characters (e.g. from shorthand to longhand), as syn. of transliterate.

Transcription is the action of making the copy, which is a transcript, or transcription. Transcript may also be used for a photocopy (or printed reproduction) of a written document.

2. Transcription is used by geneticists as a name for a process of RNA synthesis.

transduction. The carriage of genetic material from one bacterial cell to another by a bacteriophage; the phage carries part of the genome of the infected bacterium and the recipient (or transduced) cell shows characters of both bacterial parents. The phenomenon has been studied mainly among the intestinal bacteria, and the ease of transduction is taken as an indicator of relatedness.

transfer, *v.* and *n.* **1.** Used in microbiology for the act and result of making a subculture from one medium to another.

2. In taxonomy the transfer of micro-organisms within a hierarchical system may or may not involve a change of name, and is best discussed by examples.

(*a*) Moving a strain from one species to another in the same genus (as a result of segregation or of misidentification) is followed by a change in the specific epithet of the strain.

(*b*) Moving a taxon horizontally on the hierarchical scale (e.g. moving a species from one genus to another); the epithet of the species must not be changed unless it becomes a later synonym or homonym in its new position. For example on transfer of *Micrococcus aureus* to *Staphylococcus* the specific epithet remains the same, but if *Corynebacterium pyogenes* Glage 1903 were transferred to *Streptococcus*, a new specific epithet would be needed because *Streptococcus pyogenes* Rosenbach 1884 antedated the epithet pyogenes of *C. pyogenes*.

(*c*) Vertical transfer: when a subspecies is raised to a species it retains its epithet unless it becomes a later synonym or homonym (this should not happen as the same epithet is not allowed in more than one species or subspecies of a genus[1]). When a genus is raised to a family (or a family to an order) the stem of the generic name is retained and the appropriate **suffix** is added. When the name of a genus, or epithet of a lower rank does not change (i.e. is a **basionym**) in the transfer, the author citation remains that of the original author; it is shown in parentheses, and is followed by the name of the author who made the change.[2] [1, BAC.13c; BOT.24N; R24B; 64. 2, BOT.49; BAC.34b, 41a, b.]

transform, transformation, *v.* and *n.* **1.** The act and result of change produced by the addition of DNA (extracted from the cells of the donor) to recipient cells in which a small amount of genetic material is transferred to the recipient. The first – and classical – example was the experiment made by F. Griffith in which the specificity of killed pneumococci was transferred to living non-specific (R) forms. *In vitro* the ability (or competence) of bacteria to be transformed by DNA from the donor is not constant and is a state of short duration at the end of the logarithmic phase of growth.

2. In statistics, and hence some methods of numerical classification and identification, the act and result of changing original or raw data into data that are more easily manipulated in some way. A transformation may be useful to demonstrate a simple mathematical relation. For example, it is frequently desirable to represent the statistical relation between two variables as a straight line, yet plotting graphically the raw data will yield a curve; dependent on the shape of the curve, one or other (or sometimes both) sets of raw data may be transformed (e.g. by taking logarithms, or reciprocals, etc.) so that graphical plots of the transformed data will yield a straight line. LRH.

translate, translation, *v.* and *n.* Each of these words has several meanings, and may be used in many different ways.

1. The act and result of changing words or sentences from one language to another.

2. The movement of bodies (e.g. bishops, or relics of a saint) from one place to another, which, by extension to biology, applies to the movement of a taxon within a hierarchy either horizontally (e.g. from one genus to another) or vertically, a subgenus raised to a genus, or a genus to a family.

3. Geneticists use the words in a very specialized sense, viz for a 'read out' of the genetic code.

transliterate, *v.* Change the word (or letter) from one set of characters to those of another language, e.g. change from Greek to Latin characters. ZOO.[1] has a useful table which shows how a Greek word or letter is latinized; perhaps for this reason ZOO. does not regard a fault in transliteration as an 'inadvertent error' and does not permit its correction.

[1, ZOO.Ap.B.]

translucent, *adj.* Describes an object that transmits light, but not the outlines of things on the other side of the object.

tree. Dendrogram. Some authors prefer the word tree as dendrogram appears somewhat pretentious to them: does a tree cease to be a tree when chopped down and laid on its side? Dendrogram is associated with phenetic classification and an author may perfer tree when not dealing with phenetic classification. Sneath (1974, *Symp. Soc. gen. Microbiol.*, **24**, 1) refers to many kinds of trees, mostly used in connexion with evolution. LRH.

tribe (and **subtribe**). **1.** Ranks between family and genus; subtribe is not used in zoo. The names are formed in BAC.[1] and BOT.[2] by adding suffixes to the stem of the names of an included genus (type genus in bact.): *-eae* (for tribe) and *-inae* (for subtribe). In bot. the tribe (subtribe) that includes the type genus of the family name is to be based on that genus, and the tribal (subtribal) suffix added to the stem of the name of that genus; it is not followed by an author citation. In ZOO.[3] *-ini* is added to the stem of the type genus. See **suffix**.
2. Individual taxon of the rank of tribes (subtribe).
[1, BAC.9T. 2, BOT.19. 3, ZOO.R29A.]

trinomen. The three words (generic name, species name and subspecies name) that form the scientific names of a subspecies. Trinomen is used more in ZOO.[1] than in bact. or bot. The abbreviation subsp. is inserted between the specific epithet and the subspecific epithet in BAC.[2] Some bacteriologists prefer the word variety to subspecies, but this is not allowed for new names published after 1 January 1976.[3]
[1, ZOO.5. 2, BAC.13a. 3, BAC.5c.]

trivial feature. Term used in bot. for small (microscopic) morphological characters such as the sculpturing of spores or pollen grains, features that may be useful in the description or differentiation of both genera and species.

trivial name. **1.** Formerly used in zoo. for the second element of the binomial which forms the name of a species; replaced by **specific name**.
2. Incorrectly used for vernacular or common name.

-troph. Suffix attached to a term for a substance or substances used for nutrition, e.g. organotroph (using organic matter), lithotroph (using inorganic salts).

two-state character. A character that can be described or recorded unequivocally in positive and negative terms. Some characters regarded as two-state are in fact qualitative characters (such as the production of a particular end-product) which could be expressed quantitatively as a multi-state character.

type, *n.* and *v.* **1.** *n.* Frequently used as a shortened form of **nomenclatural type**, the basis of **type concept** (method). See also **holotype**, **isotype**, **lectotype**, **neotype**, **syntype**.

The type of the name of a taxon is a taxon of lower rank. But it makes dull reading and is tedious in speech to be pedantic and frequently repeat phrases such as 'the type of the name of a genus is a type species (the type species), and the type of the name of a species is a particular specimen (culture) designated as the type'. So we extend the meaning and apply it to the taxon, knowing at the back of our minds that we should not do this, excusing ourselves with the hope that we shall be better understood by saying 'the type of a genus is a species, and the type of a species is the type strain'. We do this so often that many forget that the type business is a nomenclatural device; those who know better think of types as types of taxa; and this thinking not only extends to the codes but seems to be accepted. Taxa of different rank in a hierarchical system have types; the type of a family, subfamily or tribe is a genus; the type of a genus or subgenus is a species, and that of a bact. species or subspecies is a strain[1] (isolate or culture); this, the ultimate type, is the only non-abstract form of type. In bot. and zoo. the type of a species is a specimen and cannot be a living culture. When a bacterium cannot be grown and maintained in culture, the bacterial type may be a specimen, illustration, or even a description. Types of fungi are dried specimens. The types are the name-bearers. Names of taxa above family do not have types in zoo.; in BAC.[1] and bot. some names are automatically typified by being based on generic names.[2]

2. A subdivision of a taxon, usually as a shortened form of **biotype**, **serotype**, or **phage type**.

3. *v.* Used in identification work; to find the type (serotype) of an organism, less often the biochemical type. BAC.[3] recommends serovar and biovar as preferred names, but it would be difficult to use these words as verbs.

4. *n.* The type used by printers; its size, shape, and heaviness (boldness); more correctly, **typeface**.

[1, BAC.15T. 2, BOT.(1975)10; 16. 3, BAC.Ap10BT.]

type and taxon relationship. This relationship (or lack of it) confuses many who use the terms type and taxon. In its explanation of the type concept ZOO.[1] describes the *type* as a reference standard that fixed the application of a scientific name; it becomes the nucleus of a taxon and since it is a specimen or description (in bot. and zoo.) it cannot change. In this it is contrasted with the circumscription of a taxon which may change, particularly as knowledge of the taxon increased. Each taxon theoretically has its type in a taxon of lower rank and the ultimate type is the type specimen. In bact., where the type of a species is preferably a living culture, there is greater scope for this ultimate type to show change and the taxon must be defined more broadly; for a bact. view see **representative strains**.

Although it should not be necessary, taxonomists often need reminding that a type is the type of a name.

[1, ZOO.61.]

type concept (type method). The extension of the meaning of type from the type of a name to the type of a taxon is the type concept as understood by bacteriologists and zoologists; botanists prefer the term type method and Davis & Heywood reserve type concept for **typology**. The nomenclatural device of **typification** attaches a name permanently to a specimen (strain) and the name is retained (perhaps with its ending changed) when the type is moved to another taxon of equivalent or higher rank. As an example, by Opinion A the Marburg strain of *Bacillus subtilis* was designated the type strain of the species,[1] and the species the type of the genus *Bacillus*. If any change in the taxonomic status of this species should be proposed the Marburg strain must be included in the change; if the genus *Bacillus* is divided the part that contains the Marburg strain must keep the name *Bacillus subtilis*.

When the bacterium can be grown (by culture in lifeless media, tissue culture, or by animal passage) the type will be a designated living culture (strain),[2] and in this it differs from the type of a species in mycology which is a preserved specimen or a description.[3] The type method extends beyond the type strain of species, but the types of taxa above species are taxa that are ultimately referred to the type of a species (the family Enterobacteriaceae – type genus, *Escherichia* – type species, *Escherichia coli*, – type strain NCTC 9001, ATCC 11775 – Opinion 15, 26[1]). The codes lay down procedures for the selection and designation of **nomenclatural types**.[4]

[1, BAC.Ap5. 2, BAC.18a. 3, BOT.9. 4, BAC.18; 20–22; BOT. 7–10; ZOO.70–5.]

type culture. **1.** **Strain** (isolate) chosen and designated as the nomenclatural type of a bact. species.[1] After 1 January 1976 the type must be designated by the author at the time of publication of the name of the species.[2] Before 1976 this was not required, though it was desirable, and when it was not done a type must be chosen later; when strains of the taxon isolated, studied, and named by the original author survive one of these strains must be chosen as a **lectotype**; when none of the original author's strains has survived a **neotype** must be chosen and designated.

2. Incorrectly applied to a culture in a so-called type-culture collection. Few of the strains in such collections are type strains; most, but not all, are typical (in the non-taxonomic sense) strains and can be likened to museum specimens.

3. In virology the word culture (in type culture) refers to any method of maintaining a virus in the living state, in a medium, in a host by passage, in cells or exudates, or freeze-dried material.

[1, BAC.18a–e. 2, BAC.16.]

typeface. The shape and size of characters used in printing. The typeface used is part of **journal style** but there are conventions about the printing of latinized scientific names that should be followed in writing and typing a manuscript. Binomial and trinomial names will be printed in italic and in the MS. should be underlined once; specific epithets (names) are seldom used alone but when they are they should not be underlined as they will appear in roman type.

Generic names used alone may be either in roman or italic; this is a matter of **editorial whim** or preference and even in one journal may alter with a change in editor. Names of taxa above genera are printed in roman in English journals, usually in italic in US publications. Author citations and qualifying phrases that follow a name printed in italic are made distinctive by being printed in roman or even boldface type; names in roman may be followed by a citation in italic or boldface.[1]

[1, BAC.33aN2.]

type method. Term used by botanists and mycologists for the permanent association of name and taxon; equivalent to **type concept** as used by bacteriologists. In this nomenclatural attachment of name to specimen the mycologist's type specimen cannot be a living culture; it must be a preserved specimen.[1]

[1, BOT.9.]

type-series. Zoo. term for all the specimens on which an author bases a species, excluding those he refers to as variants or excludes for some other reason.[1]

[1, ZOO.72(b).]

type species. **1.** Either the single species or one of several species included in the genus when the generic name was first validly published.[1,2,3] The species designated as the type of a genus; it is a nomenclatural type, and is the species to which the name of the genus is permanently attached. *Salmonella choleraesuis* is the type species of the genus *Salmonella* and the genus must always be so defined that it includes the species *S. choleraesuis*. If for any reason that species is excluded from the taxon, the generic name *Salmonella* is restricted to the excluded species, and a new generic name must be found for the remainder of the taxon.

2. The type of the name of a species is a designated specimen, description, or illustration, but in bact. it is preferably a **type strain**;[4] it does not need to be typical (characteristic) of the genus.

The type species may be fixed (1) by the author when he proposed the name of the genus (a holotype; type by original designation);[4,5,6,7] (2) by selection by monotypy when the genus, as originally published, had only one species;[4,7] (3) when the genus as published had more than one species, one of those described in the original publication may be chosen (a lectotype);[4,6,8] (4) when none of the original specimens on which the author based his genus has survived, a new specimen can be chosen to serve as a **neotype**,[4,6,9] unless meanwhile the generic name has been rejected.

[1, BAC.20a. 2, BOT.10. 3, ZOO.42(b); 67(a). 4, BAC.18. 5, BAC.16. 6, BOT.7. 7, ZOO.68. 8, ZOO.74. 9, ZOO.72; 75.]

type specimen. The single specimen (or other element[1]) designated as the type of the name of a species or subspecies. Type specimens must be preserved permanently; even in mycology the type specimen cannot be a living culture. If for any reason a specimen cannot be preserved the type may be a description or an illustration.

ZOO.[2] makes the surprising statement that a specimen that is the type of one species may be designated as the type of another species; which seems to encourage the creation of **objective synonyms**.

Type specimens are seldom found in bact. where the type is preferably a living culture (type strain). For an organism that cannot be cultivated the type will probably be a description, less often an illustration, and seldom a specimen.

[1, BOT.9. 2, ZOO.72(c).]

type strain. 1. The type culture; the strain to which the name of a taxon (species) is attached. It is the type of the name and only by extension becomes the type of the species. Normally the type of a bacterial species is a living culture, but for species that have not yet been grown in culture there cannot be a type strain; instead the type will be a specimen (a dead or dried preparation), an illustration, or even a description;[1] although not specifically stated in BAC. there is no reason why a description should not be supplemented by figures or photographs to show morphological features. The type of a bact. species proposed after 1 January 1976 must be designated by the author when he publishes his proposals for the taxon and its name.[2]

Older taxa may have had their type strains designated by (1) the original author (holotypy), (2) by monotypy, when only one strain was described and formed the basis of the species (an unsound taxonomic action), (3) by lectotypy, one of the strains studied and described by the original author who failed to designate a holotype, and (4) in the absence of a strain designated by the original author a **neotype** may be proposed, usually by a worker or group of workers interested in the species. If the original type of a species was an illustration or description, a new type must be designated when the species is successfully cultivated *in vitro* or maintained by passage *in vivo*. This should be done by the author who first isolates and describes the strain; if he fails to do this it should be described and designated by a subsequent author. The name of the author who isolates and describes the *culture* (as distinct from the named taxon) does not replace the author citation of the name of the species.

Because BOT.[3] refuses to accept a living culture as a type, mycologists cannot designate a type strain and must use a dried herbarium specimen (*exsiccatum*); although the spores of these specimens may remain viable for some years, cultures made from them cannot be accepted as types.

2. As the type strain is attached to the name of the species, workers will look to the strain to learn something of the species itself (i.e. the taxon). But they may well be disappointed, for the type strain need not be characteristic of the species (though it is desirable that it should be) and the seeker of information on the characters will get more satisfaction from the examination of what have been called reference (or representative) strains, which have no standing in nomenclature,[4] but are characterized and authenticated by reputable workers.
[1, BAC.18a. 2, BAC.16. 3, BOT.9. 4, BAC.19.]

-type. Suffix added to several different stems to indicate a subdivision of the category labelled by the stem. In papers on taxonomy the suffix *-type* should be used with circumspection so that the nomenclatural use of *type* can be kept pure. BAC.[1] recommends that the suffixes *-var* and *-form* should be substituted for *-type* in words such as biotype (biovar), serotype (biovar), phage type, phagotype (phagovar). But the words with the *-type* suffix are used as verbs as well as nouns, and it hardly seems possible that authors will readily agree to serovar a streptococcus, or to phagovar a staphylococcus.
[1, BAC.Ap10BT.]

typical. Characteristic; having the characters or features of something, of a living micro-organism. But here a warning is needed, the word typical has no connexion with *type* in the nomenclatural sense, for a type culture (type strain) need not be typical of the taxon to which its name is attached.

typification. 1. The application of names is determined by nomenclatural types. The main points about typification are: (1) typification applies only to the name, not to the taxon (organism). (2) The type of the name of a taxon above species is a taxon of lower rank: the type of the generic name *Escherichia* is the species *Escherichia coli*, and by extension we say that the type-species of *Escherichia* (meaning the genus) is *E. coli* (the species). (3) The name of a species is a combination, and the type of the species name (*Escherichia coli*) is the specimen (strain) to which the combination is permanently attached (strain NCTC 9001, ATCC 11775). (4) The type strain (specimen) does not need to be a good representative of the taxon (species), though in designating the type of a new species name, a strain with characters that comply with the description should be chosen.[1]

2. When a species is divided or its circumscription altered, the redefined taxon retains its name unless the type of the name has been excluded.

3. The nomenclatural type of a family name is the name of a contained genus, but not necessarily that of the genus with the oldest name;[2] it is normally the genus from whose name the family name was formed, as Brucellaceae from *Brucella*. Exceptions occur, as in Enterobacteriaceae which is not based on *Enterobacter*, but has as its type the genus *Escherichia*; exceptions such as this are usually protected by conservation. ZOO.[3] does not apply typification to names of taxa above family, but BAC.[4] extends it to class (of which the type is order) and to order and tribe, and BOT.[5] to those names that are **automatically typified** because they have a type the name of the genus on which the name of the higher taxon is based.

4. Korf & Rogers (1967, *Taxon*, **16**, 19) distinguish two means of typification: *explicit*, when an author designates the type, and *implicit* when the author fails to designate a type but by his actions indicates that only one taxon of those eligible must be the type of the next higher taxon. An example of implicit typification is typification by monotypy. These ideas can be summarized as: *explicit types*, holotype, lectotype, neotype; *implicit types*, monotype, **schizotype**.

[1, BAC.18a. 2, ZOO,64. 3, ZOO.35; 62. 4, BAC.15T. 5, BOT.(1975)10; 16.]

typify, *v.* **1.** To be characteristic of something. Seldom used in this sense in taxonomic literature.

2. To designate a nomenclatural type. In bact. the type of a species is preferably a strain (culture);[1] BOT.[2] and ZOO.[3] have specimens, illustrations, or descriptions as types. Living cultures are not acceptable as types in bot. and mycologists cannot designate a culture as a type specimen.

As typification applies only to a name and not to a taxon (except indirectly), the type designated need not be typical (in the sense of characteristic) of the taxon.

[1, BAC.18a. 2, BOT.9. 3, ZOO.72.]

typing phages. Bacteriophages used in the typing or subdivision of a bacterial taxon (usually a species but may be a serotype). Two kinds of phages may be used for this purpose: those that occur in nature and chosen because they give the sort of answer that is wanted (clear-cut and consistent lysis of the appropriate test strains), and those adapted from a parent phage to lyse other strains.

Typing phages often give a pattern of reactions and the **phage type** may be one of the patterns.

typographic(al) error. An error attributed to the compositor, but the availability of this loophole should not be made an excuse for careless proof-reading. The use of a capital for the first letter of a **specific epithet (name)** formed from the name of a person or place is regarded as a *typographical error*. Like an **orthographic error**, it should be corrected by later authors who use the name; the author citation and date of valid publication are not affected by this correction.

typography. 1. Choice of characters used in printing.

2. The characters used in printed matter. In biological nomenclature scientific names from genus to subspecies are set in a distinctive type.[1] The choice of characters to be used in printing constitutes the art of typography. In scientific journals typography is determined by the editor and printer, and in books by discussion between author and publisher.

It has become an established convention that scientific binominals (binomials) shall be printed in **italic** but the **author citation** and abbreviations for new species (sp. n.), new genus (gen. n.) or new combination (comb. n.) are in a contrasting type, usually roman. A generic name standing alone when it means 'members of the genus X' may be in italic, but many editors prefer to print a

generic name in roman. (See also **abbreviations.**) The Royal Society's *General Notes on the Preparation of Scientific Papers* (1974) recommends that the names of higher groups (family, class, order, kingdom) should be printed in roman type; if the word is in Latin form a capital is used but when in English or when used as an adjective, with a lower case first letter. American usage may differ and italic type is sometimes used for scientific names of all ranks.
[1, ZOO.Ap.E2.]

typology. According to Bock typology goes back to the teachings of Plato; it is the belief that in a natural group the members resemble an ideal type, and that any variations are not important. It is contrasted with modern population taxonomy and is interpreted more liberally. Davis & Heywood equate typology with **type concept** which term, in this dictionary, is used for the 'type method' of botanists.

typonym. A later name given to the type specimen. Absolute synonym.

typotype. The type of a type; the material (specimen) that forms the subject of the description or illustration that is the type of a name.

typus, typicus. 1. ZOO. recommends that these words, which might lead to confusion, should not be used as new specific names.[1] However, if in a newly established genus there is a species named *typus* or *typicus*, that species should be made the type species.[2]
2. BOT.[3] says that *typicus* as an infraspecific epithet is illegitimate except when it repeats the specific epithet because Art. 26 requires its use. (Art. 26 says that the infraspecific taxon that includes the type of the species must repeat the correct name (epithet) of the type as the infraspecific epithet. If the correct name of the species is *Xus typicus*, then the infraspecific epithet for the taxon that includes the type must also be *typicus*.)
3. BOT.[4] recommends the use of the word *typus* immediately before or after *the particulars* (description, diagnosis) of a nomenclatural type being designated.
[1, ZOO.D8. 2, ZOO.68(b). 3, BOT.24. 4, BOT.R37A.]

U

-u. In bact.[1] and bot.[2] a personal name ending in the letter *-u* can be made into a generic name by the addition of the letter *-a*, or into a substantival specific epithet by adding the letter *-i* to the name of a man or *-a* to the name of a woman. In ZOO.[3] the appropriate suffix will be *-us*, *-a*, or *-um* to make a generic name; to make a species name from the name of a man,[4] add *-i* or, for the name of a woman,[5] add *-a*.
[1, BAC.Ap.9BT. 2, BOT.R73B; (1975) R73C. 3, ZOO.Ap.D37. 4, ZOO.Ap.D17. 5, ZOO.R31A.]

umbonate. Term descriptive of a colony with a raised central papilla.

uniform scale of values. An idealistic view of a taxonomic hierarchy in which ranks (categories) on the same level should have the same value. Numerical taxonomists attempt to do this and it has been suggested that every taxonomist should make this ideal one of his aims, or even regard it as a rule. But a taxonomist working intensively on a particular group becomes an expert and he is likely to see differences not apparent to the less expert, and he probably becomes a **splitter** with regard to that particular group.

In bact. it is easy to name genera that are equivalent in value and others that are of different values; the genera *Escherichia*, *Shigella* and *Salmonella*, do not differ among themselves any more than the species of *Streptococcus*, and the family Enterobacteriaceae has about the same value as the genus *Streptococcus*. But it is hard to see how a uniform scale of values could ever be enforced or made the subject of a rule. In the past attempts have been made to do it by the **subordination of characters**, but it does not seem to work in practice.

uninomen, -inal. **1.** A complete name contained in one word, scientific names of taxa of the levels of genus and above (e.g. *Salmonella*, Salmonelleae) are uninominal. Generic names are in the singular, names of higher categories are in the plural; they are all treated as Latin nouns. Uninominals above genus level are usually printed in roman.

2. Uninominal systems of nomenclature have been proposed by several taxonomists, most of whom have been zoologists, but none of the schemes has seemed practicable to the majority of his fellows. Michener (1963, 1964) is one who advocated a uninominal system, but at the same time he proposed a numerical coding system and this, because it is more practical, received the greater attention.

union of taxa. The Principle[1] or Law[2] of Priority applies to the name of a taxon formed from the union of seveal taxa of the same rank. When two or more genera are united the name of the combined taxon will be that of the oldest legitimate (valid) name among the component genera.[3] If the names of the united taxa are of the same date the author who makes the union should choose one of them and his choice must be followed by other authors.[4]

The same principle holds when members of taxa of other ranks are joined; in the united taxon the name (epithet if below subgenus in bact. and genus in bot.) of the taxon with the oldest legitimate name becomes the name of the combined unit.[3,4]

[1, BAC.Pr8; BOT.PrIII. 2, ZOO.23. 3, BAC.42; BOT.57; ZOO.23(e). 4, ZOO.24(a).]

uniquely derived character. Term used by Le Quesne for a character that has changed, or evolved, from a primitive state to an advanced state only once during the evolution of the organism. It is perhaps impossible to establish this in a positive sense, but methods exist whereby characters can be shown not to be uniquely derived. LRH.

unitary designation. A one-word name applied to species should not be regarded as a generic name[1] unless it has been published as such. A bact. example of a unitary designation is the name pneumococcus, which has sometimes been printed in italic with a capital P, and treated as a combined generic and specific name. [1, BOT.20(2).]

unite. Join, combine; in taxonomy to combine two or more taxa; such a union can be made by any taxonomist, but it need not be accepted by his fellows. Classifying, uniting and splitting, do not follow any rules: the nomenclatural consequences, however, must follow the rules.

units of measurement included in descriptions should follow the SI system, but until this is better known it may be advisable to show the equivalent older units in parentheses. Microbiologists should avoid units such as micron (use micrometre, μm) and millimicron (use nanometre, nm). See **SI**.

unknown. 1. Term used for an organism that is the subject being identified. It contrasts with the 'knowns' – the organisms comprising the taxa that have been identified, characterized, and named, and with which the unknown is compared. (See **U-OTU**.)

2. Term applied to a character of a taxon for which information is not available. In tables of characters and descriptions it is often better to state that the character is unknown than to leave a blank space in a table or not to mention it in a description.

U-OTU. Unknown (unidentified) operational taxonomic unit. Cf. **OTU**.

unpublished names. 1. BOT.[1] recommends that, in forming new specific epithets, use should not be made of unpublished names found in correspondence, traveller's notes, or on herbarium labels even if the name is to be attributed to the source from whence it came.

2. Authors should not use (or even mention) unpublished names that they do not accept, especially if the original coiner of the name has not authorized publication.[2]

3. Common sources of unpublished names are those written on herbarium packets or sheets – the so-called ***herbarium names***.

[1, BOT.R23B(i). 2, BOT.R34A.]

unpublished work. When an author wishes to refer to the unpublished work of another author he should obtain the consent of the other worker. In the text he should refer to this work by inserting in parentheses the name of the author and the words 'personal communication', and perhaps the date, though this will not affect priority of a name. Personal communications should not be included in lists of references or of literature cited.

unsuitable characters. Term used by Hubálek in a paper dealing with a numerical taxonomic analysis, for characters that were negative in all strains, and thus did not have, within the group being studied, any distinguishing value. Can be applied, of course, to characters that are positive in all strains, for the same reason.

unused name. A zoologist may make out a case for the suppression of an unused name that is a senior synonym or a senior homonym of a name in current use.[1] The plea must be put before the Zoological Commission and a prima facie case will be made when (1) it can be shown that the older name is not known to have been used during the preceding 50 years, and (2) that the younger name has been used, as a presumed valid name, for a taxon by at least five different authors and in at least ten publications during the same period.[2] In making a case for the suppression of a senior homonym, the suppression abolishes the application of laws of priority and homonymy; a case to suppress a senior synonym prevents application of the Law of Priority.[3]

The rejection of an unused name can only be made by the Commission; rejection by an author before 1 January 1973 on the grounds that the name was an unused senior synonym, will not now be accepted unless the rejection has been approved by the Commission.[2]

[1, ZOO.23(a–b). 2, ZOO.79(b). 3, ZOO.79(a).]

unweighted characters. Characters that are not given special emphasis (or weight). Silvestri & Hill (1964) think that the term should be qualified by the words *a priori*, especially in a numerical taxonomy context. '"Unweighted characters" is an unfortunate term by which to describe characters used in making a classification, and that it would be better to say that characters are used in an unprejudiced manner, for if a classification (formation of taxonomic groups) is possible this will be so only because the characters will have different degrees of correlation. Highly correlated characters will be the main determinants (have the greatest weight) in creating the classification, while poorly correlated characters will have little or no influence.'

Heywood (1966) makes the point that the term is used for characters that have been selected for use (perhaps because of ease of observation), thereby weighting them *vis-à-vis* characters not selected. When used in a classification the results tell us what weight *a posteriori* should be given to particular characters.

-uric. Suffix indicating tolerance of or resistance to the substance or state that forms the main part of the word, e.g. aciduric, thermoduric.

-us. 1. When the latinized form of a personal name ends in *-us*, and it is intended to make it into a new name of a genus, the *-us* ending must be dropped before adding the appropriate suffix[1] (see Table P. 1, p. 198, and **personal name**); examples: (1) a name such as Magnus is in Latin form; the *-us* is dropped, leaving the stem ending in a consonant; therefore the letters *ia* are added to give *Magnia*; (2) from Linné, the latinized form is Linnaeus; after dropping *-us*, the stem ends in the vowel *e*, and the letter *-a* should be added to give *Linnaea*.

2. Plurals of generic names ending in *-us*. As a scientific name, the generic name is always a Latin word in the singular: as a plural it becomes a common name or vernacular word, and usually takes the plural form of the language used. English words ending in *-us* form awkward plurals, and *Fowler* says that each must be decided on its merits. Most generic names ending in *-us* are second declension Latin nouns, and the plural ending in *-i* is generally preferable to *-uses*; but while staphylococcuses is never used, viruses is the usual plural of virus. On the other hand, third declension nouns (e.g. genus) take *-era*, to form genera.

[1, BAC.Ap.9BT.]

usage of names. ZOO.[1] states that the words *existing usage* mean the most common usage, and *common usage* is defined[2] (by implication) as use by at least five different authors in at least ten publications within the last 50 years.

BAC. introduces new ideas on **revived names** and the **reuse** of names (q.v.).

[1, ZOO.80(ii). 2, ZOO.79(b).]

V

v. BOT.[1] requires the letter *v*, when it represents a vowel in a pre-1800 name, to be changed to *u*. When alternative spellings of a name or epithet exist, *u* should be used before a consonant or semivowel, and *v* before a vowel (*Taraxacum* not *Taraxacvm*).

[1, BOT.(1975)73.]

VAC system. A nomenclatural system for viruses made up of a vernacular name (V) plus a **cryptogram** (C) based on an Adansonian (A) classification. The scheme was suggested by Gibbs *et al.* (1966, *Nature*, **209**, 450) as an alternative to a non-phylogenetic scheme which used latinized binominals (**LB**) put forward for discussion by the Provisional Virus Nomenclature Committee of the International Association of Microbiological Societies.

Names in the VAC system might be in two or three parts: (1) the vernacular name would be unchanging except for translation into different languages; (2) the cryptogram or code which describes the characters of virus would be an abbreviated form of the full code which gives a unique specification of the virus. The shortened code sometimes could relate to more than one virus group, and to pin-point the particular group another element might be necessary; this part (3) could be a name, symbol, or letter specific to the group. Many such specific symbols exist for plant and insect viruses, and others could be invented.

valid name. The oldest available name of a taxon; it is the application of the Law of Priority[1] and is the zoo. equivalent of **correct name** in bact. and bot. A taxon may have many **available** names but it can have only one valid name which is usually the oldest that is in accordance with the rules. Exceptionally it is not the oldest because that has been suppressed by the Zoological Commission, or it may be a name conserved or specifically validated by the Commission.
[1, ZOO.23.]

valid publication. For a name to be validly published it must meet certain detailed requirements of BAC. and BOT. Earlier editions of BAC.[1] had a note that the word *valid* and the term *validly published* meant *with standing in nomenclature*.

The first essential is that the name must have undergone **effective publication**;[2] it must be in Latin form as required by the rules, and it must be accompanied by a description of the organism (in Latin in bot.) or by a reference to a previously and effectively published description or diagnosis.[3,4] If the last requirement had been applied stringently many bact. names would have been declared illegitimate on the grounds that they had not been validly published as they lacked descriptions good enough to distinguish them from other taxa. BOT.[5] requires the descriptions of algae to be supplemented by figures showing the essential characters, and also recommends that the description of any new taxon should indicate the ways in which it differs from *its allies*.[6]

For valid publication BAC.[3] and BOT.[7] require that a nomenclatural type should be designated when a new taxon is proposed; for bact. taxa type cultures should be designated and deposited in culture collections, which should be named in the original publication. In bact. taxa the descriptions should conform to minimal standards (yet to be set).[8] The name of a species must be shown as a combination of the generic name and specific epithet.[9]

BAC.[3] introduced new requirements for the valid publication of bact. names after 1 January 1976. The name of a new taxon or a new combination must be published in *IJSB* and be accompanied by a description or a reference to one effectively published; it need not be in Latin. The date of valid publication in *IJSB* is the important date, that of publication in any other journal or in a book will not have any significance in determining the priority of the name.

Valid publication also depends on whether the author definitely accepts his own proposals[10] (this may seem absurd, but taxonomists do much fence-sitting); sometimes names are proposed for purely theoretical groups, and these names are not acceptable. An author must indicate the rank of the taxon he names, and the name and rank of the next higher taxon.[11] The valid publication of the name of a taxon below genus is dependent on the earlier (or simultaneous) valid publication of the generic name or specific epithet in the case of an infraspecific taxon.[12]

A name is not validly published when it is merely cited as a synonym, or is mentioned incidentally or by the subordinate taxa included in it, or when it is proposed in anticipation of the future acceptance of the group concerned, or as a provisional name.[10] After 1 January 1935 a bot. name must be accompanied by a diagnosis or description in Latin;[13] this applies to fungi, and after 1 January 1958 a Latin diagnosis or description and an illustration of distinguishing features is also needed when new names for algae are proposed.[5] The name of a bact. species is not validly published when the description is based on a mixed culture.[14]

[1, BAC.(1958)12aN. 2, BAC.25a; BOT.29. 3, BAC.27. 4, BOT.32. 5, BOT.39. 6, BOT.R34B. 7, BOT.37. 8, BAC.R30a; b. 9, BAC.30; BOT.33. 10, BAC.28b; BOT.34. 11, BOT.(1975)35. 12, BOT.43. 13, BOT.36. 14, BAC.31a.]

-var. Suffix used in BAC.[1] to indicate an *infrasubspecific taxon*; these are not subject to the rules of BAC.[2] An alternative given is **-form** as biovar, chemovar, chemoform, cultivar, morphovar, phagovar, serovar.

The intention is to avoid use of the suffix -type and so preserve, in papers on taxonomic subjects, the purity of the word **type** in nomenclature.

[1, BAC.Ap.10B. 2, BAC.5d.]

variable, variability. 1. Inconstant or (of a character) unstable. To be distinguished from **variation** and **variant**. A variable character is one that, in a given strain, sometimes seems to be positive (or present) and at other times negative (absent); although symbolically stated to be + or −, the characteristic recorded may be the result of a test that would be better (and more accurately) expressed quantitatively.

2. Of a test, unreliable, not regularly repeatable; affected by extraneous factors. An example is the MR (methyl red) test, used to indicate the final pH value of a buffered glucose medium after the bacterium has grown in it for a stated time–temperature combination; pH values of 4.2 or less show a red colour, above 6.3 a yellow colour, while between the two values the indicator assumes various shades of orange. The test is influenced by the temperature and duration of incubation, and by other, less easily controlled factors.

variable spelling. Differences in the spelling of zoo. species-group names in the following list[1] are insufficient to eliminate homonymy: (1) *ae*, *oe*, or *e*; (2) *ei*, *i*, or *y*; (3) *c* or *k*; (4) the aspiration or non-aspiration of a consonant (*oxyrhynchus*, *oxyryncus*); (5) *c* or *ct*; (6) single or double consonant (*litoralis*, *littoralis*); (7) *f* or *ph*; (8) different connecting vowels; (9) transcription of the semivowel *i* as *y*, *ei*, *ej*, or *ij*; (10) *-i* or *-ii* ending of patronymic genitive; (11) *-ensis* or *-iensis* in a geographical name.

Differences in spelling (between an incorrectly spelt name and its corrected form) are not sufficient to dispose of the homonymy of two names;[2] and differences in termination due to gender in adjectival species-group names do not prevent homonymy.

[1, ZOO.58. 2, ZOO.57(b).]

variance. **1.** In statistics, the square of the standard deviation.

2. Sometimes loosely used by numerical taxonomists for **difference**.

variant. **1.** A culture that is not identical with the parent in all its characters; now largely replaced by an adapted strain or mutant. A term that can be used for any unnamed variation within a species; it avoids using the term *variety*, which has a special nomenclatural connotation.

2. Differences in spelling, especially in the use of the wrong connecting vowel in compound words are treated as orthographic variants in BAC.[1] and BOT.,[2] but this kind of variant is not regarded as an inadvertant error in ZOO.[3]

3. Orthographic variants[2,4] and inadvertant errors[3] are to be corrected.

4. Two or more names so similar that they may be confused are treated as variants;[5] when they are based on different types they are homonyms. Examples given in BOT.:[5] *Columella*, *Columellia*; *Skytanthus*, *Scytanthus*; *chinensis*, *sinensis*; *ceylanica*, *zeylanica*.

[1, BAC.Ap9A. 2, BOT.73. 3, ZOO.32(a)(ii). 4, BAC.57cN1. 5, BOT.75.]

variation. In a micro-organism, a change in the demonstrable characters of the descendants. It may be the result of an **adaptation** to the environment, or a **mutation** in which the potential to form a heritable character is suddenly lost.

variety, *abbr.* **var.** In nomenclature a category of rank below the level of species. In bact. variety was equated with subspecies,[1] but after 1975 the use of variety for *new names* will deprive those names of standing in nomenclature;[2] in bot. variety is inferior to subspecies.[3] Since 1960, variety has not been recognized in zoo., and a new name published as a variety is to be regarded as infrasubspecific,[4] not subject to ZOO., and is not an available name.[5]

BAC.[6] and BOT.[7] require the names of varieties (subspecies) to be expressed as **ternary combinations** with a word (or abbreviation) denoting the rank between the specific and subspecific epithets. When a species is divided into subspecies or varieties the one that contains the type of the species must have, as its subspecific epithet, the same epithet as the species, as *Bacillus cereus* var. *cereus*. The species *Corynebacterium diphtheriae* has been divided into varieties and one of these should be *C. diphtheriae* var. *diphtheriae*, but the usual division into three varieties, *gravis*, *intermedius*, and *mitis*, does not explore all the nomenclatural possibilities, or satisfy the requirements of BAC.[6]

[1, BAC.(1958)7N. 2, BAC.5c. 3, BOT.4. 4, ZOO.45(c); (e). 5, ZOO.15. 6, BAC.13a. 7, BOT.24.]

vegetative. **1.** Not forming spores.

2. Not forming sex organs; asexual (mycology).

vernacular name. The name of an organism or taxon expressed in the language of the country; these names do not have any status in nomenclature, and are frowned upon by botanists. In microbiology they were often preferred to scientific names when writing in English, though the fashion is passing, and they are not encouraged in American. See **common name**.

Vi antigen. A labile surface antigen found on the surface of some salmonellas, which masks O antigens and so prevents agglutination by O antisera. At one time thought (erroneously) to be associated with the virulence of the organism, hence the name Vi antigen given to it by Felix & Pitt.

vigour. A term proposed by Sneath to take into account differences between strains due to growth rate and incubation temperature on the general metabolic activity of a strain, and also the influence of the time of reading a test (or the duration of incubation) on the result. Differences in vigour are regarded as analogous to differences in size in higher organisms. Another sort of difference, **pattern 1**, analogous to shape, is found in strains of equal vigour.

VIR. Abbreviation used in this dictionary for the proposed virological code; in lower case letters used for 'in virology' or 'virological usage'.

virion. An infective virus particle; stated by **PCNV** to be the basis of the taxonomy of viruses. Four of its characteristics define the equivalent of virus families. (1) Chemical nature of the nucleic acid (RNA/DNA). (2) Symmetry of the nucleocapsid (helical/cubical/binary). (3) Presence or absence of an envelope. (4) (*a*) Helical viruses – diameter of nucleocapsid; (*b*) cubical viruses – triangulation number and number of capsomer.

virulence, virulent. The quality (and ability) to invade tissues, multiply, produce tissue response or change, and sometimes to cause death (usually of an experimental organism inoculated with a culture).

Virulence depends on distinct factors in different organisms; diphtheria bacilli are virulent because they produce an exotoxin (i.e. are toxigenic) although the organisms do not spread from the site of inoculation. Pneumococci, on the other hand, are invasive and spread from the site of inoculation to all parts of the body; the virulence of pneumococci is related to the polysaccharide capsular material.

A virulent organism is always a **pathogen** for the organism it infects or damages by its toxin, but the term virulent is not applied to all pathogens, and one does not write of virulent staphylococci. (See also **virulent phage**.)

virulent phage is a bacteriophage that multiplies in a susceptible cell, which lyses and liberates the phage particles.

viruliferous. Term used by phytopathologists to describe a vector containing a virus that can be transmitted to a plant.

-virus. Suffix added to a word or name, however arbitrarily formed, to make the generic name of a virus.[1]

[1, VR15.]

virus nomenclature. Virological nomenclature continues in a long gestation; virologists produced, but did not approve, a provisional code of nomenclature (referred to in this dictionary as VIR.). In the absence of an accepted code ICTV developed a set of rules (VR) to serve as a guide for virus workers. In drawing up these rules virologists obviously took to heart some of the lessons of **heretical taxonomy**, and tried to deal with the problems in a simple way by making simple rules. While the intention is good the rules themselves are far from clear as there is confusion between what applies to names and what to classification.

viticula. Literally, a little vine. Used by Hubálek for the corkscrew-like swirl of sediment that is created by rotating a tube of broth culture.

vogue words and phrases. Words and phrases that acquire a temporary popularity with a meaning that is slightly different from the normal. Examples of worn-out and abandoned vogue words are sophisticated (in relation to techniques); climate (referring to opinions); data (not relating to premises but to results of experiments); educated (in connexion with guesswork); moment of time, for now, and operational.

Consensus is an example of a vogue word that has retained its meaning.

vowel. 1. In nomenclature the letter *y* at the end of a surname is treated as a vowel. Generic names can be made from surnames ending in a vowel other than *-a* by adding the letter *a*, as *Noguchia* and *Gaffkya*; the letters *-ea* are added to a name ending in *-a*, as *Fallaea* from Falla; a suffix is unnecessary when the name ends in *-ea*.

2. In bact. and bot. incorrect connecting vowels may be corrected.[1,2] ZOO.,[3] however, does not accept incorrect connecting vowels as inadvertent errors and consequently they should not be corrected in zoo. names.

[1, BAC.57cN1. 2, BOT.73. 3, ZOO.32(a)(ii).]

VR. Abbreviation used in this dictionary for the Rules of Nomenclature of Viruses as amended in 1975 (Fenner, 1976).

vulgar. Of names, used for names in the language of the country. Of similar derivation to vulgate, the language of the people; the adjective vulgar is used by those not writing in their own language; English-speaking people would use vernacular or **common name**.

W

W. A consonant not found in the alphabet of classical Latin, but allowed in scientific names.[1] Seen particularly in generic names derived from the names of people (usually biologists it is intended to honour) as in *Wolbachia*, *Welchia*, and in specific epithets in *Clostridium welchii*, *Klebsiella edwardsii*.

[1, BOT.73; ZOO.11(b)(i).]

'weasel' words. Words with two or more different meanings; words that are ill-defined or of uncertain meaning.

WFCC. World Federation for Culture Collections.

weighting of characters. Attaching a greater importance to some characters than to others. For botany, Davis & Heywood list four kinds of weighting: (1) selection by ease of observation, (2) correlation (or *a posteriori*) weighting, (3) *a priori* weighting, deduced by correlation studies, and (4) rejection or residual weighting, the only characters left. (See also **unweighted characters**.) These kinds of weighting apply also in microbiology, but bacteriologists at least are trying to be more objective in their taxonomic work, and are restricting the weighting of characters to identification (or diagnostic) work.

Ernst Mayr, a zoologist and a firm opponent of numerical taxonomy, described three kinds of weighting: (1) *a priori*, (2) Adansonian non-weighting, only suitable where taxonomy is *completely immature, as in the bacteria*, (3) *a posteriori* on the *information content* of a character (the correlation between the character and a so-called natural classification).

Classifications based on unweighted characters may not be compatible with accepted identification systems, but this is to be expected since an identification scheme is artificial and only becomes practicable when the diagnostic features (i.e. the ***differentiae***) are weighted.

well-known names. See **better-known names**.

White, P. Bruce (1891–1949). The bacteriologist who did more than any other to put order into the taxonomy of salmonellas. The bases of his work were (*i*) the fundamental differences in the agglutination reactions of motile and non-motile bacteria (described in 1903 by Smith & Reagh, *Journal of Med. Res.*, **10**, 89) and (*ii*) the two kinds of variation (S → R, described by Arkwright; and diphasic variation of the flagellar antigens, described by F. W. Andrewes) that affected these reactions. The primary classification was based on the somatic antigens (O antigens of Weil & Felix) of the bacteria in the S form; secondary breakdown was made on differences in the flagellar (H) antigens.

Kauffmann extended the original classification and because of Bruce White's pioneer work the revised version was named the Kauffmann–White scheme.

Bruce White regarded the different serological combinations as stable units and was willing, on the knowledge then available, to equate them with species.

wild-type. A term used to describe the micro-organism as it exists in nature; for a pathogen it means the organism as it occurs in the tissues of the host. In microbial genetics the wild-type is said to be **prototrophic**. Variants that are less capable in synthesizing the growth factors or amino acids needed are said to be **auxotrophic.**

word. A succinct definition occurs in the description of a species-group name as *a simple word of more than one letter* or a compound word.[1] The first element of a compound species-group name may be a Latin letter used to denote a character[2] and this is joined to the main part of the name by a hyphen (the only use of a hyphen permitted by ZOO.).

[1, ZOO, 11(g)(i). 2, ZOO.26(c).]

word or phrase not intended as an epithet. In giving examples of **illegitimate epithets**, BAC.[1] quotes *Bacillus nova species*; and does not regard the *nova species* as a two-word epithet. Other examples quoted are the ordinal adjectives *primus* and *secundus*, and from this we can conclude that *Clostridium tertium* has an illegitimate epithet. BOT.[2] will not accept *words not intended as names*, or ordinal adjectives as epithets.

[1, BAC.52. 2, BOT.23.]

working strain or **type.** Useful term for a strain that, in the absence of a **holotype**, **lectotype** or **neotype**, is chosen as a good representative of the species, suitable for use as a reference strain in comparative work.

writing for publication. Papers on taxonomic subjects must, like all scientific papers, be carefully prepared, and the contents arranged in the sequence, and using the conventions for references, considered under **journal style** and **literature cited**. In microbiological taxonomy there are special points that are dealt with here.

1. *Materials used.* Under a heading such as this an author should describe exactly what he studied: specimens, specimens from which cultures were isolated (stating whether the cultures obtained were mixed or pure), cultures from other workers (who should be named), cultures from collections (accession or catalogue numbers should be stated); whenever possible the **history of a culture** should be recorded, and it is important to give any name under which a strain has appeared in earlier literature, e.g. *Staphylococcus aureus* strain Heatley, NCTC 6571. The author should indicate clearly how he will refer to the cultures (specimens) in the paper; if he intends to propose changes of scientific names it is advisable to avoid the use of latinized names in the text until 'Nomenclature' is reached in the Discussion, and it is better to refer to strains by some number or strain name (as NCTC 6571 or Heatley in the example given above).

2. *Methods* used for characterization should be given in sufficient detail to enable them to be repeated by the reader. Alternatively, a reference may be made to a previously published description of the method in a widely distributed journal or book.

3. *Results* of characterizing tests, or of statistical analyses do not need comment here, but see **symbolic notation table**.

4. *Discussion*. Under this heading the author's results will be discussed in relation to those of previous taxonomists working with the same organisms, those believed to be similar, and perhaps, organisms believed to be different. Comparisons need to be described very clearly; it is easy to confuse the reader by strings of numbers or names, and the collection of such numbers or names into groups, or into tables can be of great assistance. Organisms that seem to belong to one taxon may have been described under several different names, and here it will be important to include not only **author citations** of names, but also **reference** to workers who have made non-nomenclatural contributions to the taxonomy of the group.

In the Discussion it may be necessary to trace the history of one or more of the cultures, particularly when it is intended to show that a strain is not what it is said to be, but is a **substituted culture**.

5. *Final arrangement*. When the classification of a group is being changed in any way, the new arrangement should be clearly stated, the **taxonomic position** of the group shown, and the name of the taxon in the next higher **rank** stated.

6. *Nomenclature*. Any proposals for changes in nomenclature must be clearly stated; prevarication is not allowed by any of the nomenclatural codes. An author should know his own mind; if he does not, he should not be writing a paper for publication. He should avoid phrases such as 'The name *Camel reginae* would seem to be appropriate'; such a name would be regarded by other nomenclators as being not firmly proposed, and therefore, not validly published. If he is proposing a new taxon (whether a genus, species or subspecies) he should not neglect to designate the nomenclatural type. A paper in which the name of a new bacterial taxon is proposed must be published in *IJSB*, or the name proposed there together with a reference to the full paper if published elsewhere.[1]

[1, BAC. 24a.]

writing to be read differs from writing for publication chiefly by wrapping the meat (required by the rules and recommendations of nomenclatural codes) in a palatable pastry that not only soothes the tongue (when read aloud) but pleases the mind when read in solitude.

The author who wants his paper to be read, digested, and remembered will write simply and briefly and, when the subject is suitable, in a light-hearted manner. Papers can be overweighted with seriousness and pomposity, and made indigestible in a dull, uninspiring way; alternatively, even a serious and dull subject like nomenclature can be made palatable by a slightly eccentric approach that borders on the bantering.

Writers of scientific papers must make an effort to ease the lot of the man who is not reading his mother tongue, and should avoid long words, involved constructions, colloquial phrases, and laboratory jargon. It is essential that the summary is in simple English, and that tables and figures can be understood without reading the text.

'Trendy' words and phrases (**vogue words**) should seldom find a place in a scientific paper, but they can, occasionally, introduce a sense of relief or realism among the intricacies of the pseudo-legal phraseology of codes of nomenclature. Fortunately, with revision of the codes their wording is becoming simpler, generally clearer, but not always unambiguous, as they should be.

J. W. Howie (*Lancet*, 1957, ii, 320) contributed a masterly essay on writing to be read, and any serious author who has not read this should do so before writing his next paper.

X

xerography. Literally, dry writing, or writing without ink; a process whereby multiple dry copies are made on ordinary paper without the use of a photographic developer or heat; it is essentially a rapid process that produces large numbers of copies, and can be a substitute for **lithography**.

Xerox. Name of an industrial organization that introduced a rapid method of photocopying letters, books, or papers on ordinary, unsensitized, sheets of paper. The word Xerox is used as a verb (for the process of making copies) and as an adjective; probably derived from xerography. Xerox is a trade name.

Y

Y. 1. Although the letter *y* is not found in words of classical Latin, nomenclators are not such purists that they forbid its use in scientific names. It appears mainly in scientific names made from place names or the surnames of people, as in the generic name *Yersinia*, but it is also found in names derived from other sources, such as *Symbiotes*, *Cytophaga sylvestris*.

2. In BAC.[1] and BOT.[2] a surname ending in the letter *-y* can be made into a generic name by the addition of *-a*, as *Gaffkya* from Gaffky, or into a substantival specific epithet by adding the letter *-i*, *murrayi* from Murray.

3. In ZOO.[3] differences in spelling that involve only the letter *y* and *ei* or *i* are not sufficient to prevent the homonymity of species-group names.

[1, BAC.Ap.9BT. 2, BOT.R73B; (1975)R73C. 3, ZOO.58(2)(9).]

yeasts. Unicellular fungi which may (Endomycetales) or may not (Cryptococcaceae) produce spores. Of economic importance particularly to brewers and bakers who use them as sources of enzymes to convert sugars to alcohol and carbon dioxide.

Z

ZOO. Abbreviation used in this dictionary for Zoological Code or International Code of Zoological Nomenclature; references to ZOO. are to the edition published in 1964, and amendments published in *Bull. zoo. Nomencl.*, **29**, 168–89; zoo. is used for 'in zoology' or 'zoological practice'.

Zoological Commission. International body charged with consideration of changes in ZOO., and with making recommendations for its clarification or modification. It issues Declarations (provisional amendments), Opinions, and Directives, it compiles lists of accepted names and indexes of rejected names.[1] [1, ZOO.77,78.]

zygote. A fertilized sex cell, diploid because each gamete has contributed half the number of chromosomes that make up the normal cell which contains paired chromosomes.

zymogram. Pattern of bands detected by electrophoretic analysis of enzyme solutions. Used in taxonomic work to characterize the esterases and heat-stable catalases of mycobacteria and other organisms.

zymology. The science or study of fermentation; more specifically, the study of yeast fermentation.

References

(Note: In joint authorship of three or more authors, only the first author is given, followed by *et al.* Titles of books are given, but not of papers published in scientific journals.)

Adanson, M. (1763). *Familles des Plantes.* Paris: Vincent.

Ainsworth, G. C. (1961). *Ainsworth and Bisby's Dictionary of the Fungi.* Kew: Commonwealth Mycological Institute.

Ainsworth, G. C. (1973). *Review of Plant Pathology,* **52**, 59.

Annear, D. I. (1956). *Laboratory Practice,* **5**, 102–5.

Bacteriological Codes

(1) Buchanan, R. E. *et al.* (1948). International Bacteriological Code of Nomenclature. *Journal of Bacteriology,* **55**, 287–306. (Reprinted, 1949, *Journal of General Microbiology,* **3**, 444–62).

(2) Buchanan, R. E. *et al.* (1958). *International Code of Nomenclature of Bacteria and Viruses: Bacteriological Code.* Ames: Iowa State College Press.

(3) Editorial Board (1966). International Code of Nomenclature of Bacteria. *International Journal of Systematic Bacteriology,* **16**, 459–490.

(4) Lapage, S. P. *et al.* (1975). *International Code of Nomenclature of Bacteria: Bacteriological Code,* 1976 Revision. Washington: American Society for Microbiology.

Bergey's Manual of Determinative Bacteriology

Bergey, D. H. *et al.* (1923). First published.

Bergey, D. H. (1926). Second edition.

Bergey, D. H. (1930). Third edition.

Bergey, D. H. (1934). Fourth edition.

Bergey, D. H. *et al.* (1939). Fifth edition.

Breed, R. S. *et al.* (1948). Sixth edition.

Breed, R. S. *et al.* (1957). Seventh edition.

Buchanan, R. E. & Gibbons, N. E. (1974). Eighth edition.

Blackwelder, R. E. (1959). *Systematic Zoology,* **8**, 69.

Blackwelder, R. E. (1964). *Publications of the Systematics Association,* **6**, 17.

Blackwelder, R. E. (1967). *Systematic Zoology,* **16**, 64–72.

Botanical Code

International Code of Botanical Nomenclature (1972), adopted by the Eleventh International Botanical Congress, Seattle, 1969. Utrecht: International Bureau for Plant Taxonomy and Nomenclature.

Boulter, D. & Thurman, D. A. (1968). *In* Hawkes, J. G. (ed.) (1968). q.v.

Buchanan, R. E. (1916). *Journal of Bacteriology,* **1**, 591–6; (1917), **2**, 155–64, 347–50, 603–17; (1918), **3**, 27–61, 175–81, 301–6, 403–6, 461–74, 541–5.

Buchanan, R. E. (1925). *General Systematic Bacteriology.* Baltimore: Williams and Wilkins.

Buchanan, R. E. *et al.* (1966). *Index Bergeyana.* Baltimore: Williams and Wilkins.

Bulloch, W. (1938). *The History of Bacteriology*. London: Oxford University Press.
Cain, A. J. & Harrison, G. A. (1960). *Proceedings of the Zoological Society, London*, **135**, 1–31.
Carlile, M. J. & Skehel, J. J. (eds.) (1974). *Evolution in the Microbial World*, 24th Symposium of the Society for General Microbiology, Cambridge University Press.
Castellani, A. & Chalmers, A. J. (1919). *Manual of Tropical Medicine*, 3rd edn. New York: William Wood & Co.
Chester, F. D. (1901). *Manual of Determinative Bacteriology*. New York: MacMillan.
Cohn, F. (1872). Untersuchungen über Bakterien. I. *Beitrage zur Biologie der Pflanzen*, **1**, 187–222.
Committee (1960). *CBE Style Manual*, 3rd edn. Council of Biology Editors. Washington: American Institute of Biological Sciences.
Cowan, S. T. (1965*a*). *Advances in Applied Microbiology*, **7**, 139.
Cowan, S. T. (1965*b*). *Journal of General Microbiology*, **39**, 147.
Cowan, S. T. & Steel, K. J. (1965–74). *Manual for the Identification of Medical Bacteria*. 1st edn. 1965, 2nd edn., revised by S. T. Cowan, 1974. Cambridge University Press.
Davis, P. H. & Heywood, V. H. (1973). *Principles of Angiosperm Taxonomy*. Edinburgh: Oliver & Boyd.
DuPraw, E. J. (1965). *Systematic Zoology*, **14**, 22.
Ellis, G. (ed.) (1971). *Units, symbols and abbreviations: a guide for biological and medical editors and authors*. London: Royal Society of Medicine.
Fenner, F. (1976). See VR. (1976).
Fischer, A. (1895). *Jahrbuch fur wissenschaften Botanik. Berlin*, **27**, 1–163.
Fischer, A. (1903). *Vorlesungen ueber Bakterien*. Jena: Gustav Fischer.
Flügge, C. (1886). *Die Mikroorganismen*. Leipzig: Vogel.
Flügge, C. (1908). *Grundriss der Hygiene*.
Gram, C. (1884). *Fortschritte der Medizin*, **2**, 185.
Gyllenberg, H. G. & Niemelä, T. K. (1975). *In* Pankhurst (1975) q.v.
Haas, O. (1962). *Systematic Zoology*, **11**, 186.
Hawkes, J. G. (ed.) (1968). *Chemotaxonomy and Serotaxonomy*. London: Academic Press.
Hawksworth, D. L. (1974). *Mycologist's Handbook*. Kew: Commonwealth Mycological Institute.
Heywood, V. H. (ed.) (1966). *Modern Methods in Plant Taxonomy*. London: Academic Press.
Hugh, R. (1964). *International Bulletin of Bacteriological Nomenclature and Taxonomy*, **14**, 87–101.
Jeffrey, C. (1977). *Biological Nomenclature*, 2nd edn. London: Edward Arnold.
Jordan, E. O. & Falk, I. S. (eds.) (1928). *The Newer Knowledge of Bacteriology and Immunology*. Chicago: University of Chicago Press.
Kendall, A. I. (1902). *Proceedings of the American Public Health Annual*, **28**, 484.
Lapage, S. P. *et al.* (1975). See BAC. (1975).
Lehmann, K. B. & Neumann, R. (1896). *Atlas und Grundriss der Bakteriologie*. Munich.

Lehmann, K. B. & Neumann, R. (1901). *Bacteriology*. English Translation of 2nd German edition by G. H. Weaver.
Le Quesne, W. J. (1969). *Systematic Zoology*, **18**, 201–5.
Le Quesne, W. J. (1972). *Systematic Zoology*, **21**, 281–8.
Lysenko, O. & Sneath, P. H. A. (1959). *Journal of General Microbiology*, **20**, 284.
Mayr, E. (1965). *Systematic Zoology*, **14**, 73.
Mayr, E. (1969). *Principles of Systematic Zoology*. New York: McGraw-Hill.
Mayr, E. (1976). *Evolution and the Diversity of Life: Selected Essays*. Cambridge, Massachusetts: Harvard University Press.
Merlino, C. P. (1924). *Journal of Bacteriology*, **9**, 527.
Michener, C. D. (1963). *Systematic Zoology*, **12**, 151.
Michener, C. D. (1964). *Systematic Zoology*, **13**, 182.
Migula, W. (1894). *Arbeiten aus dem Bakteriologischen Institut der technischen Hochschule zu Karlsruhe*, **1**, 235–8.
Migula, W. (1897–1900). *System der Bakterien*. Bd. I. (1897), Bd. II. (1900). Jena: Gustav Fischer.
Morse, L. E. *et al.* (1975). *In* Pankhurst, R. J. (ed.) (1975) q.v.
O'Connor, M. & Woodford, F. P. (1975). *Writing of Scientific Papers in English*. European Life Science Editors (ELSE) & Ciba Foundation.
Orla-Jensen, S. (1909). *Zentralblatt für Bakeriologie*, II, *Abt.*, **22**, 305–46.
Pankhurst, R. J. (ed.) (1975). *Biological Identification with Computers*. London: Academic Press.
Ravin, A. W. (1963). *American Naturalist*, **97**, 307–18.
Rhodes, M. (1950). *Journal of General Microbiology*, **4**, 450–6.
Savory, T. (1962). *Naming the Living World: An Introduction to the Principles of Biological Nomenclature*. London: English Universities Press.
Silvestri, L. G. & Hill, L. R. (1964). *Systematics Association Publication*, no. 6, pp. 87–103.
Simpson, G. G. (1961). *Principles of Animal Taxonomy*. New York: Columbia University Press.
Skerman, V. B. D. (1959–67). *A Guide to the Identification of the Genera of Bacteria*, 1st edn., 1959; 2nd edn. 1967. Baltimore: Williams & Wilkins.
Skerman, V. B. D. (1969). *Abstracts of Microbiological Methods*. New York: Wiley-Interscience.
Sneath, P. H. A. (1974). *In* Carlile & Skehel (1974) q.v.
Sneath, P. H. A. & Sokal, R. R. (1973). *Numerical Taxonomy: The Principles and Practice of Numerical Classification*. San Francisco: W. H. Freeman.
Snell, J. J. S. & Lapage, S. P. (1973). *Journal of General Microbiology*, **74**, 9.
Society of American Bacteriologists. (1957). *Manual of Microbiological Methods*. New York: McGraw-Hill.
Sokal, R. R. & Sneath, P. H. A. (1963). *Principles of Numerical Taxonomy*. San Francisco: W. H. Freeman.
Stainier, R. Y. *et al.* (1966). *Journal of General Microbiology*, **43**, 159.

Viruses, Proposed Code, Rules of Nomenclature of Viruses:
(1) Wildy, P. (1971). Classification and Nomenclature of Viruses. First Report of the International Committee on Nomenclature of Viruses. *Monographs in Virology*, vol. 5. Basel: Karger.

(2) Fenner, F. (1976). Classification and Nomenclature of Viruses. Second Report of the International Committee on Taxonomy of Viruses. *Intervirology*, **7**, No. 1–2 (see also *Journal of General Virology*, **31**, 463–70).
Winslow, C.-E. A. & Rogers, A. F. (1905). *Science*, **21**, 669.
Winslow, C.-E. A. & Rogers, A. F. (1906). *Journal of Infectious Diseases*, **3**, 485.
Winslow, C.-E. A. *et al.* (1917). *Journal of Bacteriology*, **2**, 505–66.
Winslow, C.-E. A. *et al.* (1920). *Journal of Bacteriology*, **5**, 191–229.

Zoological Code
International Code of Zoological Nomenclature, adopted by the XVth International Congress of Zoology. International Commission on Zoological Nomenclature, London, 1964.
Zopf. W. (1885). *Die Spaltpilze nach dem neuesten Standpunkte bearbeitet*. Breslau.